Topics in Applied Physics   Volume 65

W0037860

Topics in Applied Physics   Founded by Helmut K. V. Lotsch

Volumes 1–56 are listed on the back inside cover

Laser Spectroscopy of Solids II

Edited by W. M. Yen

With Contributions by
D. D. Dlott   M. C. Downer   B. Henderson
C. Klingshirn   G. P. Morgan   K. P. O'Donnell
P. C. Taylor   W. M. Yen

With 144 Figures

Springer-Verlag Berlin Heidelberg GmbH

Professor *William M. Yen,* Ph. D.

Department of Physics, University of Georgia,
Athens, GA 30602, USA

ISBN 978-3-662-30929-2   ISBN 978-3-540-45939-2 (eBook)
DOI 10.1007/978-3-540-45939-2

Library of Congress Cataloging-in-Publication Data. – Laser spectroscopy of solids II. (Topics in applied physics; v. 65) Includes index. 1. Solids-Spectra. 2. Semiconductors-Spectra. 3. Laser spectroscopy. I. Yen, W. M. (William M.) II. Dlott, Dana Donald. III. Series. QC176.8.06L375 1989   530.4'1   88-24859

This work is subject to copyright. All rights are reserved, whether the whole or part of the material is concerned, specifically the rights of translation, reprinting, reuse of illustrations, recitation, broadcasting, reproduction on microfilms or in other ways, and storage in data banks. Duplication of this publication or parts thereof is only permitted under the provisions of the German Copyright Law of September 9, 1965, in its version of June 24, 1985, and a copyright fee must always be paid. Violations fall under the prosecution act of the German Copyright Law.

© Springer-Verlag Berlin Heidelberg 1989

Originally published by Springer-Verlag Berlin Heidelberg New York in 1989.
Softcover reprint of the hardcover 1st edition 1989

The use of registered names, trademarks, etc. in this publication does not imply, even in the absence of a specific statement, that such names are exempt from the relevant protective laws and regulations and therefore free for general use.

2154/3150-543210 – Printed on acid-free paper

Preface

As we noted in the second edition of the first volume of *Laser Spectroscopy of Solids* (Topics in Applied Physics, Vol. 49), we were delighted by and grateful for the reception it received from the various spectroscopy communities. We believe that the acceptance of a review volume by a critical scientific audience is simply evidence that it fulfils a current need. This was fortunately the case with the first volume. It is, of course, the hope of the contributors to this sequel volume that it also appears at an appropriate time. All indications appear to be in our favor as the activity in laser spectroscopy applied to condensed phases continues to grow in an impressive fashion and to mature into an almost routine tool in the study of solid-state optical properties.

The first volume dealt principally with optically active ordered and disordered insulating systems; this was because many of the laser spectroscopic techniques transplanted from atomic studies were first applied in this type of solid. Inevitably, as tunable lasers of various types have become readily accessible, laser spectroscopic applications have expanded to include other types of condensed systems. For example, the advent of commercial pico-second laser systems has made it possible to study the dynamics occurring in semiconductors and in semiconductor structures, studies which would not have been possible a decade ago.

In this volume, we continue to contend that conceptually all optically active materials will, with an appropriate change in some gauge, behave in analogous ways and that differences which arise are to a large extent semantic. Thus, in the following contributions, we have endeavored to minimize specialized terminology and to emphasize the common thread and methodology that bind these studies. The chapters are sufficiently tutorial that they can serve as an introduction to those wishing to learn about the subject matter. However, they also possess the depth to serve as current reviews of the understanding that the use of laser spectroscopy has brought to the phenomena which affect optically excited states in condensed phases.

The authors of each chapter have played important roles in the advances which they review. Each of them therefore has the essential perspective necessary to survey the general advances in their respective fields in a fair and comprehensive manner. This is apparent in their contributions. I am personally most thankful to them for the effort and care they took in preparing their individual chapters, all of which I believe will withstand the test of time.

Finally, I wish to acknowledge the continued support and encouragement of Dr. H. K. V. Lotsch and the editorial staff at Springer-Verlag. Their patient guidance has considerably eased my task as the organizer of this effort. Support has also been provided by the National Science Foundation, the Department of Energy, and the University of Georgia.

Athens, GA, December 1988   *William M. Yen*

Contents

Contributors

Dlott, Dana D.
   School of Chemical Sciences, University of Illinois,
   Urbana, IL 61801, USA

Downer, Michael C.
   Department of Physics, University of Texas,
   Austin, TX 78712, USA

Henderson, Brian
   University of Strathclyde, Department of Physics and Applied Physics,
   107 Rotenrow, Glasgow, G4 0NG, United Kingdom

Klingshirn, Claus
   Fachbereich Physik,
   Universität Kaiserslautern, Erwin Schrödinger Straße 1,
   D-6750 Kaiserslautern, Fed. Rep. of Germany

Morgan, Gerard P.
   Department of Physics, University College, Galway, Ireland

O'Donnell, Kevin P.
   University of Strathclyde, Department of Physics and Applied Physics,
   107 Rotenrow, Glasgow, G4 0NG, United Kingdom

Taylor, P. Craig
   Department of Physics, University of Utah,
   Salt Lake City, UT 84112, USA

Yen, William M.
   Department of Physics, University of Georgia, Athens, GA 30602, USA

1. Introduction:
Advances in the Laser Spectroscopy of Solids

William M. Yen

With 11 Figures

We are all aware of the central role spectroscopy has played in the evolution of our modern understanding of physical and chemical processes. The term spectroscopy, in its most general sense, embraces all those techniques in which an energy source is used to interact with and to probe some resonance in the atomic or material systems of interest. Advances in this area, more often than not, are intimately connected with the development of new energy sources, be they radiative or particulate, which allows greater finesse in our methodology and often leads to entirely new techniques and discoveries. The theme of this, as well as a prior volume [1.1], centers on the impact the advent of such a source, i.e., the tunable laser, has had on optical spectroscopies as they are applied to the properties of the condensed phases.

The introduction of stimulated optical devices has had a very noticeable and significant effect on spectroscopy; indeed, following a period of development, some of the techniques categorized under the heading of laser spectroscopy have become a part of the arsenal of tools used for the routine characterization of the optical properties of all types of materials. There exist many excellent and timely texts and reviews of conventional as well as laser spectroscopic methods and techniques [1.2]; consequently, it is not necessary to reiterate the contents of the existing literature. Instead, in this brief introductory overview (Chap. 1), we proceed under the assumption that the reader has been exposed to some of the basic literature, and review, briefly, the various experimental advantages which accrue from the use of laser spectroscopy. The other chapters contain material which supplements or updates subjects reviewed in [1.1]. Chapter 2 is a comprehensive and seminal review of two-photon processes in rare earth activated systems; the resurgent interest in these higher order processes is an important example of the contribution being made by laser spectroscopy. Recent advances achieved in understanding optical energy transfer in disordered systems are reviewed in Chap. 3. It should be apparent from that chapter that though some problems remain, we have made noticeable strides in obtaining a more fundamental and quantitative understanding of these processes as they occur in insulators. The discussion there centers on the application of the microscopic and macroscopic models of energy transfer which were developed in [1.1]. Chapter 4 presents a tutorial on the properties of optical color centers in insulators and reviews the experimental results obtained using spectroscopic methods based on the laser. Interest in these materials had waned until it was demonstrated that they could be used as tunable solid-state lasers. The needs of technologies based on these relatively new devices have fueled an impressive revival in this area and this activity again bears testimony to the importance of the methods advocated in these volumes. Chapter 5 surveys yet another area in which new laser-based techniques have recently made an impact, that is, the study of relaxation and transfer processes which affect vibrational excitations in molecular crystals. The remaining two chapters deal with the optical properties of crystalline and amorphous semiconductors, respectively. These chapters go a considerable way towards rectifying the omission in the first volume of chapters dealing with this important class of materials. In the general spirit of these reviews, Chap. 6 contains a comprehensive tutorial on the optical properties of semiconductors, which serves as a useful introduction to the archival reviews of spectroscopic studies appearing later in that chapter and in Chap. 7. In both cases, the focus is on advances in which laser spectroscopic methods have played a significant role; these contributions are most obvious in the area of amorphous systems and in dealing with events occurring in the subnanosecond domain.

To return to this introductory chapter, in addition to reviewing the experimental advantages of laser spectroscopy, we also comment on recent developments in laser technology and advances in experimental methodology and correlate them to selective gains made in our understanding of the optical properties of solids in general. Some of these latter developments have been mentioned in the Addendum to the second edition of [1.1], while others are dealt with later in this volume. The field of laser spectroscopy of solids, however, is presently evolving so rapidly that we can present only a sampling of the most recent advances. This chapter concludes with some general remarks regarding the directions these studies may take in the near future.

1.1 Conventional Versus Laser Spectroscopy

Generally, in conducting spectroscopic studies, the precise methods to be used are determined simply by the available resources and their convenience or ease of application. The only requirement which is placed on the results obtained with a particular technique is that they be consistent with and complementary to those obtained by other means or techniques. Thus, for example, it is possible to conduct spectroscopic measurements on the same system in the frequency and time domains; the results are simply Fourier conjugates of each other in this instance.

Conventional optical spectroscopy embraces all those techniques which employ incoherent line or blackbody radiation sources in conjunction with some form of optical analyzer or spectrograph. These techniques have been, of course, crucial to the evolution of all of modern physics and chemistry, but because of the nature of the radiation sources used, they suffer from certain well-established limitations. All of these shortcomings have been remedied by the use of lasers as the radiation source and it is now customary to identify spectroscopy which employs stimulated sources as laser spectroscopy.

Since the conception and first demonstration of the laser [1.3, 4], the intervening years have seen a rapid proliferation of stimulated devices covering an extremely broad range of frequency and temporal characteristics (Fig. 1.1). The advent of tunable lasers [1.5] is of particular importance, as is the development of pulsed laser devices which have unprecedented temporal resolutions [1.6]; these events have generalized the utility of lasers for spectroscopic applications. The virtues of laser radiation were recognized almost from the day of its discovery, so that lasers found immediate uses as alternatives to broad band, incoherent sources. More relevant to the theme of these volumes are the many other techniques which the laser made possible. A number of instances come to mind; the Raman effect in solids was first discovered in 1928 but did not become a practical analytic tool until continuous wave (cw) ion lasers became available [1.7]. Similarly, without the aid of the high power pulsed ruby laser, it would have been extremely difficult to have discovered the stimulated Raman process [1.8] and other nonlinear effects such as frequency doubling in solids [1.9]. Other early applications of the laser include the observation of two-photon absorptions in activated solids [1.10], the extension of the concept of spin echoes to the optical domain, where laser coherence plays a crucial role [1.11], and the demonstration of optically induced transparency in solids [1.12].

These early applications invariably utilized the monofrequency lasers which were then available; apart from Raman studies, this restricted the applicability of the laser-based technique to those systems in which some energy level fell in coincidence with the then accessible frequencies. For example, and as noted above, photon echoes were first observed and their properties studied in ruby; similarly, the technique of fluorescence line narrowing (FLN) in solids [1.13] was first demonstrated in the R lines of ruby. In both instances, the experiments employed temperature tuned ruby lasers [1.14]. The use of lasers for spectroscopic purposes remained limited until the introduction of the tunable dye laser [1.15], and this invention marks the origin of laser spectroscopy as we know it today.

We can also classify optical spectroscopy into two broad categories. The first, static spectroscopy, concerns itself with the structure of the spectra of the optically active solid. The features of such spectra are determined, of course, by the electronic levels of the optically active centers and by the effects of the environs on these levels. For these studies, high resolving powers in the frequency domain are often necessary.

In general, however, active centers in solids are not isolated and they may interact with each other as well as with a dynamic lattice environment. A different type of spectroscopy, i.e., dynamic spectroscopy, is required to probe into the effects of these interactions on the spectral properties. In these techniques, temporal measurements often play an important role with a concomitant requirement for high temporal resolution.

The spectra of all materials suffer from some degree of inhomogeneous broadening. This type of broadening has been discussed in detail in [1.1]. As pointed out there, the source of this broadening is to be found in some randomly fluctuating extrinsic perturbation to the atomic system. In crystals, for example, optical transitions are inhomogeneously broadened by unavoidable defects and imperfections of the lattice [1.16]. These external perturbations do not affect the intrinsic or homogeneous behavior of the optically active ion or center. However, in many cases the inhomogeneous broadening is sufficiently severe that it can obscure and conceal the intrinsic features which are of fundamental interest; this is the case, for instance, in optically activated glasses [1.17]. Laser spectroscopy has provided a convenient way by which inhomogeneous contributions are suppressed on behalf of the homogeneous features of interest; this is a central and important property of laser-based techniques. It is not that conventional spectroscopy is unable to yield the homogeneous spectra of materials [1.2], it is the experimental simplicity and ease of laser based techniques which have made them the obvious choice for such measurements.

1.2 Lasers as Spectroscopic Sources

The properties of laser radiation have been well established. The first of these to be exploited for optical experiments was the large power densities which are attainable with laser sources. In terms of spectroscopic measurements, the power simply translated into improvements in the signal-to-noise ratio, i.e., conducting spectroscopy with large effective apertures. In addition, as we have noted, the monochromatic power densities attained have allowed access to the whole spectrum of nonlinear optical properties.

The directionality and collimation of the laser beams make the steering and focusing of electromagnetic radiation a relatively simple experimental task, allowing experimentation in spatially confined areas and in small or irregular crystalline samples. In addition, the intrinsic coherent nature of laser radiation has allowed us to develop and pursue a host of techniques which find analogies in nuclear magnetic and electron spin resonance (NMR and ESR respectively) and which cannot be conducted with conventional sources. Coherent methods, such as transient or free induction decay (FID) and photon echoes, entailing time domain measurements are easiest to conduct whenever the relaxation rates encountered are long. This means the sharper a homogeneous line is in the frequency domain, the more readily accessible it is to coherent laser methodology. For solids, the slowest relaxation rate reported to data is found in the $^7F_0$ to $^5D_0$ transition of Eu^{3+} in Y_2O_3 where $T_2 = 420$ μs at 2 K [1.18]. This is equivalent to a linewidth of 760 Hz, which is well beyond the resolution limit of all conventional optical analytic instruments.

Since the publication of [1.1], laser technology has continued to advance rapidly, thus expanding our experimental capabilities in all areas of laser-based spectroscopies. For example, picosecond pulsed laser drivers were first commercialized in the late 1970s and have since become a fairly common source for spectroscopy. As an illustration, two of the reviews in this volume (Chaps. 4 and 5) present picosecond domain time-resolved studies applied to molecular vibrational dynamics and to semiconductor carrier properties, respectively. More recently, the development of colliding-pulse mode-locked (CPM) dye lasers and correlated technologies have succeeded in improving temporal measurements well into the femtosecond (fs) regime [1.19]. Here, the optical impulses entail a finite number of light oscillations, and thus fs pulses contain a broad range of Fourier frequency components centered at the laser frequency; in effect, pulses of this type can provide near white spectral coverage, a property which can be exploited experimentally [1.20].

Parallel laser developments have simultaneously enlarged the fraction of the electromagnetic spectrum over which tunable laser coverage can be obtained. Figure 1.1 summarizes schematically the regions over which laser devices can provide spectral coverage. The F-center tunable laser, which had just reached a developmental stage in the early 1980s, has expanded tunable laser sources well into the near IR [1.23]. These lasers were also instrumental in the development of soliton lasers employing optical fibers and the methodology for pulse compression. Various other solid-state systems, involving the 5d

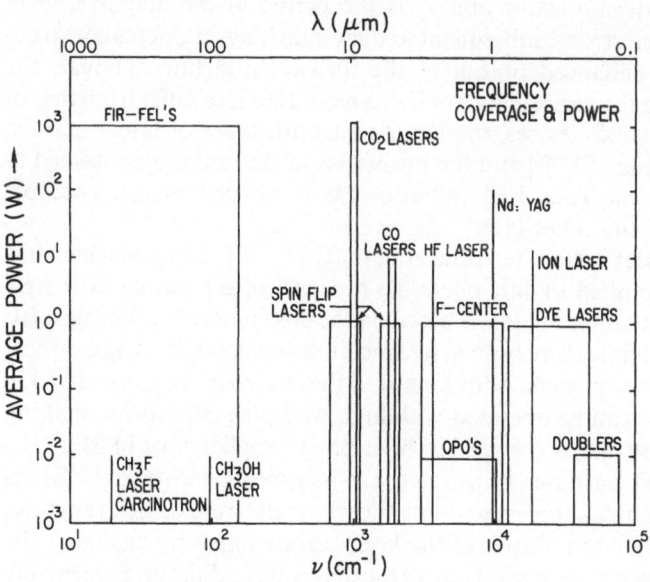

Fig. 1.1. Frequency coverage and average power outputs of a representative sampling of common monofrequency and tunable laser devices. In the FIR region, there are many molecular lines which have been used for lasing [1.21]; the Carcinotron is a traveling wave electronic device. Diode lasers operate in the general region in which F–center lasers operate; various free-electron lasers (FELs) are being attempted in the UV and VUV regions. (After [1.22])

bands of $4f$ ions [1.24] and phonon-assisted transitions of the transition metal ions [1.25], have also contributed to our arsenal of high power tunable sources predominantly in the visible and in the near IR. Semiconductor laser devices have become commercially available but they have not been fully exploited in the spectroscopic sense as yet [1.26].

The past decade has seen the successful demonstration of the principle of operation of a new class of lasers, the free-electron lasers or FELs [1.27]; these devices are unusual because they do not require a material medium for operation. As their name denotes, FELs rely on the radiation emitted by quasi-relativistic electrons as they traverse a region where they are accelerated by a periodically alternating magnetic field produced by an undulator. The principles of operation of FELs are closely related to that of traveling wave tubes and hence these devices are classical rather than quantum devices [1.28]. When the radiation emitted by the oscillating electrons is fed back, using mirrors in the case of FELs, it exerts a ponderomotive force on the electron beam, which results in beam bunching. The spatial periodicity of the electron bunches coincides with the central frequency of the radiation field. Omitting some kinematic factors, this frequency is given by

$$\omega = 1/\lambda_l \cong \gamma^2/\lambda_\mu, \tag{1.1}$$

where γ is the relativistic factor and λ_μ is the period of the magnetic field. Radiation from the electrons, subsequent to their bunching, interferes constructively, resulting in enhanced output in the forward direction. Though the principles governing the performance of FELs would seem to differ from that of conventional stimulated devices, the theory for both types of lasers may be brought into agreement [1.29] and the properties of the radiation obtained in either case are identical. Thus, FEL radiation can be utilized for spectroscopic purposes as that of any other laser.

The FEL possesses certain intrinsic advantages [1.30]. The one which has been principally exploited to date offers the prospect of generating extremely high power coherent radiation; this is based on the possibility of scaling up FEL devices, since in principle there is no active medium which might evince optical nonlinearities at higher powers. Additionally, if the electrons are recirculated in some manner, FELs can be operated with high wall-plug efficiencies, making them all the more attractive for certain high power applications [1.31].

For spectroscopic purposes, however, the most appealing feature of FELs is their potential as tunable sources spanning the whole electromagnetic range. As can be seen from (1.1), the output of the laser can be tuned by changing the energy of the electrons or the periodicity of the magnetic undulator. Present day undulators have a periodicity in the vicinity of 1–3 cm, thus with the electron accelerators available, FELs can operate well into the x-ray region [1.27].

The gain equations for FELs indicate that small signal gains are inversely proportional to the output frequency raised to the third power [1.30, 32]. Thus, the most advantageous region for FEL operation falls in the long wavelength

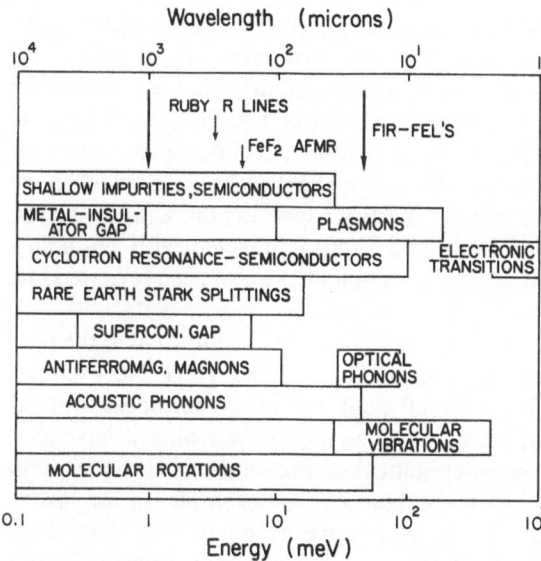

Fig. 1.2. Energies of a number of the elementary excitations in molecules and in condensed materials. The majority of the excitation which play important roles in determining the dynamical properties of optically excited states of interest to us are in the 1–100 meV (or 10–1000 cm^{-1}) region. The tuning region of the FEL [1.33] is shown by large vertical arrows which include this FIR region. (From [1.34])

portion of the spectrum, i.e., the far infrared (FIR). Though discrete frequency molecular lasers are available in the FIR [1.31], FELs offer potentially single source coverage for this whole region. Recently, a low voltage electrostatic (1–6 MeV) FEL has been demonstrated which has a tuning range from 10 to 80 cm^{-1}, outputting at several kilowatts peak power throughout the range [1.33]. The dearth of FIR sources has limited the extent to which spectroscopic investigation can be carried out in this region, yet many of the elementary excitations of solids which are of interest are to be found in this region [1.34], see Fig. 1.2. It follows that the development of FIR FELs signals the advent of laser spectroscopy in this traditionally difficult experimental region.

Visible and UV FELs have also been reported, however, invariably the range over which they can be tuned has been strictly limited by the fixed energy of the electron accelerators employed [1.30]. The optical properties of materials in the VUV and beyond still limit the development of high energy laser sources, with FELs being no exception.

More recently, efforts have been initiated to miniaturize FELs so that they can become laboratory rather than facility-sized instruments. This work centers partially in reducing the undulation period of the magnetic field. According to (1.1), a reduction in the period results in a square root reduction in the accelerator energy required for an equivalent laser frequency output. For one area of interest, i.e., the FIR, the development of appropriate micro-undulators would result in a considerable size contraction accompanied by beneficial cost reductions. Suitable micro-undulators have been constructed with periodicities of 3–4 mm, an order of magnitude reduction from undulators currently in use [1.35–37].

Insertion devices, such as the undulator, are also commonly employed with synchrotron sources to produce high intensity, high energy radiation [1.38]. The generation of radiation through the use of undulators and wigglers is the precursor to FEL operation. In the absence of broadly tunable lasers in the UV, synchrotron radiation (SR) has proven to be extremely useful as a spectroscopic source and has led to a number of SR-laser optical techniques which have helped in the elucidation of solid-state properties, especially those involving the atomic inner core of the constituents and those connected with boundary phenomena. Synchrotron radiation spectroscopies have also been reviewed in a number of places [1.39].

The performance of lasers, in general, has continued to improve in the past decade. The pulsed power of commercially available lasers, YAG : Nd, excimer, etc., driven tunable devices is now sufficient for most nonlinear optical experiments. These advances in peak as well as total power output have also made multi-laser beam spectroscopies practical. Improvements in other aspects of laser performance are also not to be ignored; for example, in cw lasers, stability of output frequencies of a few hertz have been achieved. This jitter produces an effective laser linewidth, thus, attainable laser frequency resolution is of the same order of magnitude [1.40].

The development of new laser sources allowing us access into new frequencies and into hitherto unheard of temporal resolutions, coupled with the general progress of the technical performance of all classes of laser sources, has simply resulted in a considerable increase in the experimental resources available to us and indicates the maturation of the techniques advocated in these volumes. Thus, the period since the publication of [1.1] is to be noted not so much for the pace of demonstrations of new laser-based experimental techniques as for the explosion of applications of these methods to various solid-state systems of interest.

1.3 Comments on Recent Advances of Laser Spectroscopy of Solids

The Addendum to [1.1] contains a brief update of the status of laser spectroscopic studies of solids, principally in insulating materials, up to 1985. For crystalline materials, for example, many of the conflicts surrounding optical energy transfer in the R lines of ruby [1.41] were resolved by the experiments of Gibbs et al. [1.42]. Similarly, the role of exchange in transfer and relaxation of rare earth systems [1.43] as well as the anomalous properties of the rare earth pentaphosphates [1.44] have largely been clarified. One of the gratifying results of the more recent experimental studies is that the theoretical framework for transfer and diffusion of energy developed in [1.1] has, in general, withstood the test of time [1.45]. A comprehensive update of experimental developments in transfer processes is presented in Chap. 3 of this volume and elsewhere [1.46].

Some of the other advances attained recently in our understanding of the optical properties of solids are detailed in this volume. It is notable, for example, that considerable progress has been achieved in unravelling multiphoton absorption processes in the transition metal [1.47, 48] and rare earth series [1.49, 50]. In rare earth systems, the presence of two-photon absorptions was established earlier [1.10] and explained soon thereafter in terms of the Judd-Ofelt [1.51] closure approximations by *Axe* and others [1.52]. The availability of high power tunable devices has allowed a comprehensive reexamination of these processes, resulting in a firmer understanding of this phenomenon and the resolution of a number of puzzles. The current status of this field is reviewed by one of the principal figures in these developments in Chap. 2.

In the past few years, laser spectroscopic methods have been applied to a widening assortment of materials. The fourth chapter deals with the properties of defect or color centers in various insulators; this class of optically active materials was studied extensively until the early 1970s and a basic framework for the understanding of the creation, stability and structure of the trapped electrons was established by then [1.53]. Renewed interest arose because of the development of the near IR tunable F-center lasers [1.23] and a more sophisticated foundation has evolved for these systems. As in the case of insulators, techniques such as site-selective spectroscopies used in conjunction with optically detected magnetic resonance (ODMR) have proved extremely valuable in these studies.

Similarly, both crystalline and amorphous semiconductors have also been investigated extensively with laser spectroscopy in the past decade. The application of site-selective methods has played an important role in characterizing various shallow and deep traps in ordered semiconductors and in establishing the nature of localized and delocalized states in disordered semiconductors [1.54]. Picosecond and subpicosecond laser sources have also proven to be powerful tools with which to investigate carrier dynamics in these systems and structures [1.55]. Bistable behavior of semiconductors under various optical pumping conditions has been investigated actively and has promising technical implications [1.56]. The last two chapters of this review volume detail advances in this field.

Another area in which laser-based optical methods have played a central role is in the study of elementary excitations of matter. Such excitations serve an important function in the relaxation and diffusion of optically excited states in solids and involve, more often than not, collective actions of the molecule, atomic complex or the solid lattice. Chapter 5 addresses one of these areas, i.e., vibronic relaxation processes in organic materials; as is to be noted there, the availability of short laser excitation pulses has been principally responsible for this development.

The topics selected for review in this volume are by necessity limited and, thus, no pretense is made that all significant advances in recent years are discussed. Indeed, one of the aims of this introductory overview is to provide a brief summary of happenings in the areas which are not covered or updated by these reviews. With the understanding that the following may be tainted with personal preferences, we present a synopsis of significant developments in the application of laser-based spectroscopies to the study of solid-state optical properties.

1.3.1 Relaxation and Transfer in Glasses

The observation of anomalous behavior of the excited states of centers in disordered materials as manifested through their optical linewidths has aroused a great deal of interest and activity, which has resulted in a much clearer theoretical understanding of this problem [1.57]. Experimentally, linewidth studies have been extended to include disordered organic glasses, and FLN, hole burning, and coherent methods have all been implemented in these insulating systems. The relaxation had been shown to display a seemingly universal T^2 thermal dependence down to low temperatures in the bulk of systems studied. The relaxation rate at low temperatures were also found to be anomalously large compared to the crystalline values. The latter seemed to implicate the low temperature disordered modes postulated by *Anderson* and others [1.58] as consisting of two level excitations or TLSs. Various theoretical models involving TLSs were advanced to explain the increased relaxation at low temperatures; however, in each case, the theories failed to explain the totality of linewidth results. Figure 1.3 shows the experimental results obtained to date for rare earth activated inorganic and organic glasses [1.59].

A number of experiments conducted at extremely low temperatures, i.e., 1 K and below, successfully demonstrated deviation from the higher temperature T^2 dependence. For example, in measurements conducted by *Hegarty* et al. [1.60] using photon echoes in a Nd^{3+}-doped optical fiber, a linear T dependence of the linewidth was observed, obviously implying a behavioral crossover from earlier FLN results. Subsequently, the crossover was observed in a Yb^{3+} doped phosphate glass by *Brundage* and *Yen* [1.61]. See Fig. 1.3c. FLN studies of Cr^{3+} in glasses have also contributed to the probable resolution of the linewidth problem. In these studies *Bergin* et al. [1.62] have been able to resolve the phonon sidebands associated with so-called "high field" Cr ions within the inhomogeneous glass distribution. In the high field sites, the ^2E state lies lowest and, hence, the FLN emission is mainly R-line like. Figure 1.4 illustrates the type of sideband obtained through these methods. As vibrational sideband shapes in solids can be correlated to the density of states, the figure implies that the vibrations coupling to the ^2E state depart significantly from Debye-like acoustic phonons and other optical modes existing in crystals. There is, for example, a noticeable shift of the peaks in the density towards lower energies. Similar studies have been conducted in rare-earth-doped glasses [1.63]. However, the weaker ion to lattice coupling prevented the resolution of any structure in the sidebands observed.

Huber [1.64] has subsequently shown that an altered density of vibronic states, such as those discussed above, lends an explanation to the quasi-

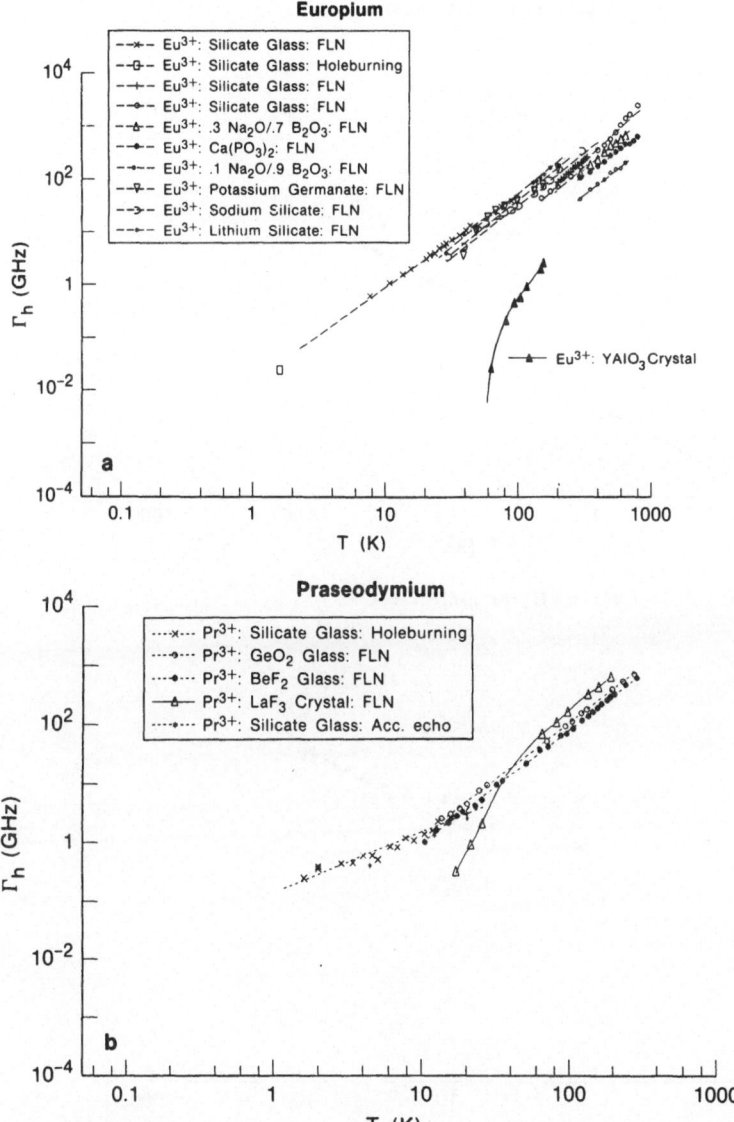

Fig. 1.3 a–d. Summary of the extant experimentally measured homogeneous optical linewidths of transitions of trivalent rare earth ions and chromium doped into inorganic glasses of varying composition as a function of temperature. Data has been obtained using a number of laser spectroscopic techniques, as denoted in the inset. The plots illustrate the striking similarity of the linewidth behavior for all the activators, especially at higher temperatures where T^2 thermal broadening is observed. Of importance is the break from this thermal dependence observed in the $^1D_2$ state of Pr^{3+}, in Nd^{3+} and in Yb^{3+}. The $^5D_0$ to $^7F_0$ states of Eu^{3+}, the $^3H_4$ to $^1D_2$ and $^3P_0$ transition in Pr^{3+}, and $^4F_{3/2}$ to ground state $^4I_{9/2}$ transition in Nd^{3+} have been investigated. For Yb^{3+} the results are for the $^2F_{5/2} \leftrightarrow ^2F_{7/2}$ fluorescence and for Cr^{3+} the results are for the $^2E R$ lines. (From [1.59])

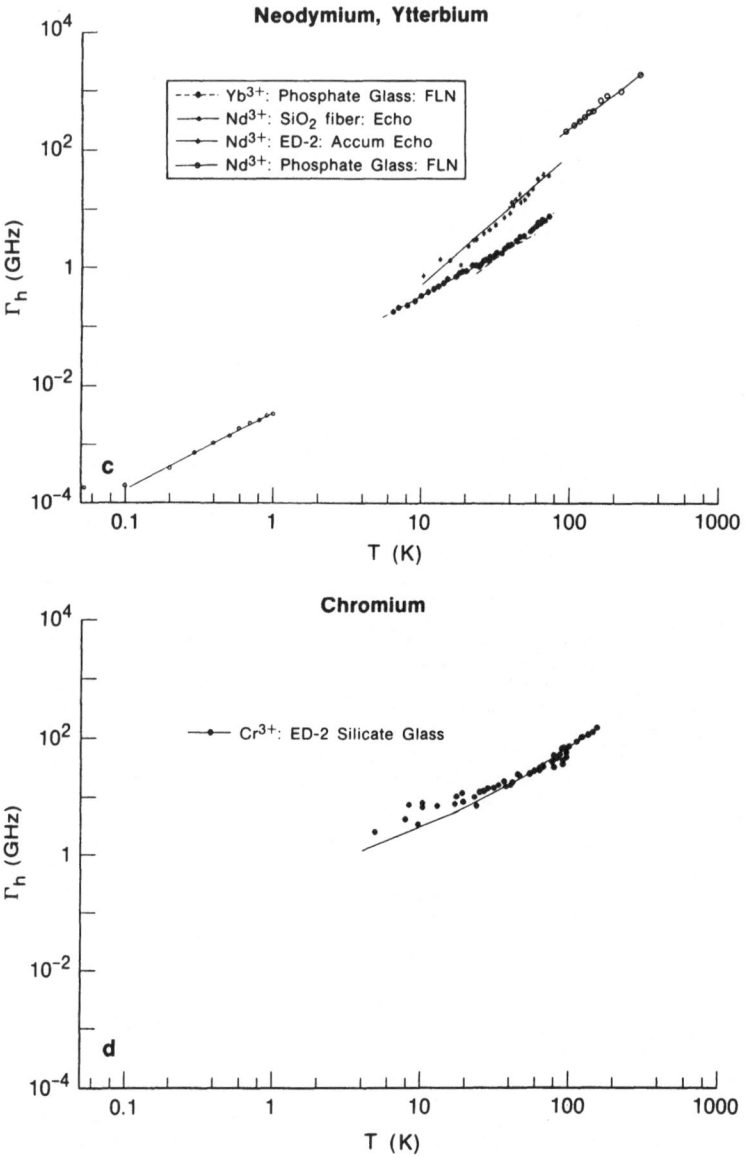

Fig. 1.3c, d. Caption see opposite side

universal T^2 behavior of the linewidth when a Raman-like two-photon scattering process is effective [1.65]. As in the case of crystals, the linewidth deviates from a T^2 dependence at lower temperatures and switches to a higher power dependence; the transition temperature occurs at approximately 1/2 the peak value of the density of states. Using the experimentally derived density of states function for the Cr-doped silicate glass, *Bergin* et al. [1.66] have been able

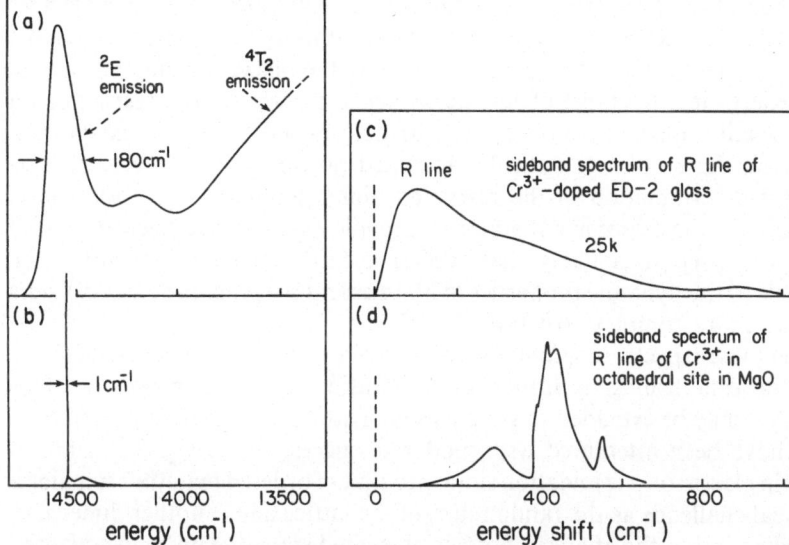

Fig. 1.4. (a) Low temperature fluorescence spectrum in the R-line region of Cr^{3+} doped into ED–2 glass, a silicate, showing an inhomogeneously broadened ^2E(R-line) emission accompanied by broad band $^4T_2$ from low field sites, i. e., sites where the crystal field is such as to make the $^4T_2$ lowest in energy and thus the emitting state. **(b)** FLN spectra when a subset of the high field or ^2E sites are excited with a moderately narrow tunable laser. Structure on the low energy side of the single R transition is evident. **(c)** Expanded view of the structure in **(b)** consiting of the vibrational assisted sideband to the R transition in ED–2. The effective phonon density of states is seen to be shifted to low energies compared to a typical sideband observed in a crystal. The latter is illustrated in **(d)**. (From [1.62])

to provide a satisfactory fit to the thermal dependence of the FLN R line in these materials. *Huber* [1.67] has also proposed a mechanism which involves the TLSs in the excited state relaxation through their elastic or vibrational fields, i.e., a phonon-mediated TLS interaction with the electronic states. His results predict that

$$\Delta V = T^{1-\mu},\tag{1.2}$$

where μ depends on the specific nature of the electronic transitions; for dipolar processes, the interaction leads to a $T^{1.3}$ dependence which is generally consistent with experimental ultra-low-temperature measurements [1.68]. Even though a number of details concerning the general applicability of the theory remain to be resolved [1.69, 70], a consistent framework has emerged which seemingly provides an acceptable answer to the linewidth problem in disordered structure.

Fractons are another type of excitation thought to exist in disordered structures where there is a restricted dimensionality [1.71]. These excitations have been invoked in the context of the linewidth problem [1.72]. In view of recent theoretical developments mentioned above, the concept of fractals does not appear to be necessary for normal organic or inorganic glasses used in the majority of these studies, but on the other hand, the relaxation and diffusion of optical energy in other types of disordered structures where some restriction on dimensionality appears may very well be determined by fractal excitations. Energy transfer studies of optically activated porous inorganic glasses have allowed a determination of the restricted dimensionality [1.73]; the effects of fractals on the transfer dynamics in organic system have been discussed elsewhere [1.74]. It is likely that similar considerations will prove to be applicable to the dynamic properties of luminescent sol preparations, polymers and other more complex structures [1.75].

Recent laser spectroscopic studies of optical energy transfer in crystals have helped in establishing a firmer understanding of these processes. This foundation may be extended to systems which are totally disordered, and such studies have been attempted with moderate success [1.76, 77]. To establish transfer in glasses on a foundation similar to that of ordered materials remains a theoretical challenge as the randomness of the lattice adds another dimension of complication to the analysis. Studies of excited state dynamics in activated transparent ceramics may prove valuable in this regard, indeed, there has been considerable recent interest in these materials [1.78].

Ceramics form a class of materials which are intermediate between ordered and disordered systems; it is possible, that is, to form nonstoichiometric mixtures in which the constituents have diverse thermal properties and one or more of them can phase separate by crystallization. Thus, with proper treatment, ceramics are made up of small crystalline areas surrounded by disordered regions. Activation of certain ceramics with common luminescent ions is possible. Depending on thermodynamic and other considerations, the activators preferentially precipitate into the crystalline or glassy phases. It is thought that this process, again under the proper preparation conditions, is very effective in confining all the activators to a specific phase [1.78]. It is this property that makes transparent ceramics interesting to study optically. Mullite ceramics, for example, are composed of Al_2O_3 in combination with SiO_2. In these materials, sapphire can be induced to crystallize into micro-crystals with dimensions ranging from a few nanometers up to micrometer sizes depending upon the melting and cooling conditions. In Cr^{3+}-activated mullites, the transition metal ion precipitates preferentially into the Al_2O_3 crystalline phase; the converse is true when rare earth dopants are introduced. The luminescence spectra of Cr-doped ceramics are generally narrower than those found in glasses but they still consist of broadened R lines (high field or ordered sites), accompanied by broad bands on the low energy side which arise from the $^4T_2$ state (low field or disordered sites). The low field emissions were believed to arise from Cr ions which find themselves in the interface between ordered and disordered phases [1.79]. In recent work, Knutson et al. [1.80] have shown that the ordered-disordered regions are not sharply defined and that a transition region exists between the Al_2O_3 and SiO_2 phases, in which the sapphire phase is quasi-ordered. These workers have also demonstrated that energy transfer processes occur between the ordered and the quasi-ordered phases [1.81]. These studies appear to establish a bridge between the dynamics which occur in totally ordered and totally random systems. Other experimental studies of ceramics have sought to characterize the existing degree of phase separation using FLN techniques and attempts have been made with various degrees of success to analyze nonradiative transfer to acceptors in the disordered phase.

1.3.2 Laser Generation and Properties of High Frequency, Narrow Band Phonons and Magnons

In the past decade, a number of laser-based techniques have allowed the generation and detection of high frequency phonons and other elementary solid-state excitations. These techniques have permitted the investigation of the relaxation and dynamical properties of these important excitations. The power of the optical method compared to ultrasonic methods lies in its ability to generate and then detect monoenergetic nonequilibrium distributions of phonons. Techniques employed for the generation of narrow band, high frequency (≥ 1 cm^{-1} or 150 GHz) phonons include excited state relaxation following selective optical pumping of an electronic state, defect-induced FIR absorption and coherent Raman scattering. The excitations can be detected with superconducting bolometers, with hot phonon sideband detectors, and more recently with a variety of coherent laser spectroscopies. These developments have been reviewed up to 1985 [1.82–84], thus, we comment only on the more recent developments.

A great deal of progress has been achieved in understanding the phonon dynamics in ruby, which has again served as the prototype system for much of this field [1.85]. It has now been established that the dynamics of the 29 cm^{-1} phonons which are created through direct process relaxation between the excited R levels of Cr^{3+} are dominated by inelastic phonon scattering. *Basun* et al. [1.86] have shown that population occurs in the presence of heat pulses. The time and power dependences of the repopulation process demonstrates that it arises from a two-phonon assisted process which couples the $\bar{E}$ and $2\bar{A}$ levels and is consistent with phonon Raman scattering from exchange-coupled Cr^{3+} pairs. Further confirmation of this hypothesis has resulted from the work of *Goosens* et al. [1.87] and from *Majetich* et al. [1.88] on the dependence of the phonon dynamics on temperature, magnetic fields, etc. Similar studies have now been conducted on alexandrite [1.88].

In other work, *van Dort* et al. [1.89] have shown that time-resolved fluorescence line narrowing (TRFLN) can be used to follow the phonon processes occurring in ruby. In these studies, a narrow subset of ions are excited into the $\bar{E}$ level of ruby by FLN techniques and are then used to detect a heat pulse. The resulting nonequilibrium luminescence of the $2\bar{A}$ state is observed to show a temporal and a spectral dependence consistent with inelastic phonon processes. Compared to Basun's work the heat pulse experiments probe exchange-coupled Cr pairs with much smaller energy shifts, i.e., weakly coupled pairs. Inelastic phonon scattering has also been studied in a ruby sample containing V^{4+} by *Happek* et al. [1.90]; in this work, a V^{4+} level is excited by FIR laser radiation and the near 29 cm^{-1} phonons generated in its relaxation are detected by R_1 luminescence from the Cr^{3+} ion.

Inelastic phonon scattering has also been identified as the principal mechanism responsible for relaxation of phonons generated in rare-earth-doped crystalline systems [1.91]. By generating phonons in one such system, $SrF_2 : Er^{3+}$, *Wietfeldt* et al. [1.92] obtained clear evidence of resonant phonon-assisted energy transfer between dimers of nonresonant Er^{3+} ions.

Following earlier work by *Hu* et al. [1.93] on 29 cm^{-1} phonons in ruby, *van Miltenburg* et al. [1.94] have demonstrated that stimulated emission can also be achieved at 1.5 cm^{-1}. In these experiments, the population inversion was obtained by pulsed optical excitation of one of the Zeeman sublevels of the $\bar{E}$ state of Cr^{3+} and stimulation was observed predominantly along the optical excitation axis. Losses, in their case, were attributed to an anharmonic three-phonon process in which two 1.5 cm^{-1} phonons interact to produce a third 3 cm^{-1} excitation, thus removing them from the stimulated frequency.

More recently, various laser spectroscopic techniques which employ optical coherence have been used to detect and to investigate the dynamics of high frequency phonons. These studies hinge on the fact that a coherently prepared emsemble of ions will undergo additional dephasing in the presence of phonons; the increased relaxation rate is proportional to the phonon occupation number and the interaction is enhanced whenever a resonance between the electronic levels and the phonons occurs. These techniques have been developed by *Meltzer* and *Macfarlane* [1.95] and have been termed phonon-induced coherence loss or PICOLO spectroscopy. PICOLO is an extremely sensitive detector of monoenergetic, nonequilibrium phonons that are resonant with some electronic feature of the active system. Work demonstrating these capabilities has been carried out on the $^1D_2$ level of $LaF_3 : Pr^{3+}$ [1.95] and on the ^2E levels of Cr^{3+} in alexandrite [1.96].

It has further been demonstrated that coherent detection techniques also can be applied to the low frequency or ultrasonic portion of the phonon spectrum. In these cases, the ultrasonic phonon waves may be created using transducers and, hence, are phase coherent themselves. When the waves interact with the coherent ensemble, instead of producing stochastic dephasing they produce a phase modulation of the coherent optical signal. An example of the results of *Boye* et al. [1.97] is shown in Fig. 1.5 where 4.45 MHz ultrasonic waves are seen to modulate the photon echoes observed in the ^2E state of alexandrite. The modulation index is found to be dependent on the relative phase between the ultrasound and the pulses which prepare the electronic $\frac{\pi}{2}$, π system. The importance of these results lies in the fact that when the echo preparation pulses are not collinear, the situation usually chosen to avoid scattered laser light, the ultrasonic wave leads to enhancement of the photon

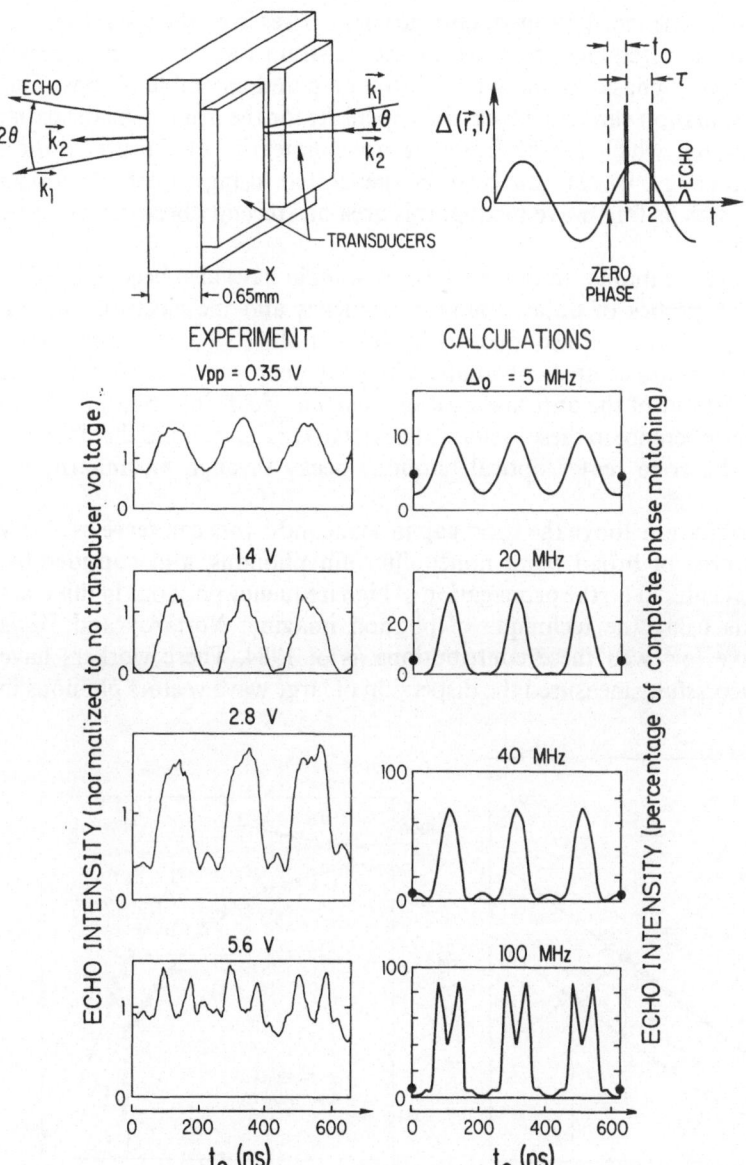

Fig. 1.5. Modulation and enhancement of photon echo intensities in alexandrite induced by ultrasonic (US) waves. The introduction of the waves produces a spatially dependent phase accumulation which offsets the phase mismatch inherent in noncollinear photon echo experimental configurations. The figure illustrates the geometry of the experiments. The transducers consist of 0.2 mm piezoelectric $BaTio_3$ attached to the sample; k_1, k_2 denote the directions of the echo preparation pulses. The graph at the top right shows the echo timing sequence in relation to the phase of the US wave; the wave produces a transition frequency shift given by Δ. Lower traces show the comparison of experimentally observed and theoretically calculated modulations of the echo intensity as a function of applied voltage on the transducers at a frequency of 4.4 MHz. See [1.97]

echo intensity. The mechanism for enhancement arises from the modulation of the ion transition frequencies by the ultrasonic strain wave which produces a spatially varying phase accumulation in the prepared ensemble of ions. This phase accumulation can be such as to compensate for the usually small spatial phase mismatch which results from the noncollinearity of the preparation pulses. Ultrasonic waves can also be generated using coherent optical techniques such as four-wave mixing; this area has recently been reviewed by *Fayer* [1.98].

Coherence techniques in the picosecond domain have also been applied to study the properties of optical phonon branches and the electron phonon generation process in semiconductors by *Bron* et al. [1.99, 100]. Zone center optical phonons are created coherently using coherent Raman scattering. The time development of the anti-Stokes Raman signal yields information on the decay of the coherence in these systems. These workers have found that for GaP and ZnSe the zone center optical phonons decay through an anharmonic process.

Laser excitation above the band gap in semiconductors can serve as a very effective source of broad band nonequilibrium phonons, and considerable activity has centered on the propagation of high frequency phonons in this class of materials using the technique of phonon imaging. *Northrop* and *Wolfe* [1.101] have reviewed these contributions as of 1984. These workers have recently successfully measured the dispersion of large wave vectors phonons in

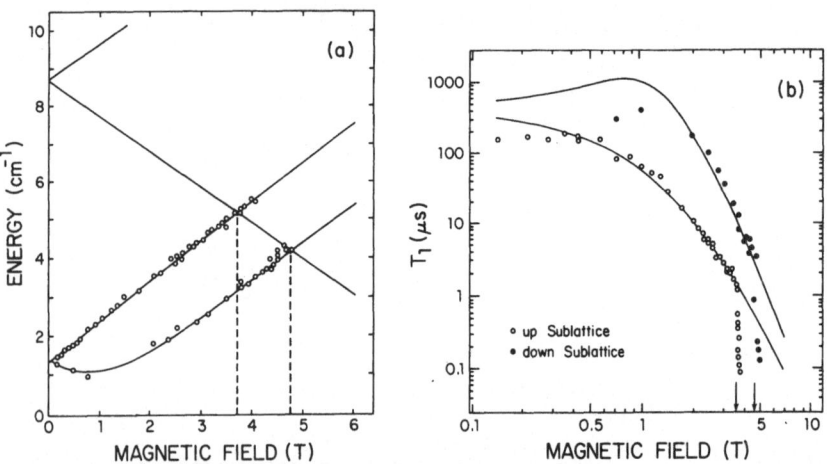

Fig. 1.6. (a) Magnetic field dependence of the splitting of the $^4F_{9/2}$ (1) Kramers doublet of Er^{3+} doped into antiferromagnetic (AF) MnF_2 with H parallel to the c axis of this tetragonal material. Also shown is the lifting of the AF sublattice magnon degeneracy by the applied field. The Er^{3+} splittings coincide with the lower magnon branch energy at 3.6 and 4.7 T. **(b)** T_1 relaxation time of $^4F_{9/2}$ (1) Er^{3+} as measured by the time evolution of the luminescence from the lowest Zeeman doublet shown as a function of field. A precipitous decrease in the relaxation rate is seen at the crossover point; this is evidence of the generation of magnons through a direct process [1.106]

InSb by using bolometers with different frequency cutoffs [1.102]. The process of cooling of hot spots on various substrates has been studied utilizing hot luminescence phonon detection by *Akimov* and his co-workers [1.103, 104].

The generation and detection in ordered materials of elementary excitations other than phonons have also been demonstrated recently. It has been established for some time that magnons in magnetic materials interact with electronic states in a manner which is totally analogous to phonons [1.105]. *Jongerden* et al. have reported the generation and detection of narrow band magnons in the vicinity of the anisotropy gap of MnF_2 doped with Er^{3+}. In their studies, the lifetimes of the transition between the lowest Kramers doublets of the $^4F_{9/2}$ state and the $^4I_{15/2}$ ground state of Er^{3+} are shown to undergo an abrupt decrease when the Zeeman splitting of the excited state doublet comes in coincidence with the magnon energy of one of the Mn^{2+} sublattices. Their results are illustrated in Fig. 1.6 [1.106]. More recently, it has been shown by *Yen* et al. [1.107] that magnons in MnF_2 can be generated throughout the Brilluoin zone directly by heat pulses or by optical pumping. These workers have observed the ballistic propagation of phonons and of magnons in this material at low temperatures using heat pulses generated by a

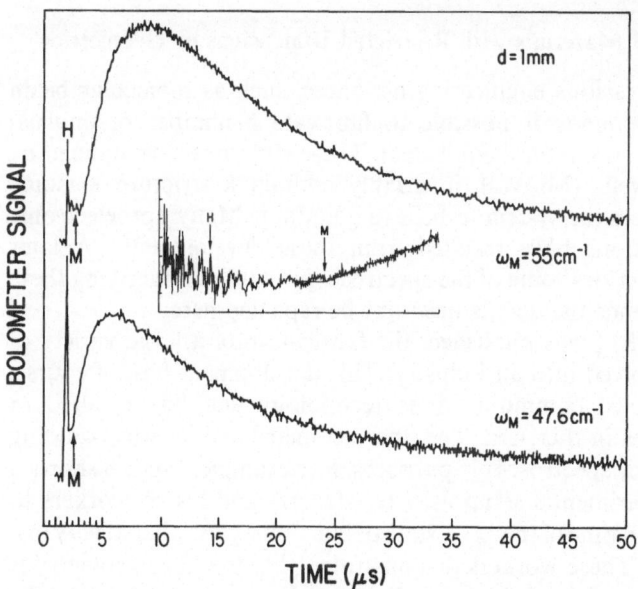

Fig. 1.7. Bolometer signals observed when various points of the two-magnon absorption of MnF_2 are pumped with a high power FIR molecular laser, at 110 and 95.2 cm^{-1} respectively. Experiments were conducted at 1.9 K, the transition temperature of the superconducting Al O_2 doped bolometer; the excitation region was approximately 1 mm from the detector. Inset shows the abrupt onset of the signal, as measured by a multichannel detector, which is characteristic of ballistic transport. The shape of the heat pulse yields information on the relaxation of the propagating magnons and on other interactions. ([1.107]; traces courtesy of W. Grill)

resistive heater and superconducting bolometer detection. By using optical excitation, magnons can be selectively created through the two-magnon FIR absorption [1.108] or through the excitation of magnon sidebands [1.109]. In the former case, a $k - k$ pair of magnons is created with no accompanying excitations other than those resulting from their relaxation. These magnons have identical energies equal to half the laser pumping frequency and reside in opposite sublattices. Figure 1.7 shows the bolometer signal observed under these conditions. The onset of the heat signal is seen to be abrupt and is characteristic of ballistic transport [1.110]. The shape of the heat pulse is determined in part by geometry but also contains information on the relaxation processes and other interactions which affect the magnon. The heat signals observed when the sidebands are pumped are generally more complex, as spin impurities in the form of excitons are created simultaneously with the propagating magnons; additionally, exciton decay products soften the abruptness of the ballistic onset. These problems are partially alleviated by destroying excitons through a biexciton annihilation process using high laser pump powers [1.111]. Though these results are preliminary, it is not difficult to foresee that with the aid of laser spectroscopic methods we will be in a position to study the dynamics of magnetic excitations in detail.

1.3.3 Spectroscopy of Materials with Restricted Dimensions or Geometries

In the past decade, various engineering advances, such as molecular beam epitaxy (MBE), have made it possible to fabricate a number of unusual structures which have no natural equivalent. These structures, exemplified by multiple quantum wells (MQWs), invariably contain a stricture in some dimension which forces the system to behave quantally. Many optoelectronic devices and applications have resulted from these developments. A later chapter (Chap. 6) describes some of the spectroscopic studies underlying these developments and hence the details need not be repeated here.

Similarly, since [1.1] was published, the fabrication of a large variety of optical fibers has evolved into an industry. This development has, of course, been driven by optical communications technology and has resulted in revolutionary changes in this area. The fiber geometry and its wave-guiding properties are ideal for spectroscopic purposes. For example, Fig. 1.8 shows a schematic of the experimental setup used by Hegarty and his co-workers to measure the relaxation time of the $^4F_{3/2}$ state of Nd^{3+} in a glass fiber at very low temperatures [1.60]. These workers demonstrated that the fiber geometry is ideal for the purposes of conducting photon echo measurements as the interaction region for the $\frac{\pi}{2}$ and π pulses can be made arbitrarily long. The pulses are introduced without the offset angle necessary in bulk samples [1.97], i.e., they are completely phase matched, and the echo signals are large and can be temporally isolated easily. In addition, these workers found that the active region of the sample could be maintained conveniently at the temperature of the cryostat simply by winding many turns of the fiber on the cold reservoir; this

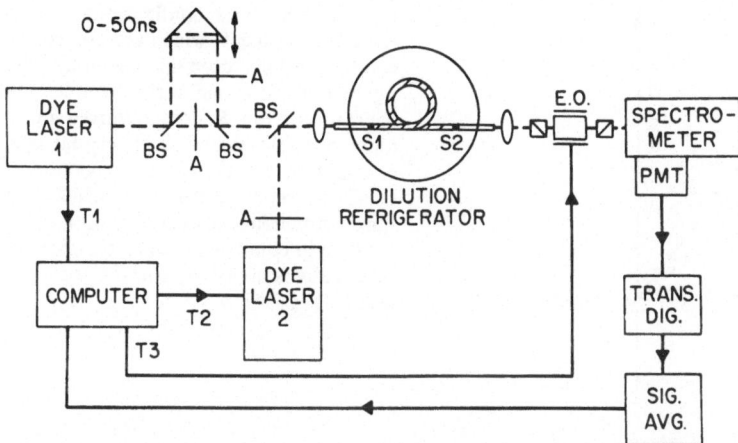

Fig. 1.8. A versatile setup to observe photon echoes at very low temperatures. The fiber is wound on the millidegree thermal bath to compensate for the notoriously poor thermal conductivity of glasses at these temperatures. The fiber configuration also allows the echo preparation pulses to be collinear while trapping the echoes in the fiber for detection outside the low temperature region. The independently driven dye lasers permit a wide range of pulse sequences. In the figure BS is a beam splitter, A an attenuator, EO an electro-optical shutter and T1–T3 are synchronizing triggers. (After [1.112])

alleviates the consequences of the poor thermal conductivity of disordered materials.

The prospect of devising on-line light amplifiers for communication purposes has led to a number of lasers designed in the fiber configuration. The majority of the activating ions used in glass fibers have been rare earths which luminesce in the near IR, i.e., ions such as Ho^{3+} or Er^{3+}. Tunable devices in this configuration have also been reported [1.113]. In these cases, single crystal fibers doped with transition elements have been employed.

The laser heated pedestal growth (LHPG) method developed by *Feigelson* [1.114] and co-workers has proven to be a very effective and versatile way of pulling single crystal, mostly multimode, fibers. *Liu* et al. [1.115] and *Yen* [1.116] have demonstrated a number of additional spectroscopic advantages accruing from the geometrical configuration. Figure 1.9 shows, for example, the tensile-stress-induced shifts observed in the R lines of Cr^{3+} in a crystalline sapphire fiber. Tensile and torsional stresses are generally hard to quantify in bulk samples, whereas the cylindrical fiber configuration lends itself nicely to these purposes.

In conducting spectroscopic studies in activated fibers, the excitation laser light is introduced parallel to the fiber axis and, hence, it will remain trapped in the fiber. Emission resulting from luminescent or other processes, such as Raman scattering, excited by the laser is emitted isotropically. It is possible to observe the light either axially, i.e., at the end of the fiber, or perpendicularly. Each configuration presents us with experimental advantages. For example, if

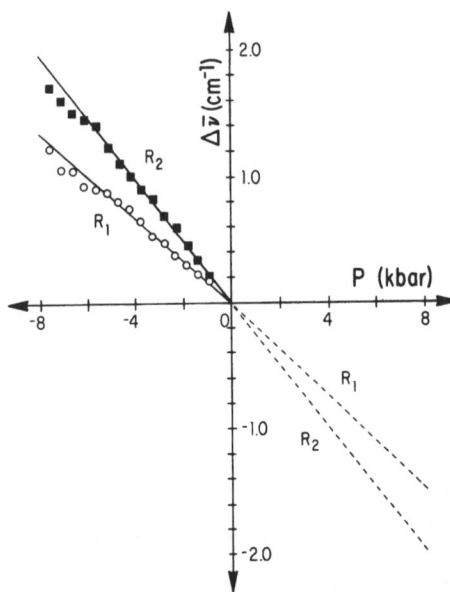

Fig. 1.9. The fiber configuration allows certain experiments which are not easy to conduct in bulk crystalline samples. The figure illustrates one such example. The behavior of the R lines observed in an unintentionally doped single crystal fiber of sapphire are shown as a function of tensile or decompressive stress achieved by simply stretching the fiber. The shifts are to the blue in contrast to red shifts observed when ruby is uniaxially compressed along its c axis [1.115]

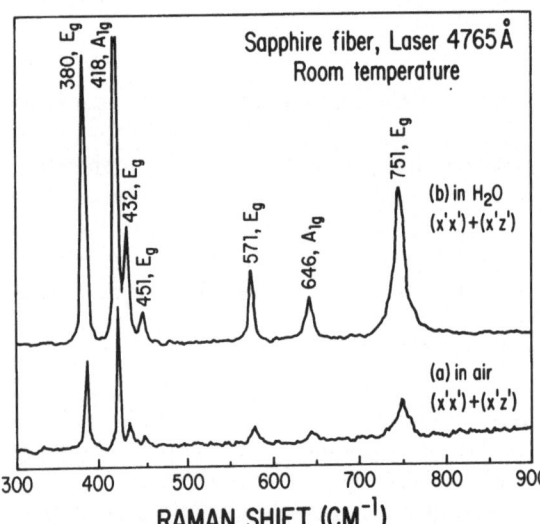

Fig. 1.10. Raman spectra of a single crystal fiber of sapphire at room temperature observed in air and when immersed in water. The index matching fluid allows isotropically emitted light to escape from the fiber while the excitation laser light remains confined because of directionality. Enhancements of nearly an order of magnitude have been observed with this experimental configuration [1.116]

the emission is analyzed perpendicular to the fiber axis, the fiber can be positioned conveniently to fill the entrance optics of an appropriate spectrograph. The axial configuration is ideal for Fabry-Perot interferometers. Figure 1.10 shows the Raman spectra of a small single crystal fiber of sapphire of 100 µm diameter observed perpendicular to the fiber axis. The figure also illustrates that by submerging the fiber in a liquid with a higher index of refraction than air the Raman signal is enhanced without appreciably increasing the excitation laser scatter. This phenomenon occurs simply because the higher external index allows more isotropic radiation to escape while confinement of the laser light is maintained because of directionality of the radiation [1.116].

Finally, it has been shown by *Jaffee* and *Yen* [1.117] that it is possible to conduct a new type of site-selective spectroscopy in very dilute glass fibers. This new method, termed dilution narrowed fluorescence spectroscopy (DNFS), has the potential of isolating a single ion in the volume excited by the laser. DNF occurs if the concentration of the active centers is so low or the physical volume of excitation is so small as to allow only a random sampling of the inhomogeneous distribution of sites. It should be clear, then, that a dilutely doped glass fiber provides the ideal medium on which to conduct DNFS. This is

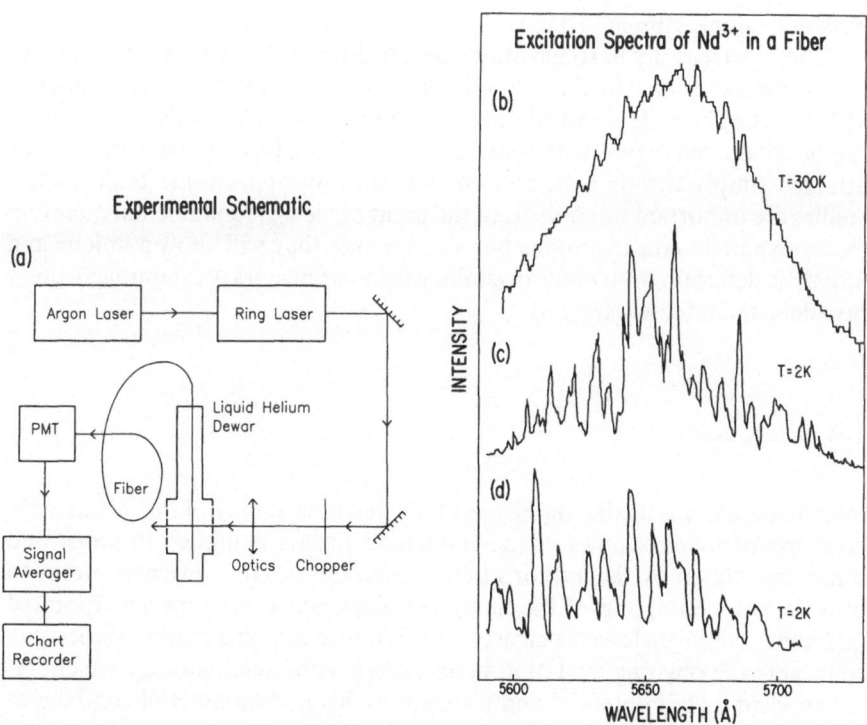

Fig. 1.11 a–d. Dilution narrowed spectra (DNS) of a Nd^{3+}-activated single mode silicate fiber. (a) The experimental setup for these observations, where a high resolution ring laser is used to excite a small volume of the fiber perpendicular to the fiber axis. The fluorescence from the $^4F_{3/2}$ state is monitored at the end of the fiber outside the temperature chamber. (b) Nonresonant excitation spectra of the Nd^{3+} at room temperature where individual sites have large thermal widths which smooth out the absorption of the $^4G_{5/2}$ state. (c) DNFS of Nd^{3+}-activated glass fiber at low temperature. The toothed spectrum represents a random sampling of the large distribution of glass sites. (d) Since DNFS samples the sites randomly, a shift in the excitation region produces a radically different excitation spectrum. These techniques open the prospect of isolating a single ion in these materials [1.117]

because the inhomogeneous distribution in glasses comprises a large number of possible crystal field configurations given approximately by the ratio

$$N(\text{sites}) = \Delta v_{\text{inh}}/\Delta v_{\text{hom}} . \tag{1.3}$$

At the same time, the excitation volume can be made very small by using single mode fibers and by employing perpendicular excitation with tight focusing of the laser. A considerable fraction of the luminescence ensuing from such excitation is trapped in the fiber and detected at the end of the fiber. The experimental configuration and preliminary results of these studies are shown in Fig. 1.11, where it is clear that DNFS has been achieved in the Nd^{3+}-doped quartz fiber. The photon count at the wings of the distribution indicates that a finite number, numbering in the tens, contribute to the recorded excitation [1.116, 117]. Similar experiments have been carried out in an extremely dilute bulk crystalline sample [1.118].

The spectroscopy of single atoms or ions localized in various magnetic or laser traps and cooled by the laser radiation are of considerable current interest [1.119]. The above demonstrations of DNFS indicate that similar experiments can be conducted in ions with specific environments. Low temperatures can be attained simply through thermal contact with an appropriate bath. DNFS studies are important not only from the point of view of fundamental quantum electrodynamics considerations but also because they will allow a unique and atomistic determination of the crystalline field parameters of inhomogeneously broadened condensed systems.

1.4 Conclusions

In concluding, it is again emphasized that this brief survey of the advances in laser spectroscopic studies of the condensed phases is limited in scope and somewhat biased by the author's preferences. Indeed, this review is meant to provide but a sampling of the many activities which are now incorporated under the heading of laser spectroscopy. It is hoped that this chapter succeeds in conveying the ongoing evolution of an experimental methodology which has not reached its full potential and which more likely than not will continue to provide us with better means with which to investigate and characterize optical materials.

It is gratifying to have been allowed the privilege of witnessing a discipline evolve from its birth, now nearly a decade and half ago. The extension of the techniques advocated in these volumes to embrace almost all facets of the study of solid-state optical properties is proof sufficient of the superiority of laser-based methods, and our expectation is that the expansion will continue as laser sources are commercialized further. For the near term, it is likely that with the aid of FELs the spectral range of laser spectroscopy studies will expand well into the millimeter range; new hybrid methods encompassing microwave techniques will be a natural by-product of this contact. Similarly, new temporal regimes have been made available to investigation, which signals prospects of a better understanding of fast nonradiative processes in liquid as well as in solids. Single ion spectroscopy, as evidenced by DNFS and by recent photo-calorimetric studies [1.120], promises to provide a means to an in-depth understanding of the optically excited state and its interaction with its environment.

Acknowledgements. The continuing support of the Division of Materials Research, National Science Foundation, and the Office of Basic Energy Studies, U.S. Department of Energy, is gratefully acknowledged. Additional support for these studies has been provided by the University of Georgia through the UGA Research Foundation.

References

1.1 W.M. Yen, P.M. Selzer (eds.): *Laser Spectroscopy of Solids*, Topics Appl. Phys., Vol. 49, 2nd ed. (Springer, Berlin, Heidelberg 1986)

1.2 J.M. Hollas: *High Resolution Spectroscopy* (Butterworths, London 1982); W. Demtröder: *Laser Spectroscopy*, Springer Ser. Chem. Phys., Vol. 5 (Springer, Berlin, Heidelberg 1981)

1.3 A.L. Schawlow, C.H. Townes: Phys. Rev. **112**, 1940 (1958)

1.4 T.H. Maiman, R.H. Hoskins, I.J. D'Haenens, C.K. Asawa, V. Evtuhov: Phys. Rev. **123**, 1151 (1961)

1.5 P.P. Sorokin, J.R. Lankard: IBM J. Res. Dev. **10**, 162 (1966)

1.6 See, e.g., P.W. Smith: Proc. IEEE **58**, 1342 (1970)

1.7 C.V. Raman, K.S. Krishnan: Nature **122**, 278 (1928); D.A. Long: *Raman Spectroscopy* (McGraw-Hill, New York 1977)

1.8 C.S. Wang: In *Quantum Electronics: A Treatise*, Vol. 1, ed. by H. Rabin, C.L. Tang (Academic, New York 1975) Chap. 7

1.9 Y.R. Shen: *Principles of Non-Linear Optics* (Wiley, New York 1984); M.D. Levenson: *Introduction to Non-Linear Optics* (Academic, New York 1982)

1.10 W. Kaiser, C.G.B. Garrett: Phys. Rev. Lett. **7**, 229 (1961)

1.11 N.A. Kurnit, I. Abella, S.R. Hartmann: Phys. Rev. Lett. **13**, 567 (1964)

1.12 S.L. McCall, E.L. Hahn: Phys. Rev. Lett. **18**, 908 (1967)

1.13 A. Szabo: Phys. Rev. Lett. **25**, 924 (1970)

1.14 D.E. McCumber, M.D. Sturge: J. Appl. Phys. **34**, 1682 (1963)

1.15 F.P. Schäfer (ed.): *Dye Lasers*, Topics Appl. Phys., Vol. 1, 2nd ed. (Springer, Berlin, Heidelberg 1977)

1.16 G.F. Imbusch, R. Kopelman: In [Ref. 1.1, Chap. 1]

1.17 M.J. Weber: In [Ref. 1.1, Chap. 6]

1.18 R.M. Macfarlane: In *Lasers, Spectroscopy and New Ideas*, ed. by W.M. Yen, M.D. Levenson, Springer Ser. Opt. Sci., Vol. 54 (Springer, Berlin, Heidelberg 1987) p. 205

1.19 J.A. Valdmanis, R.L. Fork, J.P. Gordon: Opt. Lett. **10**, 131 (1985); R.L. Fork: Opt. Lett. **11**, 629 (1986)

1.20 C.H. Brito Cruz, R.L. Fork, W.H. Knox, C.V. Shank: Chem. Phys. Lett. **132**, 341 (1986)

1.21 T.A. DeTemple: *Handbook of Molecular Lasers*, ed. by P.K. Cheo (Marcel Dekker, New York 1987) Chap. 7

1.22 Report of the FEL Subcommittee, Solid State Science Committee, National Research Council; C.K.N. Patel, chairman (National Academy of Sciences Press, Washington, DC 1982)

26   *W. M. Yen*

1.23   L.F. Mollenauer: *Laser Handbook*, Vol. 4, ed. by M.L. Stitch, M. Bass (North-Holland, Amsterdam 1985) Chap. 2
1.24   D.J. Ehrlich, P.F. Moulton, R.M. Osgood: Opt. Lett. **4**, 184 (1979)
1.25   See, e.g., P. Hammerling, A.B. Budgor, A. Pinto (eds.): *Tunable Solid State Lasers*, Springer Ser. Opt. Sci., Vol. 47 (Springer, Berlin, Heidelberg 1985)
1.26   Y. Suematsu: Phys. Today **38**, 5–32 (1985)
1.27   D.A.G. Deacon, L.R. Elias, J.M.J. Madey, G.J. Ramien, H.A. Schwettman, T.I. Smith: Phys. Rev. Lett. **38**, 892 (1977)
1.28   F.A. Hopf, P. Meystre, M.O. Scully, W.H. Louisell: Opt. Commun. **18**, 4 (1976)
1.29   W.B. Colson, S.K. Ride: *Physics of Quantum Electronics*, Vol. 7, ed. by S.F. Jacobs, H.S. Piloff, M. Sargent, M.O. Scully, R. Spitzer (Addison-Wesley, Reading, MA 1979)
1.30   C.A. Brau: Laser Focus **17**, 1–48 (1981); ibid. **23**, 2–40 (1987)
1.31   L.R. Elias: Phys. Rev. Lett. **16**, 977 (1979)
1.32   J.M.J. Madey: J. Appl. Phys. **42**, 1906 (1971)
1.33   L.R. Elias, J. Hu, G.J. Ramien: Nucl. Instrum. Methods Phys. Res. A **237**, 203 (1985)
1.34   W.M. Yen, L.R. Elias, V. Jaccarino: J. de Phys. **46**, C7–413 (1985)
1.35   R.M. White: Appl. Phys. Lett. **46**, 194 (1985)
1.36   W.M. Yen, W.M. Dennis: J. Magn. Soc. Jpn. **11**, S1–413 (1987)
1.37   J. Töpper: Diplomarbeit in Physik, Physikalischen Institut, J.W. Goethe-Universität, Frankfurt (1988) unpublished
1.38   K. Halbach: Nucl. Instrum. Methods Phys. Res. **169**, 1 (1980)
1.39   G. Margaritondo: *Introduction to Synchrotron Radiation* (Oxford University Press, New York 1988)
1.40   J.H. Hall as cited by A. Mooradian: Phys. Today **38**, 5–42 (1985)
1.41   P.M. Selzer, D.L. Huber, B.B. Barnett, W.M. Yen: Phys. Rev. B **17**, 4979 (1977)
1.42   H.M. Gibbs, S. Chu, S.L. McCall, A. Passner: *Coherence and Energy Transfer in Glasses*, ed. by P.A. Fleury, B. Golding, NATO Conf. Series VI-9 (Plenum, New York 1982) p. 373
1.43   G.P. Morgan, D.L. Huber, W.M. Yen: J. Lumin. **35**, 277 (1986)
1.44   M.M. Broer, D.L. Huber, W.M. Yen, W.K. Zwiecker: Phys. Rev. B **29**, 2382 (1984)
1.45   T. Holstein, S.K. Lyo, R. Orbach: In [Ref. 1.1, Chap. 2]; D.L. Huber: In [Ref. 1.1, Chap. 3]
1.46   W.M. Yen: *Spectroscopy of Solids Containing Rare Earth Ions*, ed. by A.A. Kaplyanskii, R.M. Macfarlane, Modern Problems in Condensed Matter Sciences, Vol. 21 (North-Holland, Amsterdam 1987) Chap. 4
1.47   R. Chien, J.M. Berg, D.S. McClure: J. Chem. Phys. **84**, 4168 (1986)
1.48   J.M. Berg, R. Chien, D.S. McClure: J. Chem. Phys. **87**, 7 (1987)
1.49   N. Bloembergen: J. Lumin. **31/32**, 23 (1984)
1.50   S.K. Gayen, D.S. Hamilton: Phys. Rev. B **28**, 3706 (1983)
1.51   B.R. Judd: Phys. Rev. **127**, 750 (1962); G.S. Ofelt: J. Chem. Phys. **37**, 511 (1962)
1.52   H. Mahr: *Quantum Electronics–Nonlinear Optics*, Vol. 1, ed. by H. Rabin, C.L. Tang (Academic, New York 1975) p. 287
1.53   W.B. Fowler: *Physics of Color Centers* (Academic, New York 1967)
1.54   B.A. Wilson: Phys. Rev. B **23**, 3102 (RC) (1981)
1.55   J. Hegarty, M.D. Sturge: J. Opt. Soc. Am. B II-2, 1143 (1985)
1.56   H.M. Gibbs: *Optical Bistability: Controlling Light with Light* (Academic, New York 1985)
1.57   M.J. Weber (ed.): *Optical Linewidths in Glasses*. J. Lumin. **36**, 179 (1987)
1.58   W.A. Phillips (ed.): *Amorphous Solids: Low Temperature Properties*, Topics Curr. Phys., Vol. 24 (Springer, Berlin, Heidelberg 1981)
1.59   R.M. Macfarlane, R.M. Shelby: J. Lumin. **36**, 179 (1987)
1.60   J. Hegarty, M.M. Broer, B. Golding, J.R. Simpson, J.B. MacChesney: Phys. Rev. Lett. **51**, 2033 (1983)
1.61   R.T. Brundage, W.M. Yen: Phys. Rev. B **33**, 4436 (1986)

1.62 F.J. Bergin, J.F. Donegan, T.J. Glynn, G.F. Imbusch: J. Lumin. **34**, 307 (1986)
1.63 D.W. Hall, S.A. Brawer, M.J. Weber: Phys. Rev. B **25**, 2828 (1982)
1.64 D.L. Huber: J. Lumin. **36**, 327 (1987)
1.65 W.M. Yen, W.C. Scott, A.L. Schawlow: Phys. Rev. **136**, A271 (1964)
1.66 F.J. Bergin, J.F. Donegan, T.J. Glynn, G.F. Imbusch: J. Lumin. **36**, 231 (1987)
1.67 D.L. Huber: J. Lumin. **36**, 307 (1987)
1.68 D.L. Huber, M.M. Broer, B. Golding: Phys. Rev. B **33**, 7789 (1986)
1.69 M. Berg, C.A. Walsh, L.R. Narasimhan, M.D. Fayer: J. Lumin. **38**, 9 (1987)
1.70 R. van den Berg, S. Volker: J. Lumin. **38**, 25 (1987)
1.71 S. Alexander, R. Orbach: J. de Phys. **43**, L625 (1982)
1.72 S.K. Lyo: Phys. Rev. Lett. **48**, 688 (1982);
      G.S. Dixon, R.C. Powell, Xu Gang: Phys. Rev. B **33**, 2713 (1986)
1.73 U. Even, K. Rademann, J. Jotner, N. Manor, R. Reisfeld: Phys. Rev. Lett. **52**, 2164 (1984)
1.74 A. Blumen, J. Klafter, G. Zumofen: *Optical Spectroscopy of Glasses*, ed. by I. Zchokke (D. Reidel, Dordrecht 1986) p. 199
1.75 J.M. Drake, P. Levitz, S.K. Sinha, J. Klafter: Chem. Phys., in press
1.76 R.T. Brundage, W.M. Yen: Phys. Rev. B **34**, 8810 (1986)
1.77 T.T. Basiev, V.A. Malyshev, A.K. Przhevuskii: *Spectroscopy of Solids Containing Rare Earth Ions*, ed. by A.A. Kaplyanskii, R.M. Macfarlane, Modern Problems in Condensed Matter Sciences, Vol. 21 (North-Holland, Amsterdam 1987) Chap. 6
1.78 G.H. Beall, D.A. Duke: *Glass – Science and Technology*, Vol. 1, ed. by D.R. Uhlmann, N.J. Kreidl (Academic, New York 1983)
1.79 L.J. Andrews, B.C. McCollum, S. Stone, D.E. Gunther, G.J. Murphy, A. Lempicki: *Development of Materials for Luminescent Solar Panels*, GTE Laboratories, Waltham, MA, Final Report DOE/ER-04996-4 (1983) unpublished
1.80 R. Knutson, H.M. Liu, W.M. Yen, T.V. Morgan: To be published
1.81 H.M. Liu, R. Knutson, W.M. Yen: Submitted to Opt. Lett.
1.82 K.F. Renk: In *Nonequilibrium Phonon Dynamics*, NATO ASI B **124**, ed. by W.E. Bron (Plenum, New York 1985) p. 59
1.83 W.E. Bron: In *Nonequilibrium Phonons in Nonmetallic Crystals*, ed. by W. Eisenmenger, A.A. Kaplyanskii, Modern Problems in Condensed Matter Sciences, Vol. 16 (North-Holland, Amsterdam 1986) p. 227
1.84 K.F. Renk: In *Nonequilibrium Phonons in Nonmetallic Crystals*, ed. by W. Eisenmenger, A.A. Kaplyanskii, Modern Problems in Condensed Matter Sciences, Vol. 16 (North-Holland, Amsterdam 1988) p. 277
1.85 A.A. Kaplyanskii, S.A. Basun: In *Nonequilibrium Phonons in Nonmetallic Crystals*, ed. by W. Eisenmenger, A.A. Kaplyanskii, Modern Problems in Condensed Matter Sciences, Vol. 16 (North-Holland, Amsterdam 1986) p. 373
1.86 S.A. Basun, A.A. Kaplyanskii, S.P. Feofilov: Sov. Phys. – Solid State **28**, 2038 (1987)
1.87 R.J.G. Goosens, J.I. Dijkhuis, H.W. de Wijn: Phys. Rev. B **32**, 7065 (1985)
1.88 S. Majetich, R.S. Meltzer, J.E. Rives: Phys. Rev., to be published (1988)
1.89 M.J. van Dort, J.I. Dijkhuis, H.W. de Wijn: J. Lumin. **38**, 217 (1987)
1.90 U. Happek, T. Holstein, K.F. Renk: Phys. Rev. Lett. **54**, 2091 (1985)
1.91 S.S. Yom, R.S. Meltzer, J.E. Rives: Phys. Rev. B **36**, 6664 (1987)
1.92 S.R. Wietfeldt, D.S. Moore, B.M. Tissue, J.C. Wright: Phys. Rev. B **33**, 5788 (1986)
1.93 P. Hu, V. Narayanamurti, M.A. Chin: Phys. Rev. Lett. **46**, 192 (1981)
1.94 J.G.M. van Miltenburg, G.J. Jongerden, J.I. Dijkhuis, H.W. de Wijn: In *Phonon Scattering in Condensed Matter*, ed. by W. Eisenmenger, K. Lassmann, S. Döttinger, Springer Ser. Solid-State Phys., Vol. 51 (Springer, Berlin, Heidelberg 1984) p. 130
1.95 R.S. Meltzer, R.M. Macfarlane: Phys. Rev. B **32**, 1248 (1985)
1.96 R.S. Meltzer: J. Lumin. **38**, 211 (1987)
1.97 D. Boye, W. Grill, J.E. Rives, R.S. Meltzer: Bull. Am. Phys. Soc. II-**33**, 800 (1988); Phys. Rev. Lett., to be published (1988)
1.98 M.D. Fayer: IEEE J. QE-**22**, 1437 (1986)
1.99 W.E. Bron, J. Kuhl, B.K. Rhee: Phys. Rev. B **34**, 6961 (1986)

1.100 B.K. Rhee, W.E. Bron: Phys. Rev. B **34**, 7107 (1986)
1.101 G.A. Northrop, J.P. Wolfe: *Nonequilibrium Phonon Dynamics*, NATO ASI B **124**, ed. by W.E. Bron (Plenum, New York 1985) p. 165
1.102 S.E. Hebboul, J.P. Wolfe: Phys. Rev. B **34**, 3948 (1986)
1.103 A.V. Akimov, A.A. Kaplyanskii, M.A. Pogarski, V.K. Tikhomirov: JETP Lett. **43**, 333 (1986)
1.104 A.V. Akimov, A.A. Kaplyanskii, E.S. Moskalenko: Sov. Phys. – Solid State **29**, 288 (1987)
1.105 W.M. Yen, G.F. Imbusch, D.L. Huber: *Optical Properties of Ions in Solids*, ed. by H.M. Crosswhite, H.W. Moos (Interscience, New York 1967) p. 301
1.106 G.J. Jongerden, A.J. Kil, J.I. Dijkhuis, A.F.M. Arts, H.W. de Wijn: J. de Phys. **46**, C7–241 (1985)
1.107 W.M. Yen, L.D. Rotter, W. Grill: J. Appl. Phys., **64**, 5470 (1988)
1.108 S.J. Allen, Jr., R. Louden, P.L. Richards: Phys. Rev. Lett. **16**, 463 (1966)
1.109 D.D. Sell, R.L. Greene, R.M. White: Phys. Rev. **158**, 489 (1987)
1.110 R.J. von Gutfeld, A.H. Nethercot: Phys. Rev. Lett. **12**, 641 (1964)
1.111 B.A. Wilson, J. Hegarty, W.M. Yen: Phys. Rev. Lett. **41**, 268 (1978)
1.112 M.M. Broer, B. Golding, W.H. Haemimerle, J.R. Simpson, D.L. Huber: Phys. Rev. B **33**, 4160 (1986)
1.113 R.J. Mears, L. Reekie, S.B. Poole, D.N. Payne: Electron. Lett. **22**, 159 (1986); M.F.J. Digonnet: IEEE J. LT-**4**, 1631 (1986) and as cited in [1.114]
1.114 R.S. Feigelson: [Ref. 1.25, p. 129]
1.115 H. Liu, K.-S. Lim, W. Jia, E. Strauss, W.M. Yen, A.M. Buoncristiani, C.E. Byvik: Opt. Lett., **13**, 931 (1988)
1.116 W.M. Yen: Proc. Conf. on Trends in Quantum Electronics, Bucharest, Aug. 1988, SPIE, to be published
1.117 S. Jaffee, W.M. Yen: Unpublished data
1.118 R. Lange, W. Grill, W. Martenssen: Europhys. Lett., **6**, 499 (1988)
1.119 See, e.g., G. Janik, W. Nagourney, H. Dehmelt: J. Opt. Soc. Am. B II-**2**, 1251 (1985)
1.120 E. Strauss, S. Walder: Europhys. Lett., **6**, 713 (1988)

2. The Puzzle of Two-Photon Rare Earth Spectra in Solids

Michael C. Downer

With 9 Figures

Ever since *Kaiser* and *Garrett* [2.1] observed two-photon absorption of ruby laser pulses in the ultraviolet absorption band of europium-doped CaF_2 in 1961, two-photon laser spectroscopy has been an important complementary technique to the linear spectroscopy of solids. Several unique characteristics of two-photon processes have contributed to their special importance in this role. Firstly, the summing of photon energies in two-photon absorption has permitted access to higher energy absorption bands then in linear absorption, while the subtraction of photon energies in Raman scattering has allowed detailed investigation of levels very near the ground state using visible frequency sources. Secondly, the presence of two independently variable polarizations provides a unique advantage over single-photon spectroscopy in determining the symmetry of the initial and final states in an optical transition. Finally, two-photon processes can explore levels of the same parity as the ground state, which are forbidden to single-photon electric dipole transitions. With these unique capabilities, two-photon spectroscopy has made important contributions to the study of excitons, biexcitons, polaritons, phonons, point impurities, and interband electronic transitions, particularly in materials such as alkali halides, cuprous halides and oxides, CdS, organic solids, and a wide variety of other semiconducting and insulating solids. For reviews of, and extensive references to, many of these applications of two-photon spectroscopy of solids, the reader is referred to [2.2–5].

This chapter focuses on major new developments since 1980 in the two-photon spectroscopy of solids doped with rare earth impurity ions. In contrast to the broad and relatively featureless *inter*-configurational transition $4f^N \rightarrow 4f^{N-1}5d$ which Kaiser and Garrett observed in $Eu^{2+}:CaF_2$, most of the recent work has focused on the sharp *intra*-configurational transitions $4f^N \rightarrow 4f^N$ within the partially filled $4f$ shell of the rare earth ion, which are parity-allowed for two-photon transitions. The new work has included extensive observations of two-photon absorption using tunable dye lasers as well as new observations of electronic Raman scattering using fixed frequency lasers. Numerous new lines and features which were inaccessible to linear spectroscopy have been observed and theoretically analyzed, particularly in solids containing divalent rare earth ions. In addition, careful observation of two-photon transition intensities and polarization anisotropies produced results which could not be explained by standard theories, a puzzle which drew the attention of several groups of theorists. New theoretical treatments of the far off-resonant intermediate states of the two-photon transitions, and their perturbation by ligand field and spin-orbit interactions, have now emerged from this analysis. At the same time, two-photon spectroscopy has emerged as a powerful probe of ligand-ion and ion-ion interactions in impurity-doped solid materials. The full consequences of these new discoveries for interpreting the one-photon spectra of rare earth systems, and the two-photon spectra of other solid materials, have not yet been deeply explored. These fundamental questions, together with the possibility of extensive new experimental observations, should keep the nonlinear spectroscopy of rare earth and related solids an active field of research for many years to come. An earlier and much briefer review of this field, which was confined to the two-photon absorption work, has been written by *Bloembergen* [2.6].

2.1 One-Photon and Two-Photon Rare Earth Spectroscopy: The Old Puzzle and the New Puzzle

J. Becquerel [2.7–9] first observed the unusually sharp absorption lines of rare earth salts as early as 1906. The narrow linewidths, which approach those observed in free atoms and molecules, stand in sharp contrast to the broad bands characteristic of most solid state spectra, and provide the principal attraction of rare earth spectra for experimentalists and theorists alike. While the optical transitions which give rise to the sharp lines are fundamentally atomic, they can also act as a sensitive probe of the interactions of the rare earth ion with its ligands and with other rare earth ions in the host crystal. Consequently rare earth spectroscopy has traditionally combined the disciplines of atomic, molecular, and solid state physics.

The precise physical meaning of these sharp line spectra, however, was a subject of intense debate for several decades after their initial observation. In a classic 1937 paper, *van Vleck* [2.10] coined the phrase "The Puzzle of Rare Earth Spectra in Solids" to describe the impasse which had arisen in understanding them. The radial wave function of the partially filled $4f$ shell in the rare earths was understood to be contracted within the filled $5s^2 5p^6$ shell [2.11], as shown in Fig. 2.1. Consequently the sharp spectral lines were widely and correctly attributed to the atomic $4f^N \rightarrow 4f^N$ transitions within the $4f$ shell [2.12, 13], since both initial and final states were protected from the braodening influences of the crystalline environment by the $5s^2 5p^6$ "shield". However, such intraconfigurational transitions have initial and final states of the same parity, and are consequently electric dipole-forbidden. Magnetic dipole and electric quadrupole transitions, while parity-allowed, obey the angular momentum selection rules $\Delta L, \Delta J \leq 1$, $\Delta L, \Delta J \leq 2$, respectively, whereas observations had already revealed numerous strong transitions with ΔL and ΔJ as large as six. The magnetic dipole and electric quadrupole couplings could therefore explain at best only a few of the observed transitions.

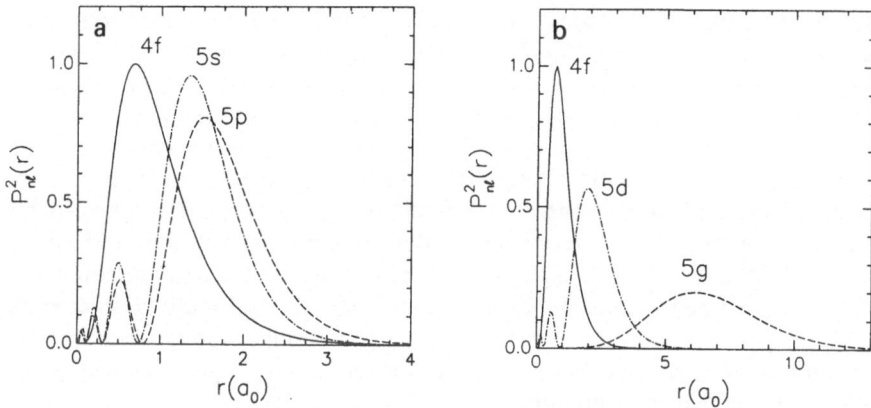

Fig. 2.1a, b. Radial distribution functions of the $4f$, $5s$, $5p$, $5d$, and $5g$ orbitals for the free Pr^{3+} ion, from Hartree-Fock calculations [2.133]. (From [2.27])

Van Vleck himself then proposed the mechanism of "forbidden electric dipole transitions" which eventually solved this puzzle. In crystals where rare earth ions occupy noncentrosymmetric sites, odd parity components of the crystal field, including both the static and vibronic components, mix states from opposite parity configurations into the $4f^N$ wave functions. In the second order, therefore, electric dipole transitions become parity-allowed. Most importantly, because the odd parity crystal field components capable of interacting with f electrons are described by tensor operators of the third and fifth ranks, electric dipole transitions with ΔL, $\Delta J \leq 6$ become possible. Nearly all the observed optical transitions in the rare earth series could therefore be accommodated. Detailed compilation of spectral data during the following decades, coupled with advances in computational methods, particularly by *Racah* [2.14–16] in the 1940s, led to quantitative fits of energy level positions throughout the lanthanide series. The extensive literature on the conventional spectroscopy of rare earth ions in condensed matter environments is summarized in the books by *Dieke* [2.17], *Wybourne* [2.18], and most recently by *Hüfner* [2.19].

The experimental observations of two-photon transitions described below created a "new puzzle of rare earth spectra", analogous in many ways to the puzzle which van Vleck faced. The angular momentum selection rules ΔL, $\Delta J \leq 2$ predicted by the lowest (second) order description of two-photon processes were again directly violated by observations, which showed strong transitions with ΔL, ΔJ as large as six. Anomalous line strength ratios and polarization anisotropics also characterized much of the data, both in two-photon absorption and in Raman scattering. The solution of the two-photon puzzle, which has been gradually assembled over the past several years, has required a more sophisticated treatment of the *intermediate* states in the two-photon transition. These intermediate states belong to high energy excited configurations, such as $4f^{N-1}d$ and $4f^{N-1}g$, which are opposite in parity to the initial and final $4f^N$ states, and extend radially far into the crystalline environment, as shown in Fig. 2.1, in sharp contrast to the localized $4f^N$ states. As with van Vleck's solution, higher order perturbative treatments have been found necessary in many cases. *Judd* and *Pooler* [2.20] showed that spin-orbit interactions among the intermediate states could drastically influence some two-photon line strengths. *Downer* et al. [2.21–24] demonstrated the role of static crystal field interactions, as well as combined static crystal field/spin-orbit interactions among the intermediate states. *Reid* and *Richardson* [2.25] have calculated that the dynamic ligand polarization contributions can be comparable in magnitude to those of the static crystal field terms. Finally, *Becker* et al. [2.26, 27] have shown that intermediate $4f^{N-1}g$ configurations are as important for some electronic Raman transitions as the $4f^{N-1}5d$ levels which are usually presumed to dominate.

2.2 Theory of Optical Transitions in Rare Earth Ions

The theory of two-photon $4f^N \rightarrow 4f^N$ processes is best considered in the context of the earlier theory of one-photon processes. There are several reasons for this. First of all, the computational methods and physical approximations involved are remarkably similar, as first pointed out by *Axe* [2.28]. Furthermore, suggestions of the revisions in two-photon theory prompted by recent experimental work can be found in earlier discussions of one-photon processes. Finally, the revisions in two-photon theory may prompt a re-examination of one-photon theory. Thus one-photon and two-photon spectroscopy of rare earths complement each other in significant ways.

2.2.1 The Judd-Ofelt Theory of One-Photon Transition Intensities

Van Vleck's concept of forbidden electric dipole one-photon transitions was developed into a quantitative theory of rare earth transition intensities in 1962 by *Judd* [2.29] and *Ofelt* [2.30]. At noncentrosymmetric crystalline sites, the perturbation of the $4f^N$ ground $\langle g|$ and final $|f\rangle$ state wave function of the rare earth ion by the odd parity components of the crystal field is described in first order perturbation theory by

$$\langle g'| = \langle g| + \sum_i E_{ig}^{-1} \langle g|H_{CF}^{odd}|i\rangle \langle i|,$$

$$|f'\rangle = |f\rangle + \sum_i E_{if}^{-1}|i\rangle \langle i|H_{CF}^{odd}|f\rangle, \qquad (2.1)$$

where the sum extends over all levels of excited configurations opposite in parity to $4f^N$. In intermediate coupling notation, g and f can be written as $\langle f^N\psi JM|$ and $|f^N\psi'J'M'\rangle$, respectively, where ψ and ψ' denote the quantum numbers other than total angular momentum J and the azimuthal component of angular momentum M needed to specify the wave functions completely. The integrated intensity, or line strength, of the single photon transition from g' to f' is then proportional to

$$\sum_{M,M'} |\langle g'|E \cdot D|f'\rangle|^2 = \sum_{M,M'} \left| -\sum_i [E_{if}^{-1}\langle g|E \cdot D|i\rangle \langle i|H_{CF}^{odd}|f\rangle \right.$$

$$\left. + E_{ig}^{-1}\langle g|H_{CF}^{odd}|i\rangle \langle i|E \cdot D|f\rangle] \right|^2 \qquad (2.2)$$

in the electric dipole approximation. Here D is the sum $\sum_j r_j$ of the position vectors for all electrons j in the f^N configuration, E is the electric field vector, and $E \cdot D$ represents the electric dipole operator. In cases where magnetic dipole and electric quadrupole contributions are important, these terms must be added to the electric dipole strength represented by (2.2). The intensity of individual Stark components is expressed by deleting the sum over M, M' and expanding the initial and final states as appropriate linear combinations of M and M' components as determined by the symmetry of the crystal field.

The actual computation of expression (2.2) as it stands is unmanageable for rare earths because of the enormous number of levels in the intermediate configurations. For example, $4f^65d$ alone contains 2725 levels [2.31]. An important simplification arises, however, since these excited configurations in the trivalent rare earth ions lie much higher in energy (typically 60000 to 200000 cm^{-1} above the ground state) than the $4f^N$ initial and final state levels. Consequently the energy denominators E_{ig} and E_{if} vary only modestly in the sum over i. Judd and Ofelt therefore introduced the *closure approximation*, in which the energy denominators are regarded as constants, thus allowing the use of the closure relation $\sum_n|n\rangle \langle n|=1$. In its most sweeping form the closure approximation regards all excited configurations as completely degenerate, leaving only a single constant energy denominator. In a more modest form, closure is performed separately within each excited configuration, each of which is assigned a different constant energy denominator. A common approximation regards the lowest energy excited configuration $4f^{N-1}5d$ as most important, and neglects the contribution of other intermediate configurations such as $4f^{N-1}ng$ and d^94f^{N+1}. As a related, though separate, approximation the energy denominators E_{if} and E_{ig} can be equated as well.

In the divalent rare earth ions the validity of the closure approximation is questionable since the excited configurations lie significantly lower in energy, because of the smaller effective nuclear charge. This fact was of little consequence for one-photon absorption, since very few sharp $4f^N \rightarrow 4f^N$ transitions are detectable anyway because of intense dipole-allowed inter-configurational transitions at the same frequencies. For the two-photon excitation of $4f^N$ levels embedded in the excited configurations, however, the closure approximation must be examined carefully, as seen below in the experimental results for Eu^{2+}.

Whatever inaccuracies the closure approximation introduces, it greatly simplifies the mathematics. The dipole operator and the crystal field operator in (2.2) unite into a single operator acting between levels of f^N. Completion of the calculation involves a series of straightforward manipulations. Operators are expressed in spherical tensor form, then the Wigner-Eckart theorem and well-known sum rules and operator techniques are applied. Details are presented in [2.29, 30]. The final result for the single photon line strength, after summing over all the components M, M' of the ground and final states and over all polarization directions is given simply by

$$S = \sum_{\lambda = 2, 4, 6} T_\lambda |\langle f^N \psi J \| U^{(\lambda)} \| f^N \psi' J' \rangle|^2 . \tag{2.3}$$

Here $U^{(\lambda)}$ is a sum of single particle unit tensor operators $\sum_n u_n^{(\lambda)}$ having the property $\langle f \| u^{(\lambda)} \| f \rangle = 1$. The reduced matrix elements $\langle f^N \psi J \| U^{(\lambda)} \| f^N \psi' J' \rangle$ involve only the angular parts of the initial and final state wave functions, and consequently can be calculated rigorously. Their values for all pairs of levels within the f^N configuration have been tabulated by *Nielson* and *Koster* [2.32] in the $L - S$ coupling scheme, and by *Carnall* et al. [2.33] in the full intermediate coupling approximation. The coefficients T_λ, on the other hand, contain radial integrals, the average energy denominator, crystal field parameters, and other numerical factors. In the approximation $E_{ig} = E_{if}$, only even values of m appear in the sum. Because of the difficulty of evaluating the T_λ accurately from first principles, it has become customary to regard them as phenomenological parameters to be determined by fitting experimental oscillator strengths. Typically one attempts to fit all observed oscillator strengths of a particular rare earth ion in a given host crystal or solution using a single set of T_λ parameters.

The oscillator strengths of numerous one-photon transitions in all trivalent rare earth ions have been analyzed according to such a fitting procedure. The most complete work is that by *Carnall* et al. [2.34, 35] on the aqueous solutions of trivalent rare earths; other analyses of rare earth oscillator strengths, including related parametrization schemes, have been reviewed by *Peacock* [2.36]. In the vast majority of cases the phenomenological treatment success- . fully accounts for the intensities of between twenty and thirty transitions in each rare earth ion, with intensities varying over as much as two orders of magnitude, with an agreement usually better than 15% of the observed oscillator strength. The overwhelming success of dozens of such fits is the chief vindication of the closure approximation and other assumptions implicit in the Judd-Ofelt analysis.

2.2.2 Later Refinements of One-Photon Transition Theory

The observation of "hypersensitive" transitions (for a compilation, see [2.36]), which could not be accommodated in the standard Judd-Ofelt formalism, prompted the proposal of another physical mechanism which could make significant contributions [2.37–40]. This mechanism involves a coupling of the $4f$ electrons with transient dipoles induced in the ligands by the radiation field, leading to an amplification of the even-parity multipolar transition amplitudes for $4f^N \rightarrow 4f^N$ radiative processes. Whereas the crystal field which appears in (2.2) reflects only the *ground state* charge distribution of the ligands, the new mechanism included the interaction of $4f$ electrons with the ligand *excited states* which are dynamically excited by the light field. *Reid* and *Richardson* [2.41] developed a parametrization scheme which clearly distinguished the influence of this "dynamic coupling" from the standard "static coupling" mechanism. They later showed that the same mechanism could be important in two-photon absorption [2.25]. In these treatments the initial intermediate, and final state wave functions in (2.2) must be regarded as products of lanthanide and ligand wave functions, with the possibility of ligand dipolar excitation explicitly included.

Another series of mechanisms for one-photon $4f^N \rightarrow 4f^N$ transitions, introduced by *Wybourne* [2.42] in 1968, is of clear interest in the light of recent two-photon absorption studies. This was the possibility of *third order contributions* involving static crystal field and spin-orbit interactions between the levels of the perturbing configurations $4f^{N-1}n'l'$. Whereas Judd and Ofelt had confined their treatment to second order contributions of the form (2.2), *Wybourne* [2.42] formally expressed the one-photon intensity as a sum of second and third order terms:

$$|\langle g'|E \cdot D|f'\rangle|^2 = \sum_{M, M'} \left| -\sum_i E_{if}^{-1} [\langle g|E \cdot D|i\rangle \langle i|H_{CF}^{odd}|f\rangle + \Leftrightarrow] \right.$$
$$\left. + \sum_{i,j} E_{if}^{-1} E_{jf}^{-1} [\langle g|E \cdot D|i\rangle \langle i|V|j\rangle \langle j|H_{CF}^{odd}|f\rangle + \Leftrightarrow] \right|^2, \quad (2.4)$$

where $\Leftrightarrow$ denotes an interchange of the electric dipole and noncentrosymmetric crystal field operators. In (2.4) V represents any part of the static Hamiltonian. Wybourne noted the potential importance of the third order terms with $V = H_{so}$, since they result in a breakdown of the spin selection rule $\Delta S = 0$, and the terms with $V = H_{CF}$, since they would bring odd rank tensor operators into play, in addition to those of even rank. Nevertheless, neither Wybourne nor subsequent investigators introduced such third order terms into the quantitative analysis of specific lanthanide transition intensity data. Indeed the overwhelming success of second order analyses virtually eliminated any motivation for doing so. Even so, the terms which Wybourne formally expressed are the direct analogs and precursors of the new contributions introduced later into two-photon formalism by *Judd* and *Pooler* [2.20] and *Downer* et al. [2.21, 22, 24]. Their clear importance for two-photon absorption should prompt re-examination of their role in one-photon lanthanide intensities. *Downer* [2.24] has re-analyzed the one-photon intensities of Gd^{3+} in aqueous solution and CaF_2 with third order contributions involving $V = H_{so}$ included. In this case, the third order contributions are modest, but significant.

For the transition $^8S_{7/2} \to {}^6P_{3/2}$ the third order contributions appear to dominate. Similar analyses for other lanthanide systems have now been performed [2.134].

A number of investigators have compared the Judd-Ofelt T_λ parameters determined from empirical fitting of intensity data to values obtained from ab initio calculations. Although it is often assumed that the $4f^{N-1}5d$ intermediate configuration makes the dominant contribution to these parameters, several notable studies have indicated otherwise. *Axe* [2.43] concluded that "*g* orbitals make an important, if indeed not a dominant, contribution to the configurational mixing responsible for electric dipole transition strengths" in europium ethylsulfate. *Krupke* [2.44], *P. J. Becker* [2.45], and *Hasunama* et al. [2.46] reached similar conclusions in analyses of optical absorption intensities in other lanthanide systems. Such conclusions are based on detailed theoretical evaluation of the Judd-Ofelt parameters, and are thus usually not evident from an empirical fit alone. Recent Raman scattering results described in Sect. 2.4 have provided the first definitive experimental confirmation of the importance of $4f^{N-1}ng$ intermediate states in lanthanide optical transitions.

2.2.3 Axe's Second Order Theory of Two-Photon Transition Intensities in Rare Earth Ions

Axe [2.28] first pointed out that the method of Judd and Ofelt applies generally to second order perturbation expressions which involve levels of much higher energy configurations as intermediate states. In particular it can be applied to second order expressions such as the following, which describes the probability of direct two-photon transitions:

$$\left| -\sum_i [\Delta_{i1}^{-1} \langle g|E_1 \cdot D|i\rangle \langle i|E_2 \cdot D|f\rangle + \Delta_{j2}^{-1} \langle g|E_2 \cdot D|j\rangle \langle j|E_1 \cdot D|f\rangle] \right|^2. \quad (2.5)$$

Goeppert-Mayer [2.47] first derived expression (2.5) from time-dependent perturbation theory carried to second order. Here $\Delta_{i1,2}$ denotes the energy $E_i - h\nu_{1,2}$ of the intermediate state above the single photon energy. According to convention, $\nu_i > 0$ for absorption of a photon, and $\nu_i < 0$ for emission of a photon. Consequently both photon frequencies are positive for two-photon absorption, whereas for Raman scattering, one is positive and one negative. Note that the second dipole operator in (2.5) plays the role of the noncentrosymmetric part of the crystal field in (2.2). Consequently explicit reference to the crystalline environment is absent in the second order description of two-photon processes in rare earths. Equation (2.5) therefore applies equally well to a free ion as to an impurity ion in a crystal.

The closure approximation is now applied to (2.5) as in the Judd-Ofelt analysis. The two dipole operators combine into a single compound operator acting between initial and final states which belong to $4f^N$. Through the use of tensor operator techniques [2.28] or second quantization techniques [2.20], this compound operator can be recoupled to yield a sum of three terms containing unit tensor operators of order $t=0$, 1, and 2 respectively:

$$\sum_{n'l'} 7(2l'+1) \begin{pmatrix} 3 & 1 & l' \\ 0 & 0 & 0 \end{pmatrix}^2 |\langle 4f|r|n'l'\rangle|^2$$

$$\times \sum_t (2t+1)^{1/2} \begin{Bmatrix} 1 & 3 & l' \\ 3 & 1 & t \end{Bmatrix} [(\Delta_{n'l'1}^{-1} + (-1)^t \Delta_{n'l'2}^{-1}] \langle E^{(1)}E^{(1)}\rangle^{(t)} \cdot U^{(t)}; \qquad (2.6)$$

$E^{(1)}$ denotes the electric field expressed in spherical tensor form. The zero order term $t=0$ contributes only to Rayleigh scattering, and can be neglected.

Two-Photon Absorption. For two-photon absorption, the term in square brackets is approximately equal to $h(v_1 - v_2)/\Delta_{n'l'}^2$ for $t=1$, where the numerical subscript on $\Delta_{n'l'}$ has been dropped on the assumption that $v_1 \approx v_2$. When the two photons have equal frequencies, as for two-photon absorption from a single beam, only the $t=2$ term is non-zero. Even for approximately equal frequencies, the $t=1$ term is much smaller than the $t=2$ term, and can usually be neglected. In this case the operator (2.6) simplifies to

$$3(2)^{3/2}(5)^{-1/2}(7)^{-1/2}\Delta_d^{-1}|\langle 4f|r|n'd\rangle|^2 \langle E^{(1)}E^{(1)}\rangle^{(2)} \cdot U^{(2)} \qquad (2.7a)$$

for intermediate states of the form $4f^{N-1}n'd$ or $4f^{N+1}n'd^9$ which involve d orbitals ($l'=2$), and to

$$(2)^{3/2}(5)^{1/2}(3)^{-1}(7)^{-1/2}\Delta_g^{-1}|\langle 4f|r|n'g\rangle|^2 \langle E^{(1)}E^{(1)}\rangle^{(2)} \cdot U^{(2)} \qquad (2.7b)$$

for intermediate states of the form $4f^{N-1}n'g$ which involve g orbitals ($l'=4$). Note that these two terms are of the same sign, and thus add constructively. Summing over all the components M, M' of the initial and final states, we find the two-photon absorption line strength in second order to be

$$S_{TPA} = A_2 [\langle E^{(1)}E^{(1)}\rangle^{(2)}]^2 |\langle f^N \psi J \| U^{(2)} \| f^N \psi J'\rangle|^2, \qquad (2.8)$$

where

$$A_2 = (8/7)\sum_{n'} [(9/25\Delta_d)|\langle 4f|r|n'd\rangle|^4 + (1/9\Delta_g)|\langle 4f|r|n'g\rangle|^4]$$

in the case where $4f^{N-1}d$ and $4f^{N-1}g$ configurations are each assigned single, but unequal, energy denominators. Equation (2.8) is analogous to the one-photon line strength (2.3). Now, however, instead of a sum of three terms, only a single term appears. Consequently the *ratio* of the line strengths of two transitions can be computed simply as the ratio of the squares of the second rank reduced matrix elements for the two transitions, without the need for phenomenological parameters. Similarly the polarization dependence of a given line strength emerges simply and rigorously from (2.8). For example, for two equivalent linearly polarized photons, $[\langle E^{(1)}E^{(1)}\rangle^{(2)}]^2 = \langle 1010|1120\rangle^2 E^4 = (2/3)E^4$, where E is the electric field amplitude. For two equivalent circularly polarized photons, on the other hand,

$$[\langle E^{(1)}E^{(1)}\rangle^{(2)}]^2 = \langle 1111|1122\rangle^2 E^4 = E^4,$$

yielding a line strength 50% greater than for linear polarization. Other polarization combinations lead to different, though readily computed, line strengths. Again no phenomenological parameters are involved. Experimental measurements of line strength ratios and polarization dependence of line strengths therefore provide rigorous tests of the second order theory of two-photon absorption.

Electronic Raman Scattering. In Raman scattering, the term in square brackets in (2.6) is equal to $2hv/\Delta^2_{n'l'}$ for $t=1$, in the approximations $v_1 \approx v_2 = v$ and $hv \ll E_{n'l'}$, and consequently is generally not negligible. This term constitutes the *antisymmetric* contribution to the Raman scattering tensor, since it changes sign upon interchange of the incident and scattered photons. For equivalent incident and scattered photon polarizations, the antisymmetric term vanishes. It is smaller than the $t=2$ term, which constitutes the *symmetric* contribution, by a factor of roughly $hv/E_{n'l'}$, which can nevertheless be as large as 0.2 in the typical case of a visible wavelength incident photon and a $4f^{N-1}5d$ intermediate state. The significant antisymmetric character of Raman scattering lends it a unique experimental importance. In particular, the antisymmetric Raman scattering operator, in contrast to the symmetric operators (2.7), has the opposite sign for intermediate states of d and g character. For d orbitals ($l'=2$), the $t=1$ term in (2.6) becomes

$$-(2)^{3/2}(3)^{1/2}(7)^{-1/2}hv\Delta_d^{-2}\langle 4f|r|n'd\rangle^2 \langle E_1^{(1)}E_2^{(1)}\rangle^{(1)} \cdot U^{(1)}, \tag{2.9a}$$

while for g orbitals ($l'=4$) it becomes

$$+(2)^{3/2}(3)^{1/2}(7)^{-1/2}hv\Delta_g^{-2}\langle 4f|r|n'g\rangle^2 \langle E_1^{(1)}E_2^{(1)}\rangle^{(1)} \cdot U^{(1)}. \tag{2.9b}$$

Expressions (2.9) were derived by applying closure separately to the intermediate states of d and g character, and recoupling the angular parts of the operators. It is not difficult to see, however, that in a full closure approximation, in which all intermediate configurations are assumed to lie at the same energy, the antisymmetric part of the Raman scattering amplitude vanishes exactly. In this approximation the antisymmetric part of (2.5) is proportional to

$$-\sum_i [\langle f|(E_1 \cdot D)(E_2 \cdot D)-(E_2 \cdot D)(E_1 \cdot D)|g\rangle]=0,$$

because the electric dipole operators commute with each other. An observed lack of Raman asymmetry thus indicates equal contributions of the various excited configurations. Similarly a measurement of the sign of the asymmetry allows the relative importance of intermediate states of d and g character to be determined. This information is not available from either one-photon or two-photon absorption from a single beam.

The Raman scattering process is usually described by a scattering tensor $\alpha_{\varrho\sigma}$, where ϱ and σ denote the polarizations of the scattered and incident photons, respectively, in either Cartesian or spherical coordinates. For describing experimental data, Cartesian coordinates are often preferred, since they can be defined with respect to crystallographic axes. For computational purposes, spherical coordinates are preferred, since $\alpha_{\varrho\sigma}$ can then be simply related to spherical tensor operators. In spherical tensor form, the scattering tensor can be expressed as a sum of an antisymmetric component $\alpha^{(1)}$ and a symmetric component $\alpha^{(2)}$, which are related to the unit tensor operators through the simple expression

$$\alpha_q^{(t)} = F_t U_q^{(t)}, \tag{2.10}$$

where the parameters F_t, originally defined by *Koningstein* and *Mortensen* [2.48], are given by

$$F_t = (-1)^t (14)(2t+1)^{1/2} \sum_{n'l'} (2l'+1)|\langle 4f|r|n'l'\rangle|^2$$

$$\times \begin{pmatrix} 3 & 1 & l' \\ 0 & 0 & 0 \end{pmatrix}^2 \begin{Bmatrix} 1 & 3 & l' \\ 3 & 1 & t \end{Bmatrix} \Delta_{n'l'}^{-1}(h\nu/\Delta_{n'l'})^{2-t} \tag{2.11}$$

for Raman transitions within f^N configurations. The ratio F_1/F_2 determines the degree of asymmetry of a transition. The spherical and Cartesian forms of the Raman scattering tensor are related by a straightforward coordinate transformation (see, for example, [2.28]). The integrated intensity of a Raman transition between two J-multiplets is proportional to

$$F_1^2|\langle f^N \psi J \| U^{(1)} \| f^N \psi J'\rangle|^2 + F_2^2|\langle f^N \psi J \| U^{(2)} \| f^N \psi J'\rangle|^2. \tag{2.12}$$

2.2.4 Third and Fourth Order Contributions to Two-Photon Processes

One of the major outcomes of experimental two-photon absorption studies in rare earths since 1980 has been the realization that the second order theory outlined above must frequently be supplemented with higher order terms in the perturbation expansion. The higher order terms are required to explain the observation of strong transitions which violate the angular momentum selection rules imposed in second order, as well as the anomalously strong intensity and polarization anisotropy of such transitions.

When second-, third-, and fourth-order terms in the perturbation series are shown explicitly, the line strength of a two-photon transition from a ground state g with components $|f^N \psi J M\rangle$ to an excited state f with components $|f^N \psi' J' M'\rangle$ is proportional to

$$\sum_{M,M'} \Bigg| -\sum_i \Delta_i^{-1} \langle |E \cdot D|i\rangle \langle i|E \cdot D|f\rangle$$

$$+ \sum_{i,j} \Delta_i^{-1} \Delta_j^{-1} \langle g|E \cdot D|i\rangle \langle i|V|j\rangle \langle j|E \cdot D|f\rangle$$

$$- \sum_{i,j,k} \Delta_i^{-1} \Delta_j^{-1} \Delta_k^{-1} \langle g|E \cdot D|i\rangle \langle i|V|j\rangle \langle j|V|k\rangle \langle k|E \cdot D|f\rangle + \dots \Bigg|^2. \tag{2.13}$$

For simplicity the case of two-photon absorption from a single beam has been assumed. The higher order terms contain extra energy denominators as well as interactions V and V' between levels of the intermediate configurations. Terms in which V and V' represent electric dipole operators describe three- and four-photon absorption processes, and are thus not relevant to our discussion of two-photon processes. Instead we are concerned with terms involving the non-radiative Hamiltonian, in particular the spin-orbit (H_{so}), crystal field (H_{CF}), central field and interelectronic Coulomb potentials. The effect of the higher order terms on two-photon selection rules is immediately evident. With $V = H_{so}$, for example, the second order spin selection rule $\Delta S = 0$ breaks down in the third order. A direct linkage of initial and final states with $\Delta S = 1$ becomes possible. Similarly, with $V = H_{CF}$, the second order selection rules $\Delta L, \Delta J \leq 2$ can be overcome in the third order. Since the part of the crystal field potential which interacts significantly with the intermediate configurations is composed of tensor components of rank 2 and 4 for levels of d character, and 2, 4, 6, and 8 for levels of g character, transitions with $\Delta L, \Delta J \leq 6$ become possible. Fourth order terms with $V = H_{so}$ and $V' = H_{CF}$ expand the selection rules even further, so as to allow a direct linkage of levels with $\Delta S = 1$ and $\Delta L, \Delta J \leq 6$. In such cases the higher order terms can rival or exceed the second order terms in magnitude despite the extra energy denominators. Eventually, however, the perturbation series converges, since beyond some finite order, no selection rules remain to be overcome for a given transition. The extra energy denominators are then no longer compensated by large numerators at higher orders.

Note that the successive terms in the pertrubation series alternate in sign, and are added together *before* squaring to determine the line strength. Interference can therefore arise, particularly when different contributions are comparable in magnitude. This interference can be constructive or destructive, depending on the relative signs of the contributions.

The expression (2.13) formally includes the dynamic coupling mechanism introduced by *Reid* and *Richardson* [2.25] if the wave functions g, i, j, g, and f are generalized to products of lanthanide and ligand wave functions. In this case the third order term in (2.13) includes contributions such as the following in which the light field transiently excites or de-excites the ligand, and electrons in the lanthanide ion and the ligand excited states interact via the Coulombic potential $V = e^2/r_{ij}$:

$$-\sum_{i,j} \Delta_{i\chi}^{-1} \Delta_{j\psi}^{-1} [\langle g_\psi g_\chi | E \cdot D | g_\psi i_\chi \rangle \langle g_\psi i_\chi | V | j_\psi g_\chi \rangle \langle j_\psi g_\chi | E \cdot D | f_\psi g_\chi \rangle$$

$$+ \langle g_\psi g_\chi | E \cdot D | i_\psi g_\chi \rangle \langle i_\psi g_\chi | V | f_\psi j_\chi \rangle \langle f_\psi j_\chi | E \cdot D | f_\psi g_\chi \rangle]. \tag{2.14}$$

The subscripts χ and ψ denote ligand and lanthanide states, respectively. Other third order terms, by contrast, assume that the ligands remain in their ground states throughout the third order coupling. The fourth order term in (2.13) can be generalized in a similar manner.

2.2.5 Computational Methods

Expressions such as (2.13) and (2.14) contain matrix elements which involve compound spherical tensor operators acting between the wave functions of multi-electron atomic configurations. *Racah* [2.14–16] developed sophisticated computational methods for evaluating such matrix elements in the 1940s, and a number of books on theoretical atomic spectroscopy, such as [2.49–51] describe such methods. Nevertheless, when matrix elements grow to the size of third and fourth order terms, these conventional methods often prove unacceptably tedious. Recent evaluations of these terms [2.20,22] have therefore employed the techniques of second quantization. This has provided elegant treatments of the many-body problem in other fields of physics, notably in the theory of superconductivity [2.52] and in that of nuclear pairing forces [2.53]. These techniques also facilitate the evaluation of complicated matrix elements in multi-electron atomic configurations, as described in detail by *Judd* [2.54].

In the second quantization method the quantum mechanical state of a system of N identical particles is specified by stating the number of particles $N_i \leq N$ possessing each possible value K_i of a complete set of single-particle dynamical variables $K = \{K_i\}$. In the case of an atomic shell nl containing N electrons, K_i refers to the quartet of quantum numbers $(nlm_{si}m_{li})$, and $N_i = 0$ or 1 because electrons, being fermions, obey the Pauli exclusion principle. Creation and annihilation operators $a_i{}^\dagger$ and a_i create and annihilate the possible states $|K_i\rangle = |nlm_{si}m_{li}\rangle$ of an nl electron. A normalized N-electron determinantal product state $\{K_1 K_2 \ldots K_N\}$ is written as a series of creation operators acting on the vacuum state: $a_1^\dagger a_2^\dagger \ldots a_N^\dagger |0\rangle = \{K_1 K_2 \ldots K_N\}$. Similarly the complex conjugate determinantal product state is formed when the corresponding annihilation operators act to the left on the vacuum state: $\langle 0|a_1 a_2 \ldots a_N = \{K_1 K_2 \ldots K_N\}^*$. From well-known properties of determinantal product states, the fundamental anti-commutation and orthonormality relations for creation and annihilation operators can be deduced. These relations are stated in the Appendix (Sect. 2.A.1).

The importance of second quantization for atomic spectroscopy is that operators which act on N particles can be represented in terms of matrix elements of single-particle operators, acting as coefficients of creation and annihilation operators. The second quantized form of a general operator $H = \sum_n h_n$, which is a sum of single-particle operators, is given in Sect. 2.A.2. The full utility of the second quantized form of operators stems from the tensor character of the creation and annihilation operators. The $(2s+1)(2l+1)$ creation operators $a_i{}^\dagger$ behave as the components of a double spherical tensor operator $\boldsymbol{a}^\dagger = \boldsymbol{a}^{\dagger(sl)}$ with rank s in spin space and rank l in orbital angular momentum space. An analogous double tensor operator $\boldsymbol{a} = \boldsymbol{a}^{(sl)}$ can be introduced for the annihilation operators. These operators can therefore be manipulated with standard tensor operator techniques [2.50,51] and can be coupled to each other to represent physically meaningful operators. In the appendix several examples are listed and an example of the detailed evaluation of a third order term with $V = H_{CF}$ is presented in Sect. 2.A.4, using the techniques of second quantization together with the well-known properties of spherical tensor operators. For an authoritative primer on the techniques of second quantization as applied to atomic spectroscopy, the reader is referred to the book by *Judd* [2.54].

2.3 Experimental Studies of Two-Photon Absorption in Rare Earths

2.3.1 Early Experimental Investigations

Kaiser and *Garrett*'s [2.1] pioneering 1961 observation of two-photon absorption in $Eu^{2+}:CaF_2$ utilized one of the earliest lasers – a pulsed, fixed frequency ruby laser – as the excitation source. Two-photon absorption was identified by monitoring subsequent ultraviolet fluorescence, which exhibited the characteristic quadratic dependence on the incident intensity. *Bayer* and *Schaak* [2.55] later investigated the polarization dependence of this same interconfigurational $4f^7 \rightarrow 4f^6 5d$ two-photon transition at several discrete excitation frequencies using a ruby laser augmented with a Raman cell. Careful measurements of polarization dependence were used to determine the symmetry of the excited states at different energies, according to the polarization rules for two-photon absorption in solids derived by *Inoue* and *Toyazawa* [2.56] and *Bader* and *Gold* [2.57].

The sharp $4f^N \rightarrow 4f^N$ transitions held a greater potential for strong two-photon absorption because of the parity selection rule and the narrower absorption linewidths. In 1966 *Singh* and *Geusic* [2.58] reported the first experimental evidence of such two-photon transitions by exciting $NdCl_3$, as well as $Nd^{3+}:LaCl_3$, $LaBr_3$, at 1.06 μm with a continuous wave neodymium laser. They observed upconverted visible fluorescence and intensity-dependent attenuation, which they attributed to a complicated excitation process involving direct two-photon excitation, radiative and nonradiative relaxation. Energy transfer processes between Nd^{3+} ions, however, may also have played a role in producing the upconverted fluorescence. Shortly afterward the Soviet group of *Ershov* et al. [2.59, 60], motivated by the possibility of a two-photon laser [2.61], used pulsed ruby and Nd^{3+}:glass laser radiation to excite three direct two-photon transitions from the $Ho^{3+}:CaF_2$ ground state: $^5I_8 \rightarrow {}^5F_4$, corresponding to $2\nu_{Nd:glass}$, $^5I_8 \rightarrow {}^5G_3$, corresponding to $2\nu_{ruby}$, and $^5I_8 \rightarrow {}^5G_3$, corresponding to $\nu_{Nd:glass} + \nu_{ruby}$. All of these observations relied on accidental coincidences between excited $4f^N$ energy levels and the summed frequencies of the excitation sources, the probability of which was enhanced by the high density of levels in the regions of interest, and by the broadening of the lines at room temperature. Another Soviet group [2.62, 63] reported further observations of $4f^N \rightarrow 4f^N$ two-photon excitation in Er^{3+}:CaF_2 and Ho^{3+}:CaF_2 using Nd:YAG laser excitation, although the observed luminescence may have been caused by stepwise excitation rather than direct two-photon absorption. *Penzkofer* and *Kaiser* [2.64] investigated the possible practical importance of two-photon absorption as a fundamental limitation on the gain of high power picosecond Nd^{3+}:glass lasers and amplifiers. Their measurements and calculations purported to show that direct two-photon absorption of intense picosecond pulses at 1.06 μm by the fully allowed $^4I_{9/2} \rightarrow ^4G_{7/2}$ transition constituted the major source of nonlinear loss in their Nd^{3+}:glass rods. Their conclusions have recently been questioned, however, by *Chase* and *Payne* [2.65] (Sect. 2.3.5).

Serious further investigations of the sharp $4f^N$ lines had to await the advent of tunable excitation sources. During the 1970s the Soviet group of *Gintoft* and *Skripko* [2.66], *Apanasevich* et al. [2.67], *Skripko* and *Gintoft* [2.68], and *Makhanek* et al. [2.69], first applied a high power pulsed tunable dye laser to two-photon excitation spectroscopy of sharp $4f^N \rightarrow 4f^N$ lines of Er^{3+}, Ho^{3+}, Eu^{3+}, Sm^{3+}, and Tb^{3+} impurities in a CaF_2 lattice. No beam polarization was reported in these studies, and crystal field components were not resolved. Since the tuning range was narrow, only one or two transitions were observed in each sample. Some of the observed luminescence signals, particularly in Ho^{3+}, arose from stepwise excitation. The authors clearly established that others, however, arose from direct two-photon absorption.

Historical hindsight, in the light of later work by *Dagenais* et al. [2.70] and by *Judd* and *Pooler* [2.20], can be applied to the observation of a strong $^7F_6 \rightarrow ^5G_6$ transition in Tb^{3+}:CaF_2 by this Soviet group [2.69]. The authors point out correctly that this transition obeys $\Delta L, \Delta J \leqq 2$ (although not $\Delta S = 0$), and that two expected neighboring transitions $^7F_6 \rightarrow ^5D_3$, $^5L_{10}$ which violate these same selection rules were not observed, in apparent accord with the second order theory. *Makhanek* et al. [2.69] fail to point out, however, that $^7F_6 \rightarrow ^5G_5$, which also obeys $\Delta L, \Delta J \leqq 2$, and also falls comfortably within their laser tuning range, was not observed either. Furthermore the square of the $U^{(2)}$ reduced matrix element for this transition is only 25% smaller than for $^7F_6 \rightarrow ^5G_6$ [2.33], while the experimental signal-to-noise ratio should have allowed observation of a signal thirty times weaker. Apparently, therefore, $^7F_6 \rightarrow ^5G_6$ exhibits an anomalously strong intensity with respect to $^7F_6 \rightarrow ^5G_5$. The former transition is the precise analog of the transition $^8S_{7/2} \rightarrow ^6P_{7/2}$ in Gd^{3+}, which *Dagenais* et al. [2.70] observed much later, in that $\Delta S = 1$, $\Delta L = 1$, and $\Delta J = 0$ in both transitions. Notably, this Gd^{3+} transition also exhibited anomalous strength, as well as anomalous polarization dependence, observations which *Dagenais* et al. [2.70] took pains to emphasize. These observations spurred *Judd* and *Pooler* [2.20] to introduce a new third order contribution with $V = H_{so}$, which their calculations revealed was unusually strong for transitions with precisely $\Delta S = 1$, $\Delta L = 1$, and $\Delta J = 0$. *Downer* [2.24] later calculated that the third order contribution with $V = H_{so}$ indeed enhances the line strength of $^7F_6 \rightarrow ^5G_6$ in Tb^{3+} one hundred-fold compared to neighboring transitions, thus explaining the intriguing observation which *Makhanek* et al. [2.69] made, but failed to appreciate fully. Corroborating measurements of the polarization anisotropy of the two-photon $^7F_6 \rightarrow {}^5G_6$ transition in Tb^{3+}, and of the neighboring transitions, would now be of great interest, but have not been made.

Fritzler and *Schaak* [2.71] made an important contribution by studying the two-photon excitation spectrum of the $4f^7 \rightarrow 4f^7$ transitions $^8S_{7/2} \rightarrow {}^6P_{7/2,\,5/2}$ in Eu^{2+}:CaF_2 and SrF_2 with a pulsed tunable dye laser. These authors demonstrated the first use of two-photon spectroscopy to observe sharp $4f^N \rightarrow 4f^N$ transitions which are overlapped by a $4f^{N-1}5d$ band, and are thus unobservable by one-photon spectroscopy. The parity selection rule suppresses the broad $4f^6 5d$ background in two-photon absorption sufficiently for the sharp parity-allowed $4f^7$ levels to emerge. *Fritzler* [2.72] analyzed the energy levels, crystal field splitting, and polarization dependent two-photon intensities of the $^6P_{7/2,\,5/2}$ multiplets. An error was made in the intensity analysis, for the data clearly showed that the $^6P_{7/2}$ line strength was much larger for linearly polarized than for circularly polarized excitation, and that the $({}^6P_{7/2})/({}^6P_{5/2})$ intensity ratio is several times larger than the corresponding ratio of the squares of the second order reduced matrix elements. Both observations contradict the second order theory of two-photon absorption. Yet *Fritzler* [2.72] stated that the agreement between calculated and observed transition probabilities was "very satisfying". *Downer* et al. [2.23] later corrected this error, and showed that a third order contribution with $V = H_{so}$, originally introduced by *Judd* and *Pooler* [2.20], was responsible for the discrepancy.

2.3.2 Two-Photon Absorption Studies of Gd^{3+} in LaF_3, Aqueous Solution, and CaF_2

Extensive observations of two-photon absorption in Gd^{3+} by *Dagenais* et al. [2.70], *Downer* et al. [2.21, 22, 24], and *Cordero-Montalvo* [2.73] starting in 1981 constituted the first systematic study of two-photon absorption in rare earths comparable in scale and thoroughness to earlier compilations of single-photon absorption data [2.17, 18]. The experiments employed pulsed, as well as continuous wave, tunable dye laser excitation with a single beam, followed by ultraviolet fluorescence detection. Polarization with respect to the crystalline axes was carefully controlled, and a systematic intensity normalization procedure was adopted. Gd^{3+} was chosen as a sample because of the high ($\sim 100\%$) UV fluorescence yield following absorption, and the lack of near resonant intermediate states at visible frequencies, thus eliminating the possibility of two-step excitation processes.

Figure 2.2 shows the 14 excited multiplets observed in these studies, and the subsequent combined nonradiative and fluorescent relaxation channels. In the crystalline samples, all the crystal field components allowed by Kramers degeneracy, apart from those of the ground state, were fully resolved. Consequently both the integrated multiplet transition intensities and the

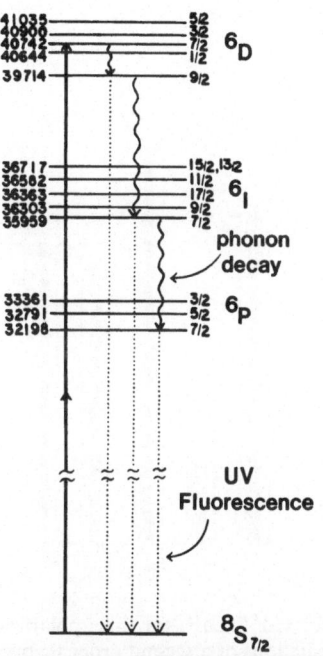

Fig. 2.2. The fourteen lowest $4f^7$ levels of Gd^{3+}:LaF_3 showing two-photon excitation of $^6D_{7/2}$ and subsequent relaxation by phonon emission and fluorescence. Crystal field splittings are not shown. Numbers at the left give the average energy of each J multiplet in cm^{-1}. The same set of levels occur in Eu^{2+}:CaF_2 at approximately 15% lower energy

individual Stark component intensities were studied in detail. In LaF_3 and aqueous solution, most of the levels studied had been observed previously in ultraviolet one-photon absorption [2.74, 75]. In CaF_2, *Cordero-Montalvo* [2.73] used two-photon absorption to observe strong lines arising from Gd^{3+} ions at cubic (centrosymmetric) sites, which are weak or absent in one-photon absorption [2.76] because of the parity selection rule in the presence of a center of inversion.

The bar graph in Fig. 2.3 presents the relative two-photon absorption line strengths for Gd^{3+}:LaF_3 on an arbitrary logarithmic scale, as compiled by *Downer* and *Bivas* [2.22]. Each vertical bar represents the integrated intensity of a J multiplet for a particular polarization of the excitation beam. The three polarizations selected – $E\perp c$, $E\|c$, and circular polarization in the c plane – were the only polarizations which could propagate through the birefringent LaF_3 crystal without change. Intensity data for Gd^{3+} in aqueous solution [2.22] and for Gd^{3+}:CaF_2 [2.73] were consistent with that shown in Fig. 2.3. In the aqueous solution, numerous Stark splittings, analogous to crystal field splittings, were observed, arising from the electric field of the quasistatic hydration complex surrounding the ion. The symmetry of these Stark components could be determined by two-photon absorption using two beams of independently variable polarizations, as shown by *Monson* and *McClain* [2.77, 78], and thus provide valuable clues to the structure of the hydration complex. The data of *Downer* and *Bivas* [2.22], however, was obtained with only a single excitation beam.

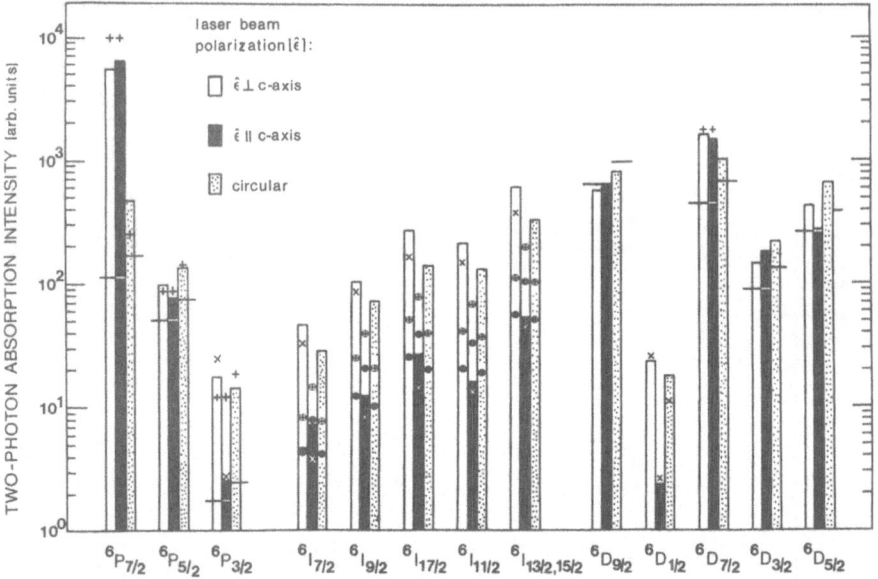

Fig. 2.3. Relative two-photon absorption line strengths for Gd^{3+}:LaF_3 for three polarizations of the excitation beam. Horizontal lines denote predictions of a second order theory. Line strengths calculated ($+$) by including third-order terms in the spin-orbit interaction; ($\bullet$) by including third-order terms in the static crystal field interaction; and ($\oplus$) by including fourth-order terms involving both spin-orbit and crystal field interactions. When J-mixing among the $4f^7$ levels is taken into account, the $\times$'s show the calculated intensities for $^6P_{3/2}$, ^6I_J, and $^6D_{1/2}$. (From [2.22])

Second Order Intensity Analysis. The horizontal lines in Fig. 2.3 denote relative intensities predicted by the second order theory of *Axe* [2.28], where the best fit was made to the four transition intensities $^8S_{7/2} \rightarrow {}^6P_{5/2}$, $^6D_{9/2, 3/2, 5/2}$, which were the only transitions which agreed both in relative intensities and polarization dependence with the Axe predictions. A measurement of the *absolute* two-photon absorption cross section of $^6P_{5/2}$ (for further discussion see Sect. 2.3.5) also agreed within experimental error with a cross section calculated using a second order theory. As further evidence of the consistency of the four transitions $^8S_{7/2} \rightarrow {}^6P_{5/2}$, $^6D_{9/2, 3/2, 5/2}$ with the second order theory, a comparison of the individual Stark component intensities with second order predictions revealed, for the most part, excellent agreement as shown in Fig. 2.4 for $^6P_{5/2}$, $^6D_{5/2}$. The theoretical values were obtained by evaluating the matrix elements of the second order operator (2.7) separately for each component M' of the excited states, and taking into account the partial mixing of M' values by the approximately D_{3h} crystal field. The M_J numbers at the base of Fig. 2.4 denote the dominant M' value for each component. For the three levels $^6P_{5/2}$, $^6D_{3/2, 5/2}$ a satisfying account of the Stark component intensities is obtained for all three polarizations.

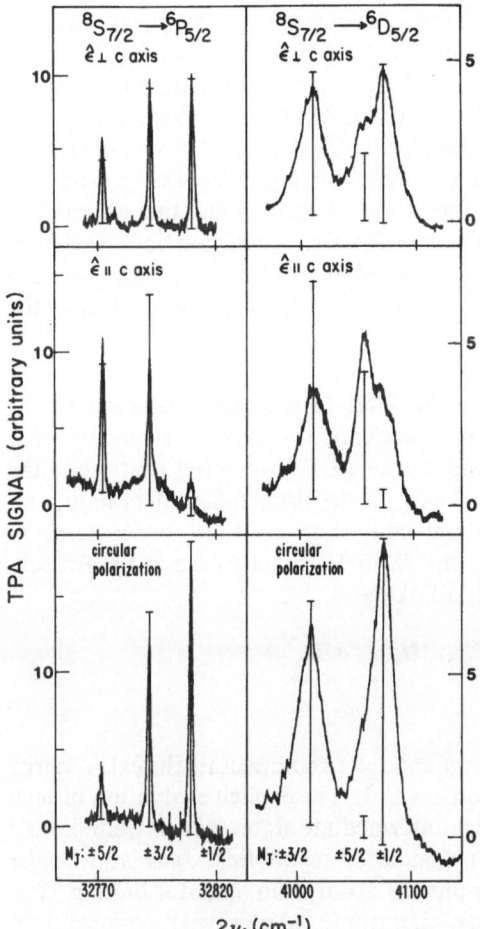

Fig. 2.4. Experimental two-photon excitation recordings of $^6P_{5/2}$ and $^6D_{1/2}$ in Gd^{3+}:LaF_3 showing polarization dependence of the individual Stark components. Vertical bars show intensities calculated with a second order theory of two-photon absorption. (From [2.22])

The one exception to full agreement with the second order theory for these four transitions is the Stark component intensities of the $^6D_{9/2}$ level. Even here the agreement for $E\perp c$ and circular polarization is excellent. Nevertheless for $E\|c$, the expected drop in intensity of the lowest two Stark components was not observed [2.22]. No satisfactory explanation of this anomaly has yet been found.

The remaining ten transitions in Gd^{3+} exhibited enormous discrepancies with the second order theory. The predictions for the ^6I_J lines fell well below the scale of Fig. 2.3, in marked disagreement with the experimental results shown. In fact, the second order theory predicts zero intensity for $^8S_{7/2} \rightarrow {}^6I_{13/2, 15/2, 17/2}$ because of the selection rule $\Delta J \leq 2$. Furthermore these "forbidden" transitions were stronger than neighboring "allowed" transitions. As a final discrepancy with second order predictions, all of the ^6I_J intensities in Gd^{3+}:LaF_3 depended strongly on the direction of linear polarization with respect to crystalline axes.

The weak $^6P_{3/2}$ and $^6D_{1/2}$ lines showed a strong anisotropy of precisely the same form. The $^6P_{7/2}$ line was unique in that the cross section for linear polarization, though isotropic, was about 12 times stronger than for circular polarization, again in direct contradiction to the Axe theory. To a much smaller extent, the same isotropic enhancement for linear polarization was evident in $^8S_{7/2} \rightarrow {}^6D_{7/2}$. Careful analysis of the effect of J-mixing of the $4f^7$ levels by the crystal field was carried out in the hope of explaining the anomalous characteristics of these transitions in terms of oscillator strength borrowed from neighboring levels through crystal field admixtures. However, such detailed analyses fell far short of explaining either the observed relative intensities or the polarization anisotropies.

Third Order Contributions in the Spin-Orbit Interaction. The case of the $^8S_{7/2} \rightarrow {}^6P_{7/2}$ transition was the first to be explained successfully by introducing a third order contribution. *Judd* and *Pooler* [2.20] observed that while the second order linkage of ^8S to ^6P was zero in the Russell-Saunders limit, and unusually small even in intermediate coupling, the third order term in expression (2.13) with $V = H_{so}$ could link ^8S to ^6P directly even in the Russell-Saunders limit. An example of such a linkage is

$$\langle f^7\,{}^8S_{7/2}|E \cdot D|f^6d\,{}^8P_{5/2}\rangle \langle f^6d\,{}^8P_{5/2}|H_{so}|f^6d\,{}^6D_{5/2}\rangle$$

$$\times \langle f^6d\,{}^6D_{5/2}|E \cdot D|f^7\,{}^6P_{7/2}\rangle, \tag{2.15}$$

which can contribute a numerator large enough to compensate the extra energy denominator which occurs in third order (2.13). The explicit evaluation of such linkages involved double closure over intermediate states, which were limited to $4f^65d$ configurations, followed by operator recoupling. *Judd* and *Pooler* [2.20] then found that the total two-photon absorption operator linking ^8S to ^6P was a sum of the usual second order operator (2.7a) plus two recoupled third order spherical tensor operators:

$$3(\tfrac{8}{35})^{1/2}\Delta_d^{-1}(E^{(1)}E^{(1)})^{(2)} \cdot U^{(2)} + (\tfrac{1}{5})(\tfrac{1}{14})^{1/2}(9\zeta_f - 4\zeta_d)\Delta_d^{-2}(E^{(1)}E^{(1)})^{(2)} \cdot W^{(11)2}$$

$$- 2(\tfrac{2}{21})^{1/2}(9\zeta_f + \zeta_d)\Delta_d^{-2}(E \cdot E)W^{(11)0}, \tag{2.16a}$$

where the third order operators $W^{(ij)k}$ represent a sum $\sum_n [w^{(ij)k}]_n$ of single-particle double tensor operators with rank i in spin space and rank j in orbital space, coupled to rank k in the space of total angular momentum [2.50], with magnitude defined by

$$\langle nl\|w^{(ij)}\|n'l'\rangle = \delta(n,n')\delta(l,l')(2i+1)^{1/2}(2j+1)^{1/2}.$$

These operators arise out of the coupling of the spin-orbit operator to each of two electric dipole operators which occur in the third order term in (2.13). In (2.16a) ζ_f and ζ_d denote spin-orbit coupling constants for f and d electrons, respectively. The corresponding operator for $4f^6 g$ intermediate states is

$$\tfrac{1}{3}(\tfrac{10}{7})^{1/2}\varDelta_g^{-1}(E^{(1)}E^{(1)})^{(2)}\cdot U^{(2)}+(\tfrac{1}{14})^{1/2}(\zeta_f-2\zeta_g)\varDelta_g^{-2}(E^{(1)}E^{(1)})^{(2)}\cdot W^{(11)2}$$
$$-(\tfrac{2}{21})^{1/2}[24\zeta_f+5\zeta_g]\varDelta_g^{-2}(E\cdot E)W^{(11)0}. \tag{2.16b}$$

The crucial operator is $W^{(11)0}$, which, being of rank zero, or scalar, can link only states of equal J, and thus contributes only to $^8S_{7/2}\to{}^6P_{7/2}$ among the ^6P lines in Gd^{3+}. Furthermore, the scalar product $E\cdot E$ vanishes for two equivalent circularly polarized photons. Finally the numerical coefficient of the third order scalar operator is twenty times larger than the other coefficients. Consequently it lends more than a hundredfold isotropic enhancement to the line strength of $^6P_{7/2}$ for linear polarization, while the line strength for circular polarization, along with that of the $^6P_{5/2,3/2}$ for all polarizations, arises solely from the weaker second rank terms in expression (2.16). These were precisely the bothersome aspects of the ^6P integrated intensities which the experimental data, as shown in Fig. 2.3, had revealed. *Downer* and *Bivas* [2.22] later showed

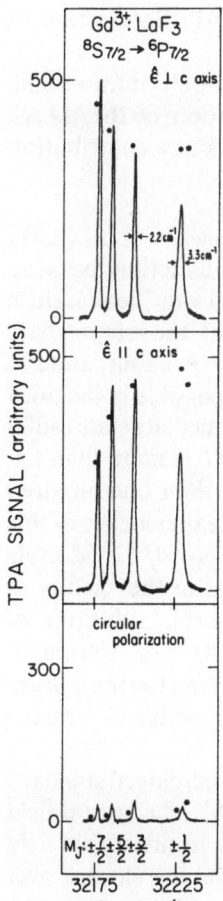

Fig. 2.5. Experimental two-photon excitation recordings of $^6P_{7/2}$ in $Gd^{3+}:LaF_3$ for three polarizations of the excitation beam. Filled circles show Stark component intensities calculated with both second-order and third-order (spin-orbit) terms included. Since the highest energy component is broader than the other components, the asterisk (*) indicates the height this component would have if it were equal in width to the other components. (From [2.22])

that the polarization dependent intensity of each Stark component of $^6P_{7/2}$ was also perfectly explained when the mutual interference among the three terms in expression (2.16a) was taken into account. While this analysis used only the operators (2.16a), inclusion of the operators (2.16b) arising from $4f^6 g$ intermediate states does not materially alter the conclusions, since the various terms enter the calculation with the same sign and similar relative magnitudes. The detailed comparison of this analysis with the observed intensities is presented in Fig. 2.5.

The third order analysis with $V = H_{so}$ also yields an operator of the form $(E^{(1)}E^{(1)})^{(2)} \cdot W^{(12)2}$ which directly links ^8S to ^6D. *Downer* and *Bivas* [2.22] showed, however, that this contribution to the ^6D line strengths was much weaker than the standard second order contribution in the intermediate coupling limit. Likewise, the contribution of the third order operator $(E^{(1)}E^{(1)})^{(2)} \cdot W^{(11)2}$ to $^6P_{5/2}$ is comparable to, but no larger than, the second order contribution. These results corroborated the earlier analysis of the four transitions $^8S_{7/2} \rightarrow {}^6P_{5/2}, {}^6D_{9/2, 3/2, 5/2}$ in terms of the second order mechanism alone. The modest enhancement in linearly polarized line strength for $^8S_{7/2} \rightarrow {}^6D_{7/2}$, shown in Fig. 2.3, was found to be a consequence of the significant admixture of $^6P_{7/2}$ in $^6D_{7/2}$ which arises from the spin-orbit interaction in intermediate coupling.

The third order contribution with $V = H_{so}$, being a purely intra-atomic scalar operator, offered no hope in explaining the violation of the $\Delta J \leq 2$ selection rule which occurred in the $^8S_{7/2} \rightarrow {}^6I_J$ transitions. A new contribution was therefore required for these transitions.

Third Order Contributions in the Static Crystal Field. *Downer* et al. [2.21] introduced a third order term involving static crystal field interactions between the intermediate states ($V = H_{CF}$). The motivation resembled van Vleck's: since the crystal field components which interact significantly with the intermediate electronic states are described by tensor operators of rank two, four, and six, $4f^N \rightarrow 4f^N$ two-photon transitions with $\Delta L, \Delta J \leq 6$ become possible in the third order. Moreover, because of the extended nature of the intermediate state radial wave functions, the crystal field interaction within $4f^{N-1}n'l'$ is more than ten times stronger than the spin-orbit interaction, and more than one hundred times stronger than crystal field interactions within $4f^N$. Clear evidence of this strength had been provided by observed crystal field splittings of $4f^{N-1}5d$ levels in rare earths [2.79, 80], which routinely showed splittings in the vicinity of 15000 cm^{-1}, and which contrasted with typical spin-orbit splittings of 1000 cm^{-1} and typical crystal field splittings within $4f^N$ of 100 cm^{-1}. Consequently the potential of this new contribution for rivalling the third order spin–orbit terms, on the one hand, and the effects of J-mixing within $4f^N$, on the other, were clear from the outset.

The actual evaluation of this new third order term again employed standard tensor operator and second quantization techniques [2.54]. The crystal field operator was expressed in spherical tensor form according to the crystal field parametrization scheme of *Wybourne* [2.18]. Following double closure over levels of the $4f^6 5d$ configuration and a series of operator recouplings, the dominant third order operator acting between $^8S_{7/2}$ and ^6I_J was found to be

$$-\tfrac{5}{7}(\tfrac{26}{35})^{1/2}\varDelta_d^{-2}(E^{(1)}(B^{(4)}E^{(1)})^{(5)})^{(6)}\cdot U^{(6)}, \tag{2.17}$$

where $B^{(4)}$ is a tensor with components equal to the coefficient of the fourth-rank component of the crystal field operator as expressed in the notation of *Wybourne* [2.18]:

$$H_{\mathrm{CF}}=\sum_n \sum_{k,q}[B_q^{(k)}C_q^{(k)}]_n.$$

The $C_q^{(k)}$ are modified spherical harmonic operators $[4\pi/(2k+1)]^{1/2}Y_{kq}$, and the $B_q^{(k)}$ coefficients are determined by the *ground state* charge distribution of the surrounding ligands and by radial integrals of f^7 electrons. The corresponding operator for $4f^6 g$ intermediate states is

$$-\tfrac{3}{77}(\tfrac{10}{91})^{1/2}\varDelta_g^{-2}(E^{(1)}(B^{(4)}E^{(1)})^{(5)})^{(6)}\cdot U^{(6)}$$

$$+\tfrac{25}{143}(\tfrac{1}{21})^{1/2}\varDelta_g^{-2}(E^{(1)}(B^{(6)}E^{(1)})^{(5)})^{(6)}\cdot U^{(6)}$$

$$+5(\tfrac{1}{3003})^{1/2}\varDelta_g^{-2}(E^{(1)}(B^{(6)}E^{(1)})^{(6)})^{(6)}\cdot U^{(6)}$$

$$+\tfrac{8}{39}(\tfrac{5}{77})^{1/2}\varDelta_g^{-2}(E^{(1)}(B^{(6)}E^{(1)})^{(7)})^{(6)}\cdot U^{(6)}. \tag{2.18}$$

Because of the smaller numerical coefficients, the contribution from the operator (2.18) can be safely neglected in this case. A completely general third order operator with $V=H_{\mathrm{CF}}$ for a two-photon transition $l^N \to l^N$ involving intermediate states of $l^{N-1}l'$ is derived in Sect. 2.A.4. The second order operator $(E^{(1)}E^{(1)})^{(2)}\cdot U^{(2)}$ has been omitted from (2.17) since its contribution is zero, or nearly zero, for all of the $^8S_{7/2}\to{}^6I_J$ transitions, even in intermediate coupling. The third order operator (2.17), on the other hand, gives a substantial linkage of $^8S_{7/2}$ to ^6I_J in intermediate coupling, although, being scalar in spin space, it gives no linkage in the Russell-Saunders limit.

The new third order contribution (2.17) was a partial success. The computed line strengths in third order were within an order of magnitude of the observed $^8S_{7/2}\to{}^6I_J$ line strengths, relative to the ^6P and ^6D lines [2.21], as indicated in Fig. 2.3. More importantly, the intensities of the ^6I_J lines relative to each other agreed extremely well with the relative magnitudes of the intermediate coupling reduced matrix elements $([^8S_{7/2}]\|U^{(6)}\|[^6I_J])^2$, as predicted by (2.17).

Nevertheless, two important shortcomings of the new explanation were evident. First, it was still somewhat too weak. Secondly, the predicted polarization anisotropy was completely wrong. The operator (2.17) predicted a roughly twofold *larger* line strength for $E\|c$ in the LaF_3 lattice, whereas the data strikingly showed a tenfold *smaller* line strength. The search therefore continued for additional contributions which could resolve the remaining anomalies.

Fourth Order Contributions. *Downer* and *Bivas* [2.22] pushed their earlier reasoning to its logical conclusion: the introduction of *fourth* order contributions involving both $V=H_{CF}$ and $V=H_{so}$ which would directly link ^8S to ^6I in the Russell-Saunders limit. Such a direct linkage is not possible in lower than fourth order. Evaluation of the fourth order term began with *triple* closure over intermediate states, which were again confined to $4f^6d$ configurations. The recoupling of the four operators which remained then led to a total two-photon absorption operator for $^8S_{7/2} \rightarrow ^6I_J$ which was a sum of the third order operator (2.17) and three new fourth order operators:

$$-\tfrac{5}{7}(\tfrac{26}{35})^{1/2}\Delta_d^{-2}(E^{(1)}(B^{(4)}E^{(1)})^{(5)})^{(6)} \cdot U^{(6)}$$
$$-\tfrac{2}{7}(\tfrac{13}{77})^{1/2}(\zeta_f-\zeta_d)\Delta_d^{-3}(E^{(1)}(B^{(4)}E^{(1)})^{(4)})^{(5)} \cdot W^{(16)5}$$
$$-\tfrac{1}{35}(\tfrac{26}{231})^{1/2}(15\zeta_f-2\zeta_d)\Delta_d^{-3}(E^{(1)}(B^{(4)}E^{(1)})^{(5)})^{(5)} \cdot W^{(16)5}$$
$$-\tfrac{5}{7}(\tfrac{2}{15})^{1/2}(3\zeta_f+2\zeta_d)\Delta_d^{-3}(E^{(1)}(B^{(4)}E^{(1)})^{(5)})^{(6)} \cdot W^{(16)6} . \tag{2.19}$$

Downer [2.24] found the contribution from the two fifth rank operators $W^{(16)5}$ to be negligible. Consequently only the two sixth rank operators remained. The final result was that the fourth order contribution approximately doubled the calculated line strength of each ^6I_J line without altering their relative intensities. The calculated line strengths, though still somewhat small, were thus brought into better agreement with the observed intensities. The calculated polarization anisotropy, however, remained the same as in third order. The picture was, therefore, still not complete.

Quantum Mechanical Interference. *Downer* [2.24] recognized that admixtures of $^6P_{7/2}$ in the ^6I_J wave functions caused by the crystal field could contribute enough intensity to interfere significantly with the above contributions to $^8S_{7/2} \rightarrow ^6I_J$, even though by themselves they could not account for the ^6I_J line strengths. *Downer* and *Bivas* [2.22] therefore evaluated the effects of J-mixing in quantitative detail.[1] From first order perturbation theory they determined that the composition of the $^6I_{7/2}$ wave function included $^6P_{7/2}$ admixtures as follows:

$$|\{^6I_J,M\}\rangle = |[^6I_J,M]\rangle + E_{IP}^{-1}\sum_q B_q^{(6)}\langle f\|C^{(6)}\|f\rangle$$
$$\times \sum_{M'}(-1)^{7/2-M'}\begin{pmatrix}\tfrac{7}{2} & 6 & J \\ -M' & q & M\end{pmatrix}\langle[^6P_{7/2}]\|U^{(6)}\|[^6I_J]\rangle |^6P_{7/2},M'\rangle. \tag{2.20}$$

Each of these wave functions, of course, has comparable admixtures of many other excited states, but only the $^6P_{7/2}$ admixtures were found to influence the

---

[1] Given the painstaking detail with which the effects of crystal field mixing within $4f^N$ were presented throughout [2.22], the baffling section of the paper by *Sztucki* and *Strek* [2.81] which addresses the "neglect" of such effects seems to reflect a misunderstanding of the earlier work

line strengths significantly. With the revised ^6I_J wave functions (2.20), there was now a third significant contribution to the line strengths: namely, the third order scalar operator of *Judd* and *Pooler* [2.20] acting on the $^6P_{7/2}$ admixture. The total two-photon absorption operator for ^6I_J then became a sum of three comparable contributions:

$$-\tfrac{5}{7}(\tfrac{26}{35})^{1/2}\Delta_d^{-2}(E^{(1)}(B^{(4)}E^{(1)})^{(5)})^{(6)}\cdot U^{(6)}$$
$$-\tfrac{5}{7}(\tfrac{2}{15})^{1/2}(3\zeta_f+2\zeta_d)\Delta_d^{-3}(E^{(1)}(B^{(4)}E^{(1)})^{(5)})^{(6)}\cdot W^{(16)6}$$
$$-2(\tfrac{2}{21})^{1/2}(9\zeta_f+\zeta_d)\Delta_d^{-2}(E\cdot E)\cdot W^{(11)0}\,. \tag{2.21}$$

The consequence of including the additional operator was a remarkable explanation of the strong polarization anisotropy of the ^6I_J lines in Gd^{3+} : LaF_3. In particular, it was found throughout the ^6I_J series that for $E\|c$, the scalar operator contributed a term comparable in magnitude, but *opposite* in sign, to the contribution from the two sixth rank operators. The result was a strong destructive interference, and a weaker line strength. For $E\perp c$, on the other hand, all three operators in (2.21) contributed terms of the *same* sign. The result was constructive interference, and an enhanced line strength. Finally, for

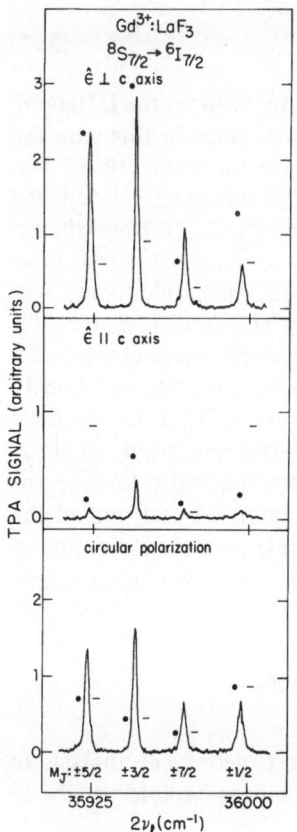

Fig. 2.6. Experimental two-photon excitation recordings of $^6I_{7/2}$ in Gd^{3+} : LaF_3 for three polarizations of the excitation beam. Closed circles show the Stark component intensities calculated by including the third-order (static crystal field) term, the fourth-order (static crystal field spin-orbit) term, and the third-order scalar (spin-orbit) term acting on the $^6P_{7/2}$ admixture in the $^6I_{7/2}$ wave function. Horizontal lines show calculated intensities when the $^6P_{7/2}$ admixture is neglected. (From [2.22])

circular polarization, the contribution from the scalar operator vanishes, which left the earlier computation unaltered. These are precisely the trends exhibited by the observed integrated intensities shown in Fig. 2.3. Figure 2.6 compares in detail the intensities calculated from the full-fledged analysis (2.21) with the observed intensities for the case of $^6I_{7/2}$. Not only were the integrated line strengths well explained, but the intensities of individual Stark components as well. A similar analysis explained the polarization anisotropy of the $^6P_{3/2}$ and $^6D_{1/2}$ lines in quantitative detail. With this final breakthrough, every major piece of the puzzle had seemingly fallen neatly into place.

Ligand Polarization. *Reid* and *Richardson* [2.25] then demonstrated that at least one piece was still missing. Following the procedures used in earlier calculations [2.40, 41] of ligand polarization effects for one-photon absorption, they deduced that the contribution of the dynamic coupling mechanism (2.14) to two-photon $^8S_{7/2} \to {}^6I_J$ transitions in $Gd^{3+}:LaF_3$ was comparable to that of the third order contribution (2.17), which involves only static crystal field perturbations. In particular, they found that following double closure over the intermediate states in the third order term (2.14), and operator recoupling, the dominant contribution to these transitions arose from a sixth-rank operator of the form

$$a\sum_L (E^{(1)}(C^{(6)}(L)E^{(1)})^{(5)})^{(6)} \cdot U^{(6)}, \qquad (2.22)$$

where $C^{(k)}(L)$ is a modified spherical harmonic tensor operator for the Lth ligand, and the summation runs over the perturbing ligands. Since in this sum the $C^{(6)}(L)$ are subject to the same symmetry restrictions as the static crystal field parameters $B^{(4)}$ which appear in (2.17), *Reid* and *Richardson* [2.25] deduced that no new polarization effects would be introduced by (2.22). Nevertheless their numerical estimate of the contribution of (2.22) to $^8S_{7/2} \to {}^6I_J$ for a *single* ligand were comparable to the contribution of (2.17). The explicit sum over ligands was not performed. Fourth order terms in which the static crystal field is replaced by dynamic ligand polarizations are also expected to give comparable contributions. Ligand polarization effects thus probably give rise to intensity contributions comparable to those from the operators (2.21), although their precise magnitude has not been calculated. The relative magnitude of static versus dynamic crystal field contributions may vary markedly in different crystalline hosts. *Sztucki* and *Strek* [2.81] have summarized and unified the third order contributions from the static and dynamic crystal field coupling mechanisms in a generalized formalism, although again no quantitative calculations were made.

2.3.3 Two-Photon Absorption Studies of Eu^{2+} in CaF_2, SrF_2, BaF_2, KCl, and KI

Downer et al. [2.23] extended their observations and theoretical analysis to Eu^{2+}, which is isoelectronic to Gd^{3+}. The motivation was twofold: firstly, to compile extensive two-photon excitation spectra of $4f^7$ levels higher in energy than the low-lying $^6P_{7/2,5/2}$ levels which *Fritzler* and *Schaak* [2.71] had observed, in particular the ^6I and ^6D groups. These levels could not be observed in one-photon absorption because of the overlapping $4f^65d$ bands. Such observations would permit a comprehensive energy level fit comparable in scope to fits of the trivalent lanthanide systems [2.33]. Secondly, to obtain a direct experimental test of the higher order theory of two-photon absorption developed from Gd^{3+} data through careful analysis of relative two-photon line strengths in Eu^{2+}. Because the separation Δ_d of the photon energy from the average energy of the intermediate $4f^65d$ states is about a factor of 3 smaller in Eu^{3+} than in Gd^{3+}, an increase in absolute two-photon line strengths was expected, and this was observed. Moreover, since the nth order contribution to two-photon absorption scales as Δ_d^{2n-2}, transitions which owed their strength primarily to third and fourth order contributions were expected to increase in intensity considerably more than those dominated by second order contributions. In short, the observations of $^8S_{7/2} \to ^6I_J$ bore out this expectation precisely; $^8S_{7/2} \to ^6P_{7/2}$, on the other hand, was one hundred times weaker than expected. The theory of the third order contribution with $V = H_{so}$, which had worked so well for Gd^{3+}, was thus thrown into disarray. Curiously the observed crystal field splitting of $^6P_{7/2}$ was also as much as five times smaller than the calculated splitting. Neither of these anomalies is understood; nor is it known whether they are related to each other.

The $4f^7$ levels which were observed by two-photon absorption in Eu^{2+} : CaF_2, SrF_2 correspond exactly to the levels studied in Gd^{3+}. The same set of J-multiplets was observed later at very similar energies in the KCl host by *Casalboni* et al. [2.82]. Fluorescence was observed from the bottom of the $4f^65d$ band following nonradiative transfer of the $4f^7$ excitation. Fluorescence yield, as in Gd^{3+}, was virtually 100%. Energy level positions for ^6P, ^6I, and ^6D lines were calculated by a comprehensive fitting program developed at Argonne National Laboratory for earlier extensive calculations of trivalent lanthanide energy levels [2.33]. The energy level fitting began with "centers of gravity" calculations of the J-multiplets, followed by detailed fits of the crystal field levels using the cubic (O_h) symmetry crystal field Hamiltonian appropriate for CaF_2 and SrF_2. The final fit yielded mean square deviations of approximately 20 cm^{-1} between observed and calculated levels, comparable to the typical quality of fits obtained with trivalent lanthanide systems. The large number of observed levels which could be incorporated into the fitting program permitted extraction of the most reliable set of atomic and crystal field parameters available to date for the Eu^{2+} systems.

Downer et al. [2.23] noted that the new ^6I and ^6D lines were considerably broader than the ^6P lines, which *Fritzler* and *Schaak* [2.71] had observed earlier. They attributed the greater broadening to the sextet character of the $4f^65d$ levels which surround ^6I and ^6D, which contrasts with the octet character of the $4f^65d$ levels which surround ^6P. Consequently the spin-preserving vibronic coupling was concluded to be more efficient for the former levels. The observations of *Casalboni* et al. [2.82] showed a similar contrast between ^6P linewidths and those of ^6I and ^6D.

Analysis of the relative two-photon intensities showed, as in Gd^{3+}, that the four transitions $^8S_{7/2} \rightarrow ^6P_{5/2}$, $^6D_{9/2,3/2,5/2}$ again agreed with second order predictions. These therefore provided a point of reference for analyzing the remaining intensities. The ^6I_J lines, relative to $^6P_{5/2}$, $^6D_{9/2,3/2,5/2}$, were an order of magnitude stronger than in Gd^{3+}, as shown in Fig. 2.7, in accord with the expected relative enhancement of the higher order contributions. In fact, detailed calculations showed that the fourth order contribution dominates in Eu^{2+}. The dynamic coupling contributions, however, were not considered. The polarization anisotropy of the ^6I_J lines was considerably less pronounced that in Gd^{3+} : LaF_3. This was determined to be the result of the higher substitution site symmetry (O_h) of CaF_2 and SrF_2 compared to LaF_3. The calculated interference and polarization anisotropy were thus much milder than for Gd^{3+} : LaF_3, in agreement with the observations. The observed polarization anisotropy, in fact, was attributed primarily to strong J-mixing among the ^6I_J levels. The intensity analysis of the ^6I_J lines in Eu^{2+} : CaF_2, SrF_2 thus in all respects experimentally confirmed the higher order nature of the transition intensities.

The greatest surprise of the Eu^{2+} results was the weakness of the third order spin-orbit contribution to $^8S_{7/2} \rightarrow ^6P_{7/2,5/2}$ and $^6D_{7/2}$. The unique enhancement of $^6P_{7/2}$ for linear polarization was again observed in Eu^{2+}, strongly suggesting that the same linkage was at work. However, this enhancement was actually *weaker* than in Gd^{3+}, rather than ten times stronger, as predicted from the $(\zeta/E_{df})^2$ scaling law and shown by the crosses in Fig. 2.7. Observations also failed to bear out similar expectations of strengthening of the third order contributions to the $^6D_{7/2}$ and $^6P_{5/2}$ lines. The later observation of a strong $^6P_{7/2}$ line in Eu^{2+} : KCl, KI by *Nunes* et al. [2.83], when the neighboring $^6P_{5/2}$ line was not observed despite a 100-to-1 signal-to-noise ratio, suggested that the strong $^6P_{7/2}$ enhancement may be present in the alkali halide hosts. However, the contrast of linear to circular polarized excitation for $^6P_{7/2}$ was not reported. Furthermore later observations by *Casalboni* et al. [2.82] confirmed the weakness of the third order spin-orbit contributions for $^8S_{7/2} \rightarrow ^6P_{7/2}$, $^6D_{7/2}$ in Eu^{3+} : KCl. The failure to observe $^6P_{5/2}$ in both of these studies may, therefore, have resulted from a strong broadening of this level, perhaps because of a lower energy onset of the sextet levels of $4f^65d$. These observations thus showed that the third order contribution with $V = H_{so}$ was much weaker than expected for all Eu^{2+} transitions in which it was expected to play a significant role.

Including the effects of $4f^6g$ intermediate states through the operator (2.16b) can only worsen the discrepancy. *Downer* et al. [2.23] argued that the problem might be partially resolved if the large spread of $4f^65d$ intermediate state energies in the divalent ion results in the closure approximation becoming invalid. They demonstrated that such a spread could cause the third order contribution to $^6D_{7/2}$ and $^6P_{5/2}$ to vanish, in reasonable agreement with the data. Their arguments, however, did not explain the data well for $^6P_{7/2}$. Further

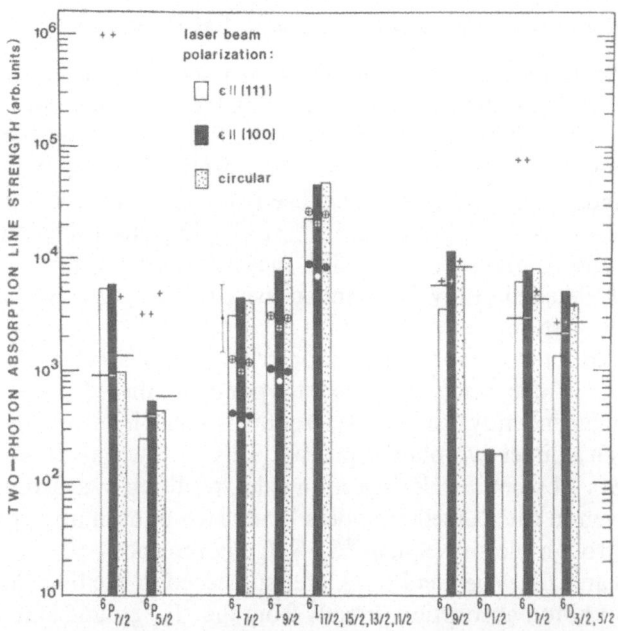

Fig. 2.7. relative two-photon absorption line strengths for Eu^{2+} : CaF_2 for three polarizations of the excitation beam. The detailed notation is the same as in Fig. 2.3. (From [2.23])

analysis along this line will require the computation of higher order terms involving the Coulomb interaction, which is primarily responsible for the spread of the intermediate levels. So far, however, the unexpected weakness of some, but not other, third order contributions to two-photon absorption in the Eu^{2+} ion remains an unsolved theoretical problem.

Nunes et al. [2.83] introduced a novel application of two-photon absorption spectroscopy to study the formation of precipitated phases of Eu^{2+} in the alkali halides KCl and KI. Using different annealing conditions, these authors identified three different structural forms for the precipitated phase of Eu^{2+}, as revealed by the Stark component structure of the $^8S_{7/2} \rightarrow {}^6P_{7/2}$ transition. The presence of a charge compensating vacancy in the neighborhood of the Eu^{2+} ion in the alkali halide hosts allows the ion to diffuse through the lattice at a rate which depends on annealing conditions. Freshly quenched doped samples of KCl and KI revealed only the four $^6P_{7/2}$ Stark components which arise from the single Eu^{2+} ion at a C_{2v} site. After annealing at various temperatures, however, new Stark components appeared which were attributed to dimers and other aggregated phases of the divalent ions. The various phases were clearly identified, and the kinetics of their formation were easily deduced using the two-photon excitation technique.

2.3.4 Two-Photon Absorption Studies of Pr^{3+} in LaF_3, $LaCl_3$, and CaF_2

Several recent studies have used two-photon absorption to access the ultraviolet $^1S_0$ level of Pr^{3+}, the highest of the 13 free ion levels which occur in the $4f^2$ configuration (Fig. 2.8). In many hosts, such as $LaCl_3$, this level had traditionally been impossible to study in one-photon absorption because of the overlapping $4f\,5d$ bands. Consequently it had been the "missing level" in parametric energy level fits for Pr^{3+} [2.84]. In Pr^{3+} : LaF_3, $^1S_0$ falls just below the onset of $4f\,5d$, and *Carnall* et al. [2.85] measured its position as $46985\ cm^{-1}$, although detailed study is inhibited even in this case by the extremely small oscillator strength.

Yen et al. [2.86] used a three-photon excitation scheme to observe $^1S_0$ in Pr^{3+}:LaF_3. One tunable dye laser was used to populate the $^1D_2$ state resonantly. A second independently tunable dye laser then excited ions from the $^1D_2$ state to $^1S_0$ through a direct two-photon process. This scheme was chosen because $^1D_2 \rightarrow {}^1S_0$ obeys all second order selection rules, resulting in a strong two-photon linkage even in the Russell-Saunders limit. Two-photon absorption was monitored through the subsequent $^1S_0 \rightarrow {}^1I_6$ fluorescence.

Cordero-Montalvo and *Bloembergen* [2.87] later succeeded in exciting $^1S_0$ in Pr^{3+} : LaF_3 by two-photon absorption directly from the $^3H_4$ ground state, despite the violation of second order selection rules. Two-photon absorption was monitored through the subsequent ultraviolet fluorescence $^1S_0 \rightarrow {}^1G_4, {}^3F_4$. This observation suggested the presence of sizeable third and fourth order linkages analogous to those which explain $^8S_{7/2} \rightarrow {}^6I_J$ transitions in Gd^{3+} and Eu^{2+}. Fourth order terms link $^3H_4$ and $^1S_0$ directly, although third order terms could contribute in intermediate coupling. The relevant third and fourth order operators take the general form

$$\sum_{k=2,4}\ \sum_{\lambda=3,4,5} [A_{k\lambda}^{(3)}(E^{(1)}(B^{(k)}E^{(1)})^{(\lambda)})^{(4)} \cdot U^{(4)} + A_{k\lambda}^{(4)}(E^{(1)}(B^{(k)}E^{(1)})^{(\lambda)})^{(4)} \cdot W^{(15)4}],$$

$$(2.23)$$

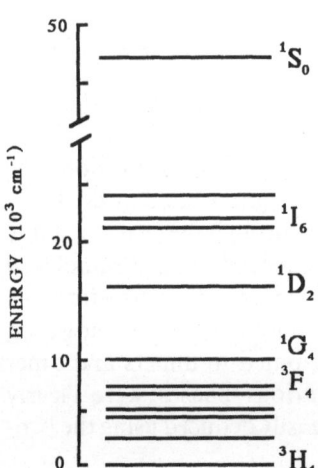

Fig. 2.8. Energy levels of the $4f^2$ configuration of Pr^{3+}. Crystal field splittings are not shown

where $A_{k\lambda}^{(n)}$ are numerical coefficients of the nth order terms. Four third order and four fourth order terms of significant magnitude arise. Detailed calculations of their relative importance, however, have not been made. Dynamic coupling contributions may also be important.

Rana et al. [2.88] extended these observations to the interesting case of $Pr^{3+}:LaCl_3$, where the $^1S_0$ had never been observed previously because of the overlapping $4f\,5d$ band. A single sharp peak resulting from direct two-photon excitation of $^3H_4 \rightarrow {}^1S_0$ was indeed observed above a featureless background. The $^1S_0$ level position was determined to be $46450.6 \pm 5\,cm^{-1}$. Thus $Pr^{3+}:LaCl_3$ became the first case where all 13 levels of the ground configuration of Pr^{3+} had been observed with a high degree of accuracy. An improved set of atomic and crystal field parameters was thus obtainable.

Other recent studies have focused on two-photon excitation of the $4f\,5d$ band in Pr^{3+}, which has attracted interest because of the potential use of transitions from this configuration in broad band tunable solid state lasers. *Lezama* and *de Araujo* [2.89] achieved efficient $4f\,5d$ excitation in 0.5 at.% Pr^{3+}, $Gd^{3+}:LaF_3$ through stepwise two-photon absorption using the $^1D_2$ intermediate level. A tunable dye laser populated the $^1D_2$ level, and the fourth harmonic of a $Nd^{3+}:YAG$ laser beam then accessed the $4f\,5d$ band through a parity-allowed single-photon transition. Strong fluorescence was then observed corresponding to the $^1S_0 \rightarrow {}^1G_4,\,{}^3F_4,\,{}^1D_2,\,{}^1I_6$ emission lines of Pr^{3+}. The authors concluded that $^1S_0$ was populated through relaxation from the $4f\,5d$ configuration. In addition the authors observed strong fluorescence from the $^6P_{7/2}$ level of Gd^{3+} which was also doped in the LaF_3 host lattice. They interpreted this latter fluorescence as evidence of an efficient energy transfer between the $4f\,5d$ configuration of Pr^{3+} and the $4f^7$ configuration of Gd^{3+}. Such energy transfer processes are treated in detail in the article by *Yen* [2.90] in this volume.

2.3.5 Absolute Two-Photon Absorption Cross Sections

The theoretical conclusions presented above have been based on *relative* intensity measurements. The difficulties of measuring *absolute* two-photon absorption cross sections with a pulsed laser source, which can have variable spatial and temporal characteristics, are well documented [2.91–93]. Nevertheless approximate measurements help to evaluate the importance of two-photon absorption in practical situations, and serve as an additional check with theoretical predictions. Consequently several measurements have been reported for $4f^N \rightarrow 4f^N$ transitions. All have involved significant experimental uncertainty, although the recent measurement by *Chase* and *Payne* [2.65] of the fully allowed $^4I_{9/2} \rightarrow {}^4G_{7/2}$ transition in $Nd^{3+}:YAG$ and $Nd^{3+}:YLF$ is the most thorough. This transition occurs near half the wavelength of Nd^{3+} lasers (1.06 µm), and is thus a potential source of nonlinear loss in such lasers, as originally suggested by *Penzkofer* and *Kaiser* [2.64].

The measurement of *Chase* and *Payne* [2.65] compared the luminescence signal induced by two-photon absorption with that induced by a one-photon transition of known absorption cross section. In Nd^{3+}:YAG, peak Stark component cross sections between 1.3 and 18×10^{-52} cm^4 s were measured for linewidths of 1–3 cm^{-1}. These values are comparable to typical cross sections for allowed two-photon transitions with off-resonant intermediate states in other materials [2.2]. The spectrally integrated cross section for $^4I_{9/2} \rightarrow {}^4G_{7/2}$ in Nd^{3+}:YAG was 1.2×10^{-40} cm^4, which was nearly an order of magnitude larger than in Nd^{3+}:YLF. These values were in order-of-magnitude agreement with calculations from a second order theory, although the strong host dependence was not well understood. *Chase* and *Payne* [2.65] concluded that two-photon absorption was not a significant loss mechanism in Nd^{3+} lasers, and that *Penzkofer* and *Kaiser* [2.64] had erred in their earlier assessment of its importance (Sect. 2.3.1). They also concluded that recent suggestions by *Altshuler* et al. [2.94, 95] for cancelling the nonlinear refractive index n_2 of a host lattice with the anomalous frequency-dependent n_2 of rare earth impurities was probably not feasible for low impurity concentrations.

The cross section measurement of the middle Stark component of $^8S_{7/2} \rightarrow {}^6P_{5/2}$ in Gd^{3+} : LaF_3 by *Downer* and *Bivas* [2.22] is notable in being the only one to use a well-characterized *continuous wave* TEM_{00} beam for excitation, thus greatly reducing the experimental uncertainties arising from spatial and temporal variations in a pulsed excitation source. With a linewidth of 3 cm^{-1} a peak cross section of 2×10^{-55} cm^4 s was measured. The corresponding cross section for the strongest Stark component of $^6P_{7/2}$ was 10^{-53} cm^4 s. These smaller values reflect the partially spin-forbidden nature of these transitions, and are also in satisfactory agreement with the appropriate second and third order cross section calculations. *Gayen* and *Hamilton* [2.96] measured a cross section of 5×10^{-54} cm^4 s for the parity-forbidden two-photon zero-phonon transition $4f \rightarrow 5d$ in Ce^{3+} : CaF_2. *Chase* and *Payne* [2.65] have critically reviewed a number of earlier measurements of absolute two-photon absorption cross sections for $4f^N \rightarrow 4f^N$ transitions.

2.3.6 Cooperative Two-Photon Absorption by Rare Earth Ion Pairs

At sufficiently high dopant concentrations, impurity ions begin to interact significantly with each other. Energy transfer processes, which are discussed in the article by *Yen* [2.90] in this volume, are one consequence of such interactions. In addition ion pairs may act cooperatively as dimers in the absorption and emission of light. Such processes closely resemble "laser-induced collisions" [2.97, 98] in gas phase experiments. Recently several investigators have shown that two-photon absorption spectroscopy offers a promising new direction in the study of cooperative absorption and of ion–ion interactions.

Leite and *de Araujo* [2.99] showed theoretically that a distinct asymmetric lineshape similar to a Fano profile [2.100], arising from interference between the transition amplitudes via intermediate states localized on each ion in the pair, is expected in the vicinity of a two-photon cooperative absorption resonance. The two-photon absorption lineshape could thus yield information on the ion–ion interaction potential. Analysis was carried out for Pr^{3+} ion pairs, but the effect has not been observed experimentally. *Vial* and *Buisson* [2.101] have experimentally studied Pr^{3+} ion pairs in LaF_3 by a two-photon absorption technique. For strongly coupled ions the energy levels for single ion excitation become significantly nondegenerate. Consequently resonant double excitation can only be achieved through absorption of two photons of different frequencies. The frequency difference was a measure of the strength of ion–ion coupling. The signature of double excitation was observation of fluorescence at frequencies higher than either excitation frequency. Different classes of ion pairs were studied, as well as the dynamics of the upconverted fluorescence formation and decay.

Expanding on this same work, *Barthem* et al. [2.102] demonstrated experimentally that two-photon absorption by Nd^{3+} ion pairs in $LiYF_4$ could serve as a technique for measuring a narrow homogeneous linewidth within a broad inhomogeneous profile. The first laser selects a class of crystal strains and the second laser, at a slightly different frequency, is absorbed only by pairs already singly excited. An observed narrowing of linewidths in the excitation spectrum demonstrated that the two ions of the pair experienced nearly the same crystalline strain. This technique is most closely related to the "absorption line narrowing" observed in gases [2.103, 104], where a class of atomic velocities is selected instead of crystalline strains. It also resembles the "fluorescence line narrowing" technique [2.105] in which homogeneous linewidths can be observed in resonant fluorescence.

2.3.7 Other Recent Two-Photon Absorption Studies in the Rare Earths

Gayen and *Hamilton* [2.96] studied the two-photon excitation of the $5d$ level from the $^2F_{5/2}$ ground state of $Ce^{3+}:CaF_2$. The parity-forbidden $4f \to 5d$ two-photon transition owes its intensity to configuration mixing by the noncentrosymmetric crystal field of C_{4v} site symmetry, analogous to dipole forbidden single-photon $4f^N \to 4f^N$ transitions. The sharp zero-phonon line and the portion of the sideband caused by single-phonon assistance displayed a similar polarization anisotropy, while the portion arising from multi-photon assistance was isotropic with respect to polarization. *Kramer* and *Boyd* [2.106] have used three-photon $4f^3 \to 4f^2 5d$ transitions to access levels of the excited configuration in $Nd^{3+}:YAG$.

Bleijenberg et al. [2.107] used stepwise two-photon excitation of Tm^{3+} in La_2O_2S to excite preferentially Tm^{3+} levels above the absorption band of the host lattice. This experiment identified a Franck-Condon shifted state above this absorption band for the first time, thus explaining previously observed thermal quenching of $^1I_6$ emissions as thermally promoted crossovers to this Franck-Condon shifted state. *Rao* et al. [2.108] have reported observations of both direct and stepwise two-photon excitation processes in Ho^{3+} in alkaline earth fluoride lattices. These observations explored higher energy two-photon transitions then the earlier work in Ho^{3+} by *Apanasevich* et al. [2.67]. *Munir* et al. [2.109] demonstrated the use of optoacoustic detection of two-photon absorption in rare-earth-doped glasses, which contrasts with and complements the more popular method of fluorescence detection.

2.4 Experimental Studies of Electronic Raman Scattering in Rare Earths

2.4.1 Early Investigations

The first two-photon process to be detected experimentally was Raman scattering, observed in the 1920s using conventional light sources. *Rasetti* [2.110] first observed an electronic Raman transition in the NO molecule in 1932. *Elliott* and *Loudon* [2.111] predicted in 1963 that electronic Raman scattering could be observed in transition metal or lanthanide impurity ions in crystals. Shortly afterward, *Hougen* and *Singh* [2.112,113] observed the electronic Raman transitions $^3H_4 \rightarrow \, ^3F_{2,3,4}$, $^3H_{4,5,6}$ in $Pr^{3+}:LaCl_3$ using a mercury lamp excitation source. All the transitions which they observed obeyed ΔL, $\Delta J \leq 2$, and $\Delta S = 0$, and the reported transition intensities agreed with a second order theory. With the advent of the laser, the potential of Raman spectroscopy was greatly increased. Extensive laser spectroscopic studies of the electronic Raman transitions of the rare earths were conducted by *Koningstein* [2.48, 114–121] and co-workers. Among the more notable achievements of these studies was the clear identification of the antisymmetric contribution to Raman scattering. A striking example was the observation of a strong $^7F_0 \rightarrow \, ^7F_1$ transition in Eu^{3+}, for which the symmetric term proportional to $\langle [^7F_0] \| U^{(2)} \| [^7F_1] \rangle^2$ vanishes, leaving only the antisymmetric contribution [2.116,117]. Other electronic Raman studies in rare earths were conducted by *Wadsack* et al. [2.122], and in transition metal ion impurities by *Christie* and *Lockwood* [2.123,124], *Hoff* et al. [2.125], and *Jones* and *Kuok* [2.126]. This early Raman work has been reviewed by *Clark* and *Dines* [2.5].

2.4.2 Recent Electronic Raman Scattering Studies of Tm^{3+}, Er^{3+}, and Ho^{3+} in Tetragonal Phosphate Host Crystals

Becker et al. [2.26, 27, 127, 128] have recently carried out systematic, quantitative measurements of the intensities and asymmetries of electronic Raman transitions in Tm^{3+}, Er^{3+}, and Ho^{3+}, and have compared these in detail to calculations based on the second order theory of *Axe* [2.28]. In earlier Raman studies such intensity comparisons, when present, had been qualitative, as in the studies of $PrCl_3$ by *Hougen* and *Singh* [2.112,113], *Axe* [2.43], and

Koningstein [2.129], and the study of dysprosium garnets by *Wadsack* et al. [2.122]. The insight provided, and the questions raised, by the new Raman results have been quite different from those which arose out of two-photon absorption work. For example, strong evidence for the importance, and perhaps dominance, of g electrons as intermediate states in both two-photon and one-photon processes has emerged. In addition, strong discrepancies with the second order theory have been found, particularly in Tm^{3+}, even for Raman transitions which obeyed the second order selection rules $\Delta S = 0$, ΔL, $\Delta J \leq 2$. In such cases, the higher order contributions introduced by *Judd* and *Pooler* [2.20], *Downer* et al. [2.21, 22, 24], and *Reid* and *Richardson* [2.25] to explain partially forbidden two-photon transitions are unlikely to play a significant role. The new Raman results have thus set the stage for important new theoretical work. Clearly a complete solution to the puzzle of two-photon rare earth spectra in solids will require careful attention to both Raman scattering and two-photon absorption results.

Experimental Results. The Raman measurements of *Becker* et al. [2.26, 27, 127, 128] were performed both with pure lanthanide phosphate crystals and with dilute lanthanide-doped $LuPO_4$ and YPO_4 crystals at liquid helium temperature. Various lines of an argon-ion laser provided polarized excitation, and polarization-analyzed scattered light was frequency-analyzed in a double monochromator equipped with a photon counting detection system. The measurements were made with the incident light propagating along the crystallographic X axis, with either Y or Z polarization, and with scattered light collected along the Y axis, with either X or Z polarization analysis. The lanthanide ions occupied sites of D_{2d} symmetry.

Guha [2.130] first measured the low temperature Raman spectrum of $TmPO_4$. Figure 2.9 depicts polarized Raman spectra obtained by *Becker* et al. [2.127]. Raman transitions from the lowest crystal field level of the ground $^3H_6$ multiplet to excited levels of $^3H_6$ are shown, although transitions to $^3F_4$ were also observed. Each crystal field level is labelled by an irreducible representation of D_{2d}, as determined by an earlier crystal field analysis of these low-lying levels observed in absorption [2.131]. An asymmetry with respect to interchange of the polarization of incident and scattered photons is expected in the ground multiplet only for transitions of Γ_5 symmetry (in this case $\Gamma_1 \rightarrow \Gamma_5$), of which three were expected and observed. Of these, the weakest one at 30 cm^{-1} showed some asymmetry, whereas the stronger ones at 138 and 280 cm^{-1} appeared symmetric within experimental error. Asymmetry was measured by comparing the $Y(XZ)X$ and the $Y(ZY)X$ spectra, where a notation such as $Y(XZ)X$ refers to scattering geometry as follows: propagation direction of scattered photon (polarization of scattered photon, polarization of incident photon) propagation direction of incident photon. This contrasted with another common notation due to *Damen* et al. [2.132]. Transitions to the $^3F_4$ multiplet, for which $\Delta J = 2$, were symmetric, since only the second rank symmetric Raman tensor (2.6) contributes to the scattering intensity.

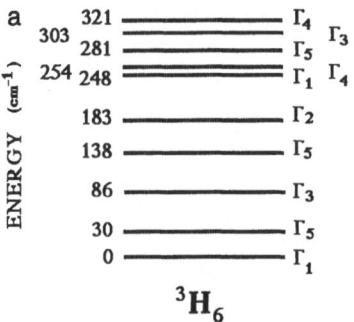

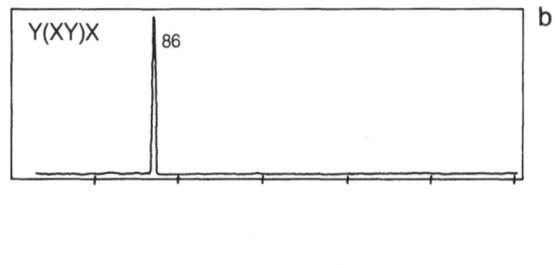

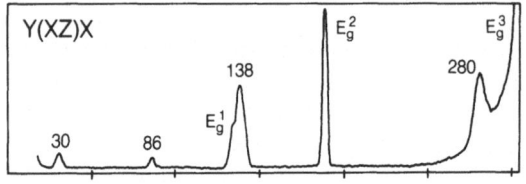

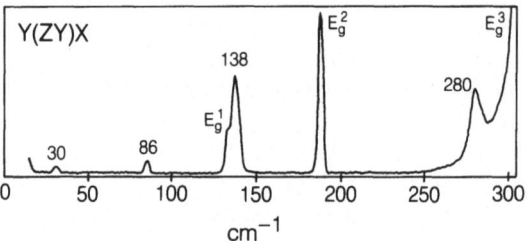

Fig. 2.9. (a) Crystal field levels of the ground $^3H_6$ multiplet of Tm^{3+} in the D_{2d} site of $TmPO_4$. **(b)** 0–300 cm^{-1} Raman scans of $TmPO_4$, taken at approximately 10 K with the 514.5 line of an Ar^+ laser. Lines e_1 through e_4 correspond to electronic transitions to excited crystal field levels of the ground multiplet at 30, 86, 138, and 280 cm^{-1}. Other features are vibronic. Full scale is 25000 counts/s for the XY scan, and 25000 counts/s for the bottom two scans. (From [2.27])

In addition to the Γ_5 transitions, a strong electronic Raman line at 86 cm^{-1} and several vibrational Raman lines were observed. Lines corresponding to crystal field levels at 183, 248, 254, 303, and 321 cm^{-1}, however, were not observed. Spectra obtained with dilute doped crystals showed no essential differences, suggesting the absence of cooperative effects among the lanthanide ions. Similarly, none of the Raman spectra for Tm^{3+} samples depended on excitation wavelength, indicating the absence of resonant enhancement effects.

Becker et al. [2.27, 127] also reported similar polarized Raman spectra for $ErPO_4$ and $HoPO_4$. In $ErPO_4$ strong resonant enhancement effects were observed [2.128] for an argon-ion laser excitation wavelength of 488.0 nm, which is nearly coincident with the one-photon transition $^4I_{15/2} \rightarrow {}^4F_{7/2}$. All other spectra, however, were nonresonant. Of the four nonresonant electronic Raman transitions observed among the crystal field levels of the $^4I_{15/2}$ ground multiplet of Er^{3+}, the three at 33, 53, and 145 cm^{-1} showed a pronounced polarization asymmetry. The exceedingly weak Raman lines observed in Ho^{3+} represented the first observation of electronic Raman scattering in this ion. Several of these lines exhibited the strongest asymmetry observed in any of the ions studied.

Raman Intensity Analysis. *Becker* [2.26, 27, 127] analyzed the polarized Raman intensities in three separate ways: (1) comparisons of calculated and observed asymmetries for the transitions of Γ_5 symmetry; (2) comparisons of calculated and observed relative intensities of transitions to individual Stark components; and (3) comparisons of calculated and observed integrated multiplet→multiplet intensities. We consider each of these analyses in turn.

The degree of asymmetry $I_{XZ,YZ}/I_{ZX,ZY}$ of a Raman transition is calculated from the ratio F_1/F_2 as given by (2.11). When intermediate configurations other than $4f^{N-1}d$ are neglected, and an average energy denominator $E_{df} \approx 100000 \text{ cm}^{-1}$ is assumed, (2.11) yields $F_1/F_2 \approx +0.25$. However, *Becker* et al. [2.26, 27, 127] found that Raman asymmetries calculated from this ratio represented the observed asymmetries poorly, as shown by the comparison in Table 2.1. Most notable are the large discrepancies for the Γ_5 transitions of $TmPO_4$, where in all cases the calculated asymmetry was opposite to that observed. In unpublished data for $HoPO_4$, *Becker* [2.27] found still larger discrepancies. By choosing significantly smaller values of the ratio F_1/F_2, on the other hand, *Becker* et al. [2.26, 27] found a great improvement in the agreement between calculated and observed asymmetries, also shown in Table 2.1. In fact, for both $TmPO_4$ and $ErPO_4$, values of F_1/F_2 near zero yielded the best agreement. As discussed in Sect. 2.2.3, a ratio $F_1/F_2 \approx 0$ signifies the approximately equal importance of $4f^{N-1}g$ and $4f^{N-1}d$ intermediate states. In the case of $HoPO_4$, a strongly *negative* ratio $F_1/F_2 = -0.22$ actually yielded the best fit, signifying the *dominance* of g intermediate states over those of d character. These conclusions were quite surprising in light of the several-fold higher

Table 2.1. Observed and calculated asymmetries of electronic Raman transitions of Γ_5 symmetry. (From [2.27])

Ion	Transition [cm^{-1}]	Observed asymmetry	Calculated asymmetry	
			$(4f^{N-1}nd$ intermediate states only)	$(4f^{N-1}nd$ and $4f^{N-1}ng$ intermediate states)
Tm^{3+}	30	3.0	0.1[a]	3.1[b]
	138	0.9	5.8[a]	0.8[b]
	280	1.2	0.5[a]	1.1[b]
Er^{3+}	33	5.3	3.5[a]	5.2[c]
	53	0.2	0.4[a]	0.5[c]
	145	0.6	1.9[a]	1.1[c]
Ho^{3+}	68	∞	0.0049[a]	210[d]
	82	∞	0.0061[a]	430[d]
	193	0.15	230[a]	0.1[d]
	251	0.75	0.0049[a]	29000[d]

[a] $F_1/F_2 = +0.25$; [b] $F_1/F_2 = -0.03$; [c] $F_1/F_2 = +0.03$; [d] $F_1/F_2 = -0.22$.

energy of the $4f^{N-1}g$ orbitals compared to $4f^{N-1}d$ orbitals in the free lanthanide ions.

As further support for their argument, *Becker* et al. [2.26, 27] calculated the relative intensities and polarization anisotropies of *all* Raman transitions studied, based on assumptions of either $4f^{N-1}d$ or combined $4f^{N-1}g$ and $4f^{N-1}d$ intermediate states. As the comparison in Table 2.2 shows, the latter

Table 2.2. Observed and calculated intensities of electronic Raman transitions. (From [2.27])

Ion	Transition	Polarization	Observed intensity	Calculated intensity	
				$(4f^{N-1}nd$ intermediate states)[a]	$(4f^{N-1}nd$ and ng intermediate states)[b]
Tm^{3+}	30 cm^{-1}	XZ, YZ	3.7	13.2	13.0
		ZX, ZY	1.2	86.4	4.2
	86	XY, YX	228.0	228.0	228.0
	138	XZ, YZ	21.6	51.6	22.0
		ZX, ZY	24.0	7.4	27.0
	183	XY, YX	Not observed	11.1	0.2
	248	XX, YY	Not observed	10.2	10.2
		Z, Z	Not observed	41.0	41.0
	254	XX, YY	Not observed	47.6	47.6
	280	XZ, YZ	23.4	11.2	17.6
		ZX, ZY	20.3	23.4	16.0
	303	XY, YX	Not observed	0.4	0.4
	321	XX, YY	Not observed	7.8	7.8
Er^{3+}	33	XX, YY	Not available	0.6	0.6
		XY, YX	15.2	15.2	15.2
		XZ, YZ	3.0	46.6	5.0
		ZX, ZY	0.6	13.1	0.9
	53	XX, YY	Not available	0.04	0.04
		ZZ	Not observed	0.2	0.2
		XY, YX	0.9	14.6	0.2
		XZ, YZ	0.9	1.8	4.5
		ZX, ZY	6.1	42.9	9.4
	105	XX, YY	Not available	2.0	2.0
		ZZ	1.5	7.8	7.8
		XY, YX	Not observed	1.7	0.02
		XZ, YZ	Not observed	0.6	0.5
		ZX, ZY	Not observed	0.4	0.5
	145	XX, YY	Not available	0.2	0.2
		XY, YX	1.8	8.4	8.4
		XZ, YZ	0.6	4.9	3.8
		ZX, ZY	0.9	2.5	3.5

[a] $F_1/F_2 = +0.25$; [b] $F_1/F_2 = -0.03$ for Tm^{3+}, $+0.03$ for Er^{3+}.

Table 2.3. Multiplet→multiplet electronic Raman scattering intensities for Tm^{3+}, Er^{3+}, and Ho^{3+}. (From [2.27])

Ion	Transition	Observed intensity	Calculated intensity
Tm^{3+}	$^3H_6 \rightarrow {}^3H_6$	100	100
	$^3H_6 \rightarrow {}^3F_4$	12	67
Er^{3+}	$^4I_{15/2} \rightarrow {}^4I_{15/2}$	39	18
Ho^{3+}	$^5I_8 \rightarrow {}^5I_8$	4	51

assumption, corresponding to small values of F_1/F_2, again yielded a superior overall fit to the observed relative intensities. To be sure, significant discrepancies still remained, most notably the total absence of the 248 cm^{-1} and 254 cm^{-1} lines of $TmPO_4$ in light of their strong predicted intensities. Nevertheless the assumption of a significant contribution from $4f^{N-1}g$ intermediate states greatly improved the representation of the Raman data in all respects. This conclusion was the most significant finding of the new Raman results.

Relative multiplet→multiplet Raman scattering intensities were also measured and calculated for $TmPO_4$, $ErPO_4$, and $HoPO_4$. Intensities observed from the spectra of different ions were calibrated by reference to a common scattering intensity for the E_g^5 phonon in each phosphate crystal. Integrated intensities were calculated using (2.12) with values of F_1/F_2 determined from the previous fits. Table 2.3 shows the results of the comparison, in which the calculated and observed integrated intensities of $^3H_6 \rightarrow {}^3H_6$ in Tm^{3+} were set arbitrarily equal. While the agreement was fair, discrepancies of factors of 6 and 12, respectively, for $^3H_6 \rightarrow {}^3F_4$ in Tm^{3+} and $^5I_8 \rightarrow {}^5I_8$ in Ho^{3+} were clearly evident.

The Importance of g Electrons in Lanthanide Transition Intensities. The Raman asymmetry and relative intensity measurements of *Becker* et al. [2.26, 27, 127] provided the most direct experimental evidence to date of the importance of orbitals of g character in lanthanide spectroscopy. These results should prompt a re-examination of both one-photon and two-photon absorption intensity data, where the influence of g electrons could well be equally important, even if less directly observable. Indeed, as mentioned in Sect. 2.2.2, intimations of the importance of g electrons appear in several earlier studies of lanthanide transition intensities. In the case of two-photon absorption, the importance of intermediate states of g character may be felt in third and fourth order contributions involving the crystal field operator, since in these cases the sixth-rank components $\Sigma_q B_q^{(6)} C_q^{(6)}$, which act on g but not d electrons, come into play in addition to the second- and fourth-rank components, as seen in (2.18). A comprehensive picture can emerge only from complementary study of Raman, one-photon and two-photon absorption data in the same samples.

The microscopic physical nature of the intermediate states of g character poses a separate and equally challenging problem. While the free ion picture lends little support to the importance of $4f^{N-1}g$ intermediate states, such states are clearly altered greatly in a crystalline environment. A molecular orbital description in which lanthanide nd and ng orbitals are hybridized with ligand valence orbitals must therefore represent the actual intermediate states more accurately. Linear combinations of such orbitals which transform like the components of a g-electron wave function would constitute effective intermediate states of g character. Nevertheless a general microscopic theory of such states and their role in lanthanide optical transitions is lacking. A change in crystalline environment should strongly influence the symmetry of such molecular orbital states. Thus from an experimental point of view, complementary Raman data for lanthanides in different crystalline hosts would be extremely valuable. Measurements of direct sum frequency absorption from sources at different frequencies would also provide valuable asymmetry data analogous to that obtained in electronic Raman scattering.

Remaining Discrepancies in Electronic Raman Scattering Results. While the introduction of $4f^{N-1}g$ intermediate states accounted for many of the apparent anomalies in the recent Raman data, other discrepancies remained unexplained. The most important of these were the missing 248 cm^{-1} and 254 cm^{-1} lines in $TmPO_4$, and the significant discrepancies in predicting integrated multiplet→multiplet intensities. Measurement error resulting from such causes as polarization leakage, accidental resonant enhancement, or cooperative ion–ion effects was carefully analyzed, and cannot account for discrepancies of this magnitude. The "missing" Raman levels may actually be severely broadened, perhaps because of accidental degeneracy with a phonon energy. Such an effect should be sensitive to the crystalline environment. Clearly a combination of new experimental work and theoretical analysis will be necessary to unravel this new puzzle.

2.5 Future Directions

Two-photon spectroscopy has brought about a much deeper understanding of the optical properties of rare-earth-doped solids. The most significant findings have been the strong influence of spin-orbit and crystal field interactions among intermediate states upon intra-$4f^N$ transitions, the importance of ligand polarization effects, the importance of intermediate states of g character, and the extensive observation of sharp $4f^N$ levels embedded in broad $4f^{N-1}5d$ bands. While the basic theoretical framework for the two-photon transitions is now in place, significant details – the weakness of third order spin-orbit contributions in Eu^{2+} (Sect. 2.3.3), the puzzling behavior of the $^8S_{7/2} \rightarrow {}^6D_{9/2}$

transition in $Gd^{3+}:LaF_3$ (Sect. 2.3.2), the missing 248 cm^{-1} and 254 cm^{-1} Raman lines of $TmPO_4$ (Sect. 2.4.2) – still demand theoretical attention.

In a larger context, the re-analysis of single photon data in the light of the findings of two-photon spectroscopy, particularly the importance of third order contributions and g orbital contributions, remains largely unexplored. Furthermore, application of two-photon spectroscopy to related solids, such as transition-metal- and actinide-doped lattices, offers the promise of further experimental discoveries. These, along with yet unforeseen problems, should continue to stimulate theoretical and experimental interest in the two-photon spectroscopy of solids.

2.A Appendix

2.A.1 Commutation and Orthonormality Relations for Creation and Annihilation Operators

$$a_i^+ a_j^+ + a_j^+ a_i^+ = 0,$$

$$a_i a_j + a_j a_i = 0,$$

$$a_i a_j^+ + a_j^+ a_i = \delta_{ij}.$$

2.A.2 Spherical Tensor Operators Expressed in Second Quantized Form

$$\sum_n h_n = \sum_{i,j} a_i^+ \langle i|h|j \rangle a_j$$

(general form for sums of single-particle operators)

$$\sum_n (C^{(k)})_n = (-1)^l (2)^{1/2} (2l+1)(2k+1)^{-1/2} \begin{pmatrix} l & k & l \\ 0 & 0 & 0 \end{pmatrix} (a^+ a)^{(0k)}$$

(modified spherical harmonic operator, k even)

$$\sum_n s_n \cdot l_n = [l(l+1)(2l+1)/2]^{1/2} (a^+ a)^{(11)0}$$

(spin-orbit operator)

$$E \cdot \sum_n r_n = (-1)^{l'} [2(2l+1)(2l'+1)/3]^{1/2} \begin{pmatrix} l & 1 & l' \\ 0 & 0 & 0 \end{pmatrix} \langle nl|r|n'l' \rangle$$
$$\times E^{(01)} \cdot [(a^+ b)^{(01)} - (b^+ a)^{(01)}]$$

(electric dipole operator)

a, a^+ annihilate and create the $2(2l+1)$ states of the l electron; b, b^+ annihilate and create the $2(2l'+1)$ states of an opposite parity l' electron.

2.A.3 Relationship Between Coupled Creation and Annihilation Operators

$$(aa^+)^{(kk')} + (-1)^{2l+2s-k-k'}(a^+a)^{(kk')} = \delta(k,0)\delta(k',0)(2s+1)^{1/2}(2l+1)^{1/2}.$$

2.A.4 Derivation of Third-Order Operator Involving the Crystal Field Interaction

We consider a two-photon transition between two states belonging to the configuration l^N with intermediate states belonging to $l^{N-1}l'$. We let a^+ and a create and annihilate the $2(2l+1)$ states of the l electron, while b^+ and b do the same for the $2(2l'+1)$ states of the l' electron. The crystal field interacts much more strongly with the l' electron, so we neglect its interaction with l electrons. The crystal field operator can then be written

$$H_{CF} = \sum_n \sum_k (B^{(k)} \cdot C^{(k)})_n$$
$$= \sum_n \sum_{i,j} b_i^+ \langle n'l'm_{l'i}|B^{(k)} \cdot C^{(k)}|n'l'm_{l'j}\rangle b_j. \qquad (2.A.1)$$

The matrix element is evaluated by the Wigner-Eckart theorem (Eq. 5.4.1 of [2.51]), and with the help of Eqs. 3.7.3 and 5.4.6 of [2.51] we convert (2.A.1) to

$$H_{CF} = (-1)^{l'}(2)^{1/2}(2l'+1)\sum_k (2k+1)^{-1/2}\begin{pmatrix} l' & k & l' \\ 0 & 0 & 0 \end{pmatrix} B^{(0k)} \cdot (b^+b)^{(0k)}. \qquad (2.A.2)$$

Following double closure over the levels of $l^{N-1}l'$, the third order two photon absorption operator which must be re-coupled is

$$(-1)^l[2(2l+1)(2l'+1)^{2/3}](2)^{1/2}\begin{pmatrix} l & l & l' \\ 0 & 0 & 0 \end{pmatrix}^2$$
$$\times \Delta_{ll'}^{-2}\sum_k (2k+1)^{-1/2}\begin{pmatrix} l' & k & l' \\ 0 & 0 & 0 \end{pmatrix}$$
$$\times [E^{(01)} \cdot (a^+b)^{(01)}B^{(0k)} \cdot (b^+b)^{(0k)}E^{(01)} \cdot (b^+a)^{(01)}], \qquad (2.A.3)$$

where $\Delta_{ll'}$ is the average energy denominator. We rewrite the crystal field part of the operator as

$$B^{(0k)} \cdot (b^+b)^{(0k)} = (2k+1)^{1/2}(B^{(0k)}(b^+b)^{(0k)})^{(00)}$$
$$= (2k+1)^{1/2}\langle(1/2\ 1/2)0\ 0\ 0|1/2(1/2\ 0)1/2\ 0\rangle$$
$$\times \langle(l'l')k\ k\ 0|l'(l'\ k)l'\ 0\rangle((B^{(0k)}b^+)^{(1/2\ l')}b)^{(00)}. \qquad (2.A.4)$$

The re-coupling coefficients are converted to $6-j$ symbols via Eq. 6.1.5 of [2.51], and are both equal to one. In an exactly analogous manner, we convert the electric dipole operator to

$$E^{(01)} \cdot (b^+a)^{(01)} = -(3)^{1/2}((E^{(01)}a)^{(1/2\ l')}b^+)^{(00)}, \qquad (2.A.5a)$$

$$E^{(01)} \cdot (a^+b)^{(01)} = -(3)^{1/2}((E^{(01)}a^+)^{(1/2\ l')}b)^{(00)}. \qquad (2.A.5b)$$

We now substitute these results for the second two operators in the brackets in (2.A.3), then re-couple again to obtain

$$-(3)^{1/2}(2k+1)^{1/2}((B^{(0k)}b^+)^{(1/2\ l')}b)^{(00)}((E^{(01)}a)^{(1/2\ l')}b^+)^{(00)}$$

$$= +(3)^{1/2}(2k+1)^{1/2}\langle(1/2\ 1/2)0(1/2\ 1/2)0\ 0|(1/2\ 1/2)0(1/2\ 1/2)0\ 0\rangle$$

$$\times \langle(l'l')0(l'l')0\ 0|(l'l')0(l'l')0\ 0\rangle$$

$$\times ((B^{(0k)}b^+)^{(1/2\ l')}(E^{(01)}a)^{(1/2\ l')})^{(00)}(bb^+)^{(00)} + Z.\qquad(2.A.6)$$

The minus sign has appeared because an anticommutation occurs in passing b to the right of a. The symbol Z denotes a sum of terms proportional to $(bb^+)^{(mn)}$ for which $(mn) \neq (00)$. From Sect. 2.A.3, $(bb^+)^{(mn)} = (-1)^{m+n}(b^+b)^{(mn)}$. Since the final state in the transition contains no l' electrons, $b|f\rangle = 0$, so the matrix element of Z between levels of l^N vanishes. As for the remaining term, we use $(bb^+)^{(00)}|f\rangle = (2)^{1/2}(2l'+1)^{1/2}|f\rangle$, and convert the re-coupling coefficients via Eqs. 6.4.2, 6.4.3, and 6.4.14 of [2.15] to $6-j$ symbols, which are found to be equal to $1/2$ and $(2l'+1)^{-1}$, respectively. An additional re-coupling isolates b^+ from the other operators:

$$(B^{(0k)}b^+)^{(1/2\ l')}(E^{(01)}a)^{1/2\ l')})^{(00)}$$

$$= -\langle(1/2\ 0)1/2\ 1/2\ 0|1/2(0\ 1/2)1/2\ 0\rangle$$

$$\times \langle(l'\ k)l'\ l'\ 0|l'(k\ l')l'\ 0\rangle \times (((B^{(0k)}(E^{(01)}a)^{(1/2\ l')})^{(1/2\ l')})b^+)^{(00)},$$

where the re-coupling coefficients are both equal to one. Using these results, we now rewrit (2.A.6) and combine it with the left-hand electric dipole operator [rewritten as 2.A.5b)], so that the bracketed operator in (2.A.3) becomes

$$+3(1/2)^{1/2}(2k+1)^{1/2}(2l'+1)^{-1/2}$$

$$\times ((E^{(01)}a^+)^{(1/2\ l')}b)^{(00)}((B^{(0k)}(E^{(01)}a)^{(1/2\ l')})^{(1/2\ l')}b^+)^{(00)}.\qquad(2.A.7)$$

Now in close analogy to (2.A.6), we unite b and b^+ on the right-hand side by rewriting (2.A.7) as

$$-3(1/2)^{1/2}(2k+1)^{1/2}(2l'+1)^{1/2}$$

$$\times \langle(1/2\ 1/2)0(1/2\ 1/2)0\ 0|(1/2\ 1/2)0(1/2\ 1/2)0\ 0\rangle$$

$$\times \langle(l'\ l')0(l'\ l')0\ 0|(l'\ l')0(l'\ l')0\ 0\rangle$$

$$\times ((E^{(01)}a^+)^{(1/2\ l')}(B^{(0k)}(E^{(01)}a)^{(1/2\ l')})^{(1/2\ l')})^{(00)}(bb^+)^{(00)} + Z$$

$$= -(3/2)(2k+1)^{1/2}(2l'+1)^{-1}((E^{(01)}a^+)^{(1/2\ l')}(B^{(0k)}(E^{(01)}a)^{(1/2\ l')})^{(1/2\ l')})^{(00)}.$$

$$(2.A.8)$$

The comments following (2.A.6) apply here as well. We now isolate a by rewriting the right-hand operator as

$$(B^{(0k)}(E^{(01)}a)^{(1/2\ l')})^{(1/2\ l')}$$

$$=\sum_t \langle (1/2\ 0)\,1/2\ 0\ 1/2\,|\,1/2\,(0\ 0)0\ 1/2\rangle\langle(1\ l)l'\ k\ l'|l\,(1\,k)t\ l'\rangle$$

$$\times\,((B^{(0k)}E^{(01)})^{(0t)}a)^{(1/2\ l')}$$

$$=(2l'+1)^{1/2}\sum_t(2t+1)^{1/2}\begin{Bmatrix}l&1&l'\\k&l'&t\end{Bmatrix}((B^{(0k)}E^{(01)})^{(0t)}a)^{(1/2\ l')}\ . \qquad (2.A.9)$$

The operators a^+ and a can then be united through the equation

$$((E^{(01)}a^+)^{(1/2\ l')}((B^{(0k)}E^{(01)})^{(0t)}a)^{(1/2\ l')})^{(00)}$$

$$=\sum_n \langle(0\ 1/2)\,1/2\,(0\ 1/2)\,1/2\ 0\,|\,(0\ 0)0\,(1/2\ 1/2)0\ 0\rangle$$

$$\times\,\langle(1\ l)l'\,(t\ l)l'\ 0\,|\,(1\ t)n\,(l\ l)n\ 0\rangle$$

$$\times\,((E^{(01)}(B^{(0k)}E^{(01)})^{(0t)})^{(0n)}(a^+a)^{(0n)})^{(00)}$$

$$=(-1)^t((2l'+1)/2)^{1/2}\sum_n(2n+1)^{1/2}\begin{Bmatrix}1&l&l'\\l&t&n\end{Bmatrix}(E^{(1)}(B^{(k)}E^{(1)})^{(t)})^{(n)}\cdot U^{(n)}\,, \qquad (2.A.10)$$

where we have made use of the relation $(a^+a)^{(n)}=-[(1/2)(2n+1)]^{1/2}U^{(n)}$. Applying (2.A.9) and (2.A.10) to (2.A.8), the third order two-photon absorption operator (2.A.3) becomes

$$(-1)^{l+1}(2l+1)(2l'+1)^2\begin{pmatrix}l&1&l'\\0&0&0\end{pmatrix}^2 \varDelta_{ll'}^{-2}$$

$$\times\sum_{k,t,n}(-1)^t(2t+1)^{1/2}(2n+1)^{1/2}\begin{pmatrix}l'&k&l'\\0&0&0\end{pmatrix}\begin{Bmatrix}l&1&l'\\k&l'&t\end{Bmatrix}\begin{Bmatrix}1&l&l'\\l&t&n\end{Bmatrix}$$

$$\times(E^{(1)}(B^{(k)}E^{(1)})^{(t)})^{(n)}\cdot U^{(n)}\,, \qquad (2.A.11)$$

which is the desired general result. To obtain the specific result (2.17) for f^N, with intermediate states $f^{N-1}d$, substitute $l=3$ and $l'=2$. To obtain the result (2.18) for f^N, with intermediate states $f^{N-1}g$, substitute $l=3$ and $l'=4$.

Acknowledgements. I would like to acknowledge the financial support of the Robert A. Welch Foundation during preparation of this manuscript. It is also a pleasure to acknowledge the contributions of my friends and colleagues over the years which made this chapter possible, especially Nicolaas Bloembergen, Brian Judd, Hannah and Hank Crosswhite, Cesar Cordero-Montalvo, Albert Bivas, Mario Dagenais, Reinhard Neumann, Philippe Becker, and Bill Yen.

References

2.1   W. Kaiser, C.G.B. Garrett: Phys. Rev. Lett. **7**, 229 (1961)
2.2   H. Mahr: "Two-Photon Absorption Spectroscopy", in *Quantum Electronics: Nonlinear Optics*, Vol. 1, ed. by H. Rabin, C.L. Tang (Academic, New York 1975) p. 287
2.3   D.S. Chemla, A. Maruani: Prog. Quantum Electron. **8**, 1 (1982)
2.4   J.A. Koningstein, O.S. Mortensen: "Electronic Raman Transitions", in *The Raman Effect*, Vol. 2, ed. by A. Anderson (Dekker, New York 1973) p. 519
2.5   R.J.H. Clark, T.J. Dines: "Electronic Raman Spectroscopy", in *Advances in Infrared and Raman Spectroscopy*, Vol. 9, ed. by R.J.H. Clark, R.E. Hester (Heyden, London 1982) p. 282
2.6   N. Bloembergen: J. Lumin. **31/32**, 23 (1984)
2.7   J. Becquerel: C.R. Acad. Sci. **142**, 775 (1906)
2.8   J. Becquerel: Le Radium **4**, 328 (1907)
2.9   J. Becquerel, H. Kammerlingh Onnes: Proc. Acad. Amsterdam **10**, 592 (1908)
2.10   J.H. van Vleck: J. Phys. Chem. **41**, 67 (1937)
2.11   M. Mayer: Phys. Rev. **60**, 184 (1941)
2.12   R. Tomaschek: Phys. Z. **33**, 878 (1932); Nature **130**, 740 (1932)
2.13   S. Freed: Phys. Rev. **38**, 2122 (1931)
2.14   G. Racah: Phys. Rev. **62**, 438 (1942)
2.15   G. Racah: Phys. Rev. **63**, 367 (1943)
2.16   G. Racah: Phys. Rev. **76**, 1352 (1949)
2.17   G.H. Dieke: *Spectra and Energy Levels of Rare Earth Ions in Crystals* (Wiley, New York 1968)
2.18   B.G. Wybourne: *Spectroscopic Properties of Rare Earths* (Wiley, New York 1965)
2.19   S. Hüfner: *Optical Spectra of Transparent Rare Earth Compounds* (Academic, New York 1978)
2.20   B.R. Judd, D.R. Pooler: J. Phys. C **15**, 591 (1982)
2.21   M.C. Downer, A. Bivas, N. Bloembergen: Opt. Commun. **41**, 335 (1982)
2.22   M.C. Downer, A. Bivas: Phys. Rev. B **28**, 3677 (1983)
2.23   M.C. Downer, C.D. Cordero-Montalvo, H. Crosswhite: Phys. Rev. B **28**, 4931 (1983)
2.24   M.C. Downer: "Two-Photon Spectroscopy of Rare Earth Ions in Condensed Matter Environments"; Ph. D. Thesis, Harvard University (1983)
2.25   M.F. Reid, F.S. Richardson: Phys. Rev. B **29**, 2830 (1984)
2.26   P.C. Becker, N. Edelstein, B.R. Judd, R.C. Leavitt, G.M.S. Lister: J. Phys. C **18**, L 1063 (1985)
2.27   P.C. Becker: "Electronic Raman Scattering in Rare Earth Phosphate Crystals"; Ph. D. Thesis, Lawrence Berkeley Laboratory, University of California, Berkeley (1986)
2.28   J.D. Axe, Jr.: Phys. Rev. **136**, A42 (1964)
2.29   B.R. Judd: Phys. Rev. **127**, 750 (1962)
2.30   G.S. Ofelt: J. Chem. Phys. **37**, 511 (1962)
2.31   G.H. Dieke, H.M. Crosswhite: Appl. Opt. **2**, 675 (1963)
2.32   C.W. Nielson, G.F. Koster: *Spectroscopic Coefficients for the p^N, d^N, and f^N Configurations* (MIT Press, Cambridge, MA 1964)
2.33   W.T. Carnall, H. Crosswhite, H.M. Crosswhite: "Energy Level Structure and Transition Probabilities of Trivalent Lanthanides in LaF_3"; Argonne National Laboratory Report (1977)
2.34   W.T. Carnall, P.R. Fields, B.G. Wybourne: J. Chem. Phys. **42**, 3797 (1965)
2.35   W.T. Carnall, P.R. Fields, K. Rajnak: J. Chem. Phys. **49**, 4412 (1968)
2.36   R.D. Peacock: Struc. Bonding **22**, 83 (1975)
2.37   C.K. Jorgensen, B.R. Judd: Mol. Phys. **8**, 281 (1964)
2.38   S.F. Mason, R.D. Peacock, B. Stewart: Chem. Phys. Lett. **29**, 149 (1974)
2.39   S.F. Mason, R.D. Peacock, B. Stewart: Mol. Phys. **30**, 1829 (1975)
2.40   B.R. Judd: J. Chem. Phys. **70**, 4830 (1979)

2.41   M.F. Reid, F.S. Richardson: J. Chem. Phys. **79**, 5735 (1983)
2.42   B.G. Wybourne: J. Chem. Phys. **48**, 2596 (1968)
2.43   J.D. Axe, Jr.: J. Chem. Phys. **39**, 1154 (1963)
2.44   W.F. Krupke: Phys. Rev. **145**, 325 (1966)
2.45   P.J. Becker: Phys. Status Solidi (b) **43**, 583 (1971)
2.46   M. Hasunama, K. Okada, Y. Kato: Bull. Chem. Soc. Jpn. **57**, 3036 (1984)
2.47   M. Goeppert-Mayer: Ann. Phys. (Leipzig) **9**, 273 (1931)
2.48   J.A. Koningstein, O.S. Mortensen: Phys. Rev. **168**, 75 (1968)
2.49   J.C. Slater: *Quantum Theory of Atomic Structure* (McGraw-Hill, New York 1960)
2.50   B.R. Judd: *Operator Techniques in Atomic Spectroscopy* (McGraw-Hill, New York 1963)
2.51   A.R. Edmonds: *Angular Momentum in Quantum Mechanics* (Princeton University Press, Princeton 1960)
2.52   J.R. Schrieffer: *Theory of Superconductivity* (Benjamin, New York 1964)
2.53   A.M. Lane: *Nuclear Theory* (Benjamin, New York 1964)
2.54   B.R. Judd: *Second Quantization in Atomic Spectroscopy* (Johns Hopkins Press, Baltimore 1967)
2.55   E. Bayer, G. Schaak: Phys. Status Solidi **41**, 827 (1970)
2.56   M. Inoue, Y. Toyazawa: J. Phys. Soc. Jpn. **20**, 363 (1965)
2.57   T.R. Bader, A. Gold: Phys. Rev. **171**, 997 (1968)
2.58   S. Singh, I.E. Geusic: Phys. Rev. Lett. **17**, 865 (1966)
2.59   B.V. Ershov, Y.P. Pimenov, A.M. Prokhorov, V.B. Federov: Dokl. Akad. Nauk SSSR **172**, 309 (1967) [Sov. Phys. – Dokl. **12**, 47 (1967)]
2.60   B.V. Ershov, Y.P. Pimenov, A.M. Prokhorov, V.B. Federov: Sov. Phys. – Solid State **10**, 1300 (1968)
2.61   P.P. Sorokin, N. Braslau: IBM J. Res. Dev. **8**, 177 (1964)
2.62   R.I. Gintoft, A.G. Makhanek: J. Appl. Spectrosc. (USSR) **14**, 406 (1971)
2.63   P.A. Apanasevich, R.I. Gintoft, A.G. Makhanek: J. Appl. Spectrosc. (USSR) **16**, 323 (1972)
2.64   A. Penzkofer, W. Kaiser: Appl. Phys. Lett. **21**, 427 (1972)
2.65   L.L. Chase, S.A. Payne: Phys. Rev. B **34**, 8883 (1986)
2.66   R.I. Gintoft, G.A. Skripko: J. Appl. Spectrosc. (USSR) **17**, 1480 (1972)
2.67   P.A. Apanasevich, R.I. Gintoft, V.S. Korolkov, A.G. Makhanek, G.A. Skripko: Phys. Status Solidi (b) **58**, 745 (1973)
2.68   G.A. Skripko, R.I. Gintoft: J. Appl. Spectrosc. (USSR) **24**, 245 (1977)
2.69   A.G. Makhanek, G.A. Skripko: Phys. Status Solidi (a) **53**, 243 (1979)
2.70   M. Dagenais, M.C. Downer, R. Neumann, N. Bloembergen: Phys. Rev. Lett. **46**, 561 (1981)
2.71   U. Fritzler, G. Schaak: J. Phys. C **9**, L 23 (1976)
2.72   U. Fritzler: Z. Phys. B **27**, 289 (1977)
2.73   C.D. Cordero-Montalvo: Phys. Rev. B **31**, 5433 (1985)
2.74   R.L. Schiesow, H.M. Crosswhite: J. Opt. Soc. Am. **59**, 602 (1969)
2.75   W.T. Carnall, P.R. Fields, R. Sarup: J. Chem. Phys. **54**, 1476 (1971)
2.76   J. Makovsky: J. Chem. Phys. **46**, 390 (1967)
2.77   P.R. Monson, W.M. McClain: J. Chem. Phys. **53**, 29 (1970)
2.78   W.M. McClain: J. Chem. Phys. **55**, 2789 (1971)
2.79   W.S. Heaps, L.R. Elias, W.M. Yen: Phys. Rev. B **13**, 94 (1975)
2.80   E. Loh: Phys. Rev. **184**, 348 (1969)
2.81   J. Sztucki, W. Strek: Phys. Rev. B **34**, 3120 (1986)
2.82   M. Casalboni, R. Francini, U.M. Grassano, R. Pizzoferrato: Phys. Rev. B **34**, 2936 (1986)
2.83   L.A.O. Nunes, F.M. Matinaga, J.C. Castro: Phys. Rev. B **32**, 8356 (1985)
2.84   R.S. Rana, F.W. Kaseta: J. Chem. Phys. **79**, 5280 (1983)
2.85   W.T. Carnall, P.R. Fields, R. Sarup: J. Chem. Phys. **51**, 2587 (1969)
2.86   W.M. Yen, C.G. Levey, S. Huang, S.T. Lai: J. Lumin. **24/25**, 6597 (1981)
2.87   C.D. Cordero-Montalvo, N. Bloembergen: Phys. Rev. B **30**, 438 (1984)

2.88   R.S. Rana, C.D. Cordero-Montalvo, N. Bloembergen: J. Chem. Phys. **81**, 2951 (1984)
2.89   A. Lezama, C.B. de Araujo: Phys. Rev. B **34**, 126 (1986)
2.90   G.P. Morgan, W.M. Yen: "Optical Energy Transfer in Insulators", in this volume
2.91   L. Swofford, W.M. McClain: Chem. Phys. Lett. **34**, 455 (1975)
2.92   J.P. Hermann, J. Ducuing: Phys. Rev. A **5**, 2557 (1972)
2.93   G. Koren, C. Cohen, W. Low: Solid State Commun. **16**, 257 (1975)
2.94   G.B. Altshuler, V.B. Karasev, S.A. Kozlov, A.V. Ovchinnikov: Pis'ma zh. Tekh. Fiz. **9**, 799 (1983) [Sov. Tech. Phys. Lett. **9**, 344 (1983)]
2.95   G.B. Altshuler, S.A. Kozlov: Kvantovaya Elektron. (Moscow) **12**, 698 (1985) [Sov. J. Quantum Electron. **15**, 459 (1985)]
2.96   S.K. Gayen, D.S. Hamilton: Phys. Rev. B **28**, 3706 (1983)
2.97   S.E. Harris, D.B. Lidow: Phys. Rev. Lett. **33**, 674 (1974)
2.98   P. Cahuzac, P.E. Toschek: Phys. Rev. Lett. **40**, 1087 (1978)
2.99   J.R. Rios Leite, C.B. de Araujo: Chem. Phys. Lett. **73**, 71 (1980)
2.100 U. Fano, J.W. Cooper: Phys. Rev. **137**, A 1364 (1965)
2.101 J.C. Vial, R. Buisson: J. de Phys., Lett. **43**, 339 (1982)
2.102 R.B. Barthem, J.C. Vial, F. Madeore: J. Lumin. **34**, 47 (1985)
2.103 C. Delsart, J.C. Keller: Opt. Commun. **15**, 91 (1975)
2.104 C. Delsart, J.C. Keller: Opt. Commun. **16**, 388 (1976)
2.105 A. Szabo: Phys. Rev. Lett. **27**, 323 (1975)
2.106 M.A. Kramer, R.W. Boyd: Phys. Rev. B **23**, 986 (1981)
2.107 K.C. Bleijenberg, F.A. Kellendonk, C.W. Struck: J. Chem. Phys. **73**, 3586 (1980)
2.108 D. Narayana Rao, J. Prasad, P.N. Prasad: Phys. Rev. B **28**, 20 (1983)
2.109 Q. Munir, E. Wintner, A.J. Schmidt: Opt. Commun. **36**, 467 (1981)
2.110 F. Rasetti: Z. Phys. **66**, 646 (1930)
2.111 R.J. Elliott, R. Loudon: Phys. Lett. **3**, 189 (1963)
2.112 J. Hougen, S. Singh: Phys. Rev. Lett. **10**, 406 (1963)
2.113 J. Hougen, S. Singh: Proc. R. Soc. London A **277**, 193 (1964)
2.114 J.A. Koningstein: J. Opt. Soc. Am. **56**, 1405 (1966)
2.115 J.A. Koningstein: J. Chem. Phys. **46**, 2811 (1967)
2.116 J.A. Koningstein, O.S. Mortensen: Phys. Rev. Lett. **18**, 831 (1967)
2.117 O.S. Mortensen, J.A. Koningstein: J. Chem. Phys. **48**, 3971 (1968)
2.118 J.A. Koningstein, O.S. Mortensen: Nature **217**, 445 (1968)
2.119 J.A. Koningstein: Can. J. Chem. **49**, 2336 (1971)
2.120 J.A. Koningstein: *Introduction to the Theory of the Raman Effect* (Reidel, Dordrecht 1972)
2.121 D. Nicollin, J.A. Koningstein: Chem. Phys. **49**, 377 (1980)
2.122 R.L. Wadsack, J.L. Lewis, B.E. Argyle, R.K. Chang: Phys. Rev. **3**, 4342 (1971)
2.123 J.H. Christie, D.J. Lockwood: Chem. Phys. Lett. **8**, 120 (1971)
2.124 D.J. Lockwood: "Light Scattering from Electronic and Magnetic Excitations in Transition-Metal Halides", in *Light Scattering in Solids III*, ed. by M. Cardona, G. Güntherodt, Topics Appl. Phys., Vol. 51 (Springer, Berlin, Heidelberg 1982) p. 59
2.125 J.T. Hoff, P. Grunberg, J.A. Koningstein: Appl. Phys. Lett. **20**, 358 (1972)
2.126 G.D. Jones, M.H. Kuok: J. Phys. C **12**, 715 (1979)
2.127 P.C. Becker, N. Edelstein, G.M. Williams, J.J. Bucher, R.E. Russo, J.A. Koningstein, L.A. Boatner, M.M. Abraham: Phys. Rev. B **31**, 8102 (1985)
2.128 P.C. Becker, G.M. Williams, R.E. Russo, N. Edelstein, J.A. Koningstein, L.A. Boatner, M.M. Abraham: Opt. Lett. **11**, 282 (1986)
2.129 J.A. Koningstein: J. Chem. Phys. **51**, 1163 (1969)
2.130 S. Guha: Phys. Rev. B **23**, 6790 (1981)
2.131 P.C. Becker: J. Chem. Phys. **81**, 2872 (1984)
2.132 T.C. Damen, S.P.S. Porto, S.B. Bell: Phys. Rev. **142**, 570 (1966)
2.133 N.E. Edelstein: Private Communication
2.134 M.C. Downer, G.W. Burdick, D.K. Sardar: J. Chem. Phys. **87**, 1787 (1988)

3. Optical Energy Transfer in Insulators

Gerard P. Morgan and William M. Yen

With 27 Figures

The transfer of optical energy in inorganic insulating solids has been studied for many years and continues to be an active area of fruitful research in physics. There are many technological and fundamental reasons why a good understanding of the processes which affect the transfer of energy is important. On the technical side, for example, there appears to be a never ending demand for new and improved phosphors for such devices as televisions and fluorescent lamps. The development of luminescent solar collectors has been advanced significantly by using transition metal ion doped glass ceramic materials. Transition metal ion and rare earth doped solids form an important class of materials which can be used in applications as diverse as lasers and IR photon counters. An improved understanding of the processes which affect the storage and transfer of optical energy in ordered or partially disordered crystalline systems is necessary if major advances are to be continued. At a more fundamental level, the transfer of optical energy is just one aspect of the more general area of transport phenomena. A better understanding of these mechanisms is needed to describe excitation transport in amorphous materials.

The transfer of energy in solid phosphors has been studied since early in this century [3.1, 2]. However, it was not until the late 1940s that a basic theoretical framework was developed by *Förster* [3.3] and *Dexter* [3.4] to describe the process of sensitized luminescence. Their work provided the background necessary for further quantitative studies in the area, which were intensified in the 1960s by the search for efficient solid-state laser materials [3.5]. Much of the early work dealt with a study of phosphors which were commonly activated by transition metal ions and it has only been in the last 20 years that trivalent rare earths have received significant attention. The first instance of energy transfer in a rare earth doped solid was reported by *Tomaschek* [3.6], who observed an activation of Sm^{3+} by Gd^{3+} in SrS. Further evidence of $4f$ ion-ion interactions was reported by *Pringsheim* [3.1] and *Bolden* and *Kroger* [3.7]. However, the main interest in rare earth transfer dynamics was initiated by the work of *Van Uitert* et al. [3.8]. The later development of this area can be partially traced to the fact that the energy levels of rare earth ions in crystals were not fully understood. This changed with the weak crystalline field approximations reviewed by *Dieke* [3.9] and *Hüfner* [3.10]. The optically active $4f$ electrons are partially shielded by the other valence electrons and so are only weakly perturbed by the electric fields of neighboring ions. Thus, their optical spectra consist of sharp lines very similar to those of their free ion counterparts making them ideal probes for other important weak interactions. In contrast, the unshielded $3d$ level ions are affected by strong lattice effects.

The development of narrowband tunable laser sources has contributed to significant theoretical and experimental advances in recent years. These new laser spectroscopic techniques permit much greater refinement in experimental detail, in both the frequency and temporal domains. Most experimental investigations have involved materials doped with rare earth ions, and so our treatment will reflect this fact. However, reference will be made to transition metal ion systems as appropriate.

Our main interest in this chapter will be to outline recent advances in our understanding of ion-ion interactions which can lead to various types of optical energy transfer. In Sect. 3.1 the advances in the field of energy transfer prior to the introduction of laser spectroscopic techniques will be outlined. Theoretical developments and experimental results which probe the microscopic aspects of energy transfer will be reviewed in Sect. 3.2 and macroscopic spectral dynamics of intraline energy transfer in Sect. 3.3. Experiments which have highlighted our understanding of transfer processes involving distinct optical centers will be presented in Sect. 3.4 and finally the topic of resonant energy transfer will be reviewed in Sect. 3.5.

3.1 Preliminary Remarks

Energy migration among optically active rare earth ions in solids has been reviewed extensively in the past and has covered a diverse range of processes [3.8, 11–13]. Fundamentally, the transfer process involves the initial excitation of a set of ions termed donors (D) by light, followed by a transfer of energy to other ions, termed acceptors (A). This is quite a general description, covering numerous possible processes, some of which are illustrated in Fig. 3.1. These include (i) resonant and (ii) nonresonant transfer, (iii) cross relaxation and (iv) upconversion.

A complete description of any of the above processes involves two distinct considerations. Firstly, the microscopic ion-ion interaction responsible for the transfer must be understood and then used to predict the resultant macroscopic observable behavior. *Förster* [3.3] initially proposed that the mechanism responsible for the transfer of energy was an interionic electric dipole–electric dipole interaction. *Dexter* [3.4] subsequently developed the treatment independently to include other multipolar interactions. These processes were considered to occur resonantly and so depended on the overlap of the emission and absorption spectra of the donor and acceptor ions, respectively. In the absence of any A–D back transfer, the transfer probability can be written in the form

$$W_{12} = \frac{2\pi J^2}{\hbar^2} \int g_1(\omega) g_1(\omega) d\omega, \tag{3.1}$$

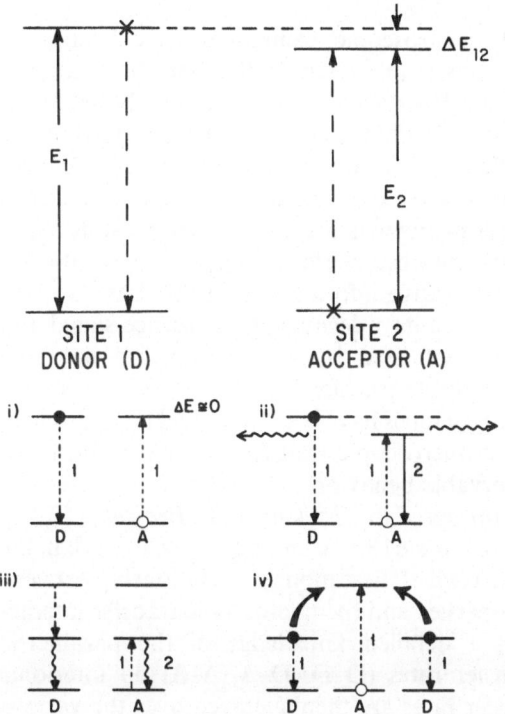

Fig. 3.1. Schematic illustration of energy transfer between donor and acceptor ions. Typical situations encountered in energy transfer studies in solids are illustrated: (*i*) resonant D–A transfer; (*ii*) nonresonant transfer; (*iii*) cross relaxation; and (*iv*) upconversion

where $J = \langle 1^*, 2|H_{\mathrm{int}}|1, 2^* \rangle$, and the asterisk (*) denotes the excited ion; H_{int} is the interaction Hamiltonian for the appropriate multipolar interaction which couples the ions 1 and 2; J depends on the nature of the optical excitation and includes any parametric dependencies, such as the interionic separation. The energy mismatch between donor and acceptor (ΔE_{12} in Fig. 3.1) does not appear explicitly, but is incorporated into the lineshape overlap integral, since the transfer is assumed to be resonant.

This model was useful for describing systems with, for example, transition metal ions where the emission and absorption lines can be broad and one can often expect a large overlap to occur. However, in the case of rare earth impurities where the spectral linewidths are much narrower than ΔE_{12}, resulting mostly in well-isolated lines, this simple process involving resonant energy transfer was found to be inadequate. The theory was extended to include various energy conversion processes [3.14, 15]. Ultimately, it was proposed that a resonant transfer of energy through the overlap of the donor phonon sideband with the acceptor's pure electronic transition was involved. This approach still suffered from some drawbacks. For example, the coupling of rare earth ions to the vibrating host lattice is weak, resulting in negligible phonon sideband structure in $4f$ emission spectra. In addition, for the case of a small mismatch, ΔE_{12}, the small density of acoustic phonon states should result in an extremely slow transfer rate.

Two types of conventional spectroscopic techniques have been used experimentally to study energy transfer processes. In the first, the excitation spectrum of the acceptor is measured. If certain features can be attributed as due to donor absorption then it is assumed that the acceptors are being fed by the donor system. Although it is difficult to extract quantitative details from this technique, it was used extensively in early experiments to verify the existence of energy transfer phenomena. Experiments which study the temporal development of either the donor or acceptor fluorescence following pulsed excitation of the donor system yield more quantitative information. In this latter experimental approach, the donor system emits a decaying fluorescence signal and the acceptor system may be characterized by both rising and decaying fluorescence components which are due to transfer dynamics as well as intrinsic decay processes. The fluorescence dynamics have been described by a system of rate equations [3.16] which yield microscopic transfer parameters from the analysis of the macroscopic observable behavior.

Various models have been proposed by *Inokuti* and *Hirayama* [3.17], *Yokota* and *Tanimoto* [3.18], and *Burshtein* [3.19], among others, to explain the observed fluorescence behavior. In each of these models, a relationship between the macroscopic fluorescence properties and the microscopic transfer interactions is established. Therefore, a detailed knowledge of the parametric dependence of the various transfer rates (D–D, D–A, A–A) on interionic separation is required. The transfer rates are then averaged over the various possible donor and acceptor spatial distributions. In such calculations most quantities are treated as random variables and are adjusted to provide a best fit to the experimental data. However, conventional spectroscopy fails to provide any direct information on the transfer parameters between equivalent states of similar ions (i.e., D–D or A–A), so various assumptions as to the character and strength of such processes must be made. Because of these shortcomings, many of the conclusions drawn from the various models have been questioned in recent years in the light of the newer techniques involving laser spectroscopy.

Broadly tunable laser sources have become available in the past 10–15 years and have been used to develop new spectroscopic techniques which probe at a more fundamental level both the static and dynamic processes involving optically active centers. Fluorescence line narrowing (FLN) techniques and time-resolved fluorescence line narrowing (TRFLN) have proven to be especially powerful for studying the dynamics of optical energy transfer. Fluorescence line narrowing was first observed in solids by *Szabo* [3.20] in 1970, and later TRFLN was first applied to rare earth doped solids by *Flach* et al. [3.21]. The characteristic linewidths (homogeneous) of $4f$ and many $3d$ ions in solids can be narrow at low temperatures. In these cases, the effects of random crystalline strains result in an inhomogeneously broadened transition which is a convolution of the homogeneous linewidths from ions in different strained environments [3.22, 23]. Pulsed laser sources have been developed with frequency bandwidths which are much less than the inhomogeneous transition profiles of the optically active ions. Thus it is possible to excite and study a

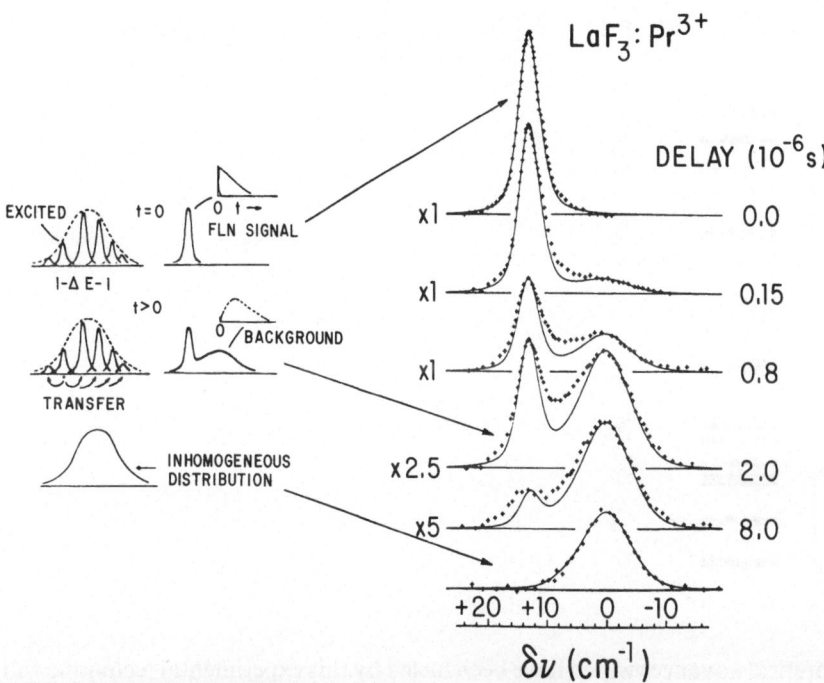

Fig. 3.2. Schematic representation of FLN (*left*) and TRFLN (*right*) in an inhomogeneously broadened distribution. Experimental traces are from a study of the $^3P_0$ state of a 20 at. % sample of Pr^{3+} in LaF_3 at 10 K. Transfer within the inhomogeneous line of the "accidentally degenerate" $^3P_0 - (^3H_6)_1$ fluorescent transition is illustrated. Solid lines represent theoretical fits to the time evolution [3.24]

subset of ions and to monitor the transfer of optical excitation to other ions within the inhomogeneous feature, as illustrated in Fig. 3.2 [3.24]. This spectral diffusion process requires some type of energy shifting mechanism among ions which are alike in all respects except for small (1 part in 10^5) variations in local crystalline fields. The study of like-ion dynamics by TRFLN permits measurements of D–D and A–A mechanisms which were not possible with more conventional spectroscopic techniques. It should be apparent from Fig. 3.2 that TRFLN does not give any information on spatial transfer processes, nor does it allow a study of any coherent resonant energy dynamics. Such information can be obtained by other techniques such as four-wave mixing and hole-burning spectroscopy and will be discussed in Sect. 3.5.

3.2 Time-Resolved Fluorescence Line Narrowing Studies

In this section we will discuss recent advances in our understanding of ion-ion transfer processes using the technique of TRFLN. Some of the recent

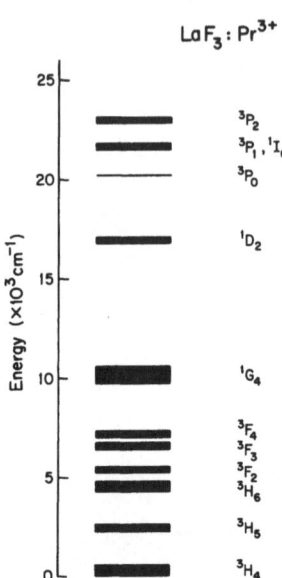

LaF$_3$: Pr^{3+}

Fig. 3.3. Energy level diagram of an isolated Pr^{3+} ion in LaF$_3$. (After [3.26])

theoretical advances which have been fueled by this experimental technique will also be outlined. However, the reader is directed to *Holstein* et al. [3.25] for a complete discussion of current models.

The bulk of the TRFLN experiments have been performed on the material La$_{(1-x)}$Pr$_x$F$_3$, and because of the wealth of information accumulated it continues to be a model system for testing many other experimental techniques. The host material, LaF$_3$, is a nonhygroscopic crystal which can be doped with Pr^{3+} up to its stoichiometric limit PrF$_3$. The energies of some of the low-lying states are illustrated in Fig. 3.3 where the thick lines represent the spread of the $(2J+1)$ components for each J manifold due to the crystalline field. A detailed description of the energy levels for this system has been reported by *Carnall* [3.26]. When the Pr ions are excited, it is found that only three states emit fluorescence. They are $^1S_0$ (~ 50000 cm^{-1}), $^3P_0$ and $^1D_2$ levels. For dilute samples at low temperatures these states have lifetimes of 700 ns, 51 μs and 500 μs, respectively. Significant shortening of the lifetimes of the $^3P_0$ and $^1D_2$ states occur at higher temperatures or higher dopant concentrations, due to self-quenching effects.

3.2.1 Observation of Spectral Transfer in LaF$_3$:Pr^{3+}

TRFLN was first demonstrated by *Flach* et al. [3.21] in the LaF$_3$ crystalline system doped with 5 at.% Pr^{3+}. The $(^3H_4)_1$–$^3P_0$ transition was excited and the resultant resonant fluorescence was observed. This transition, which has an inhomogeneously broadened linewidth of ~ 4 cm^{-1} at low temperatures, was excited at the line center with a pulsed (5 μs) narrowband (1.5 GHz) laser. The

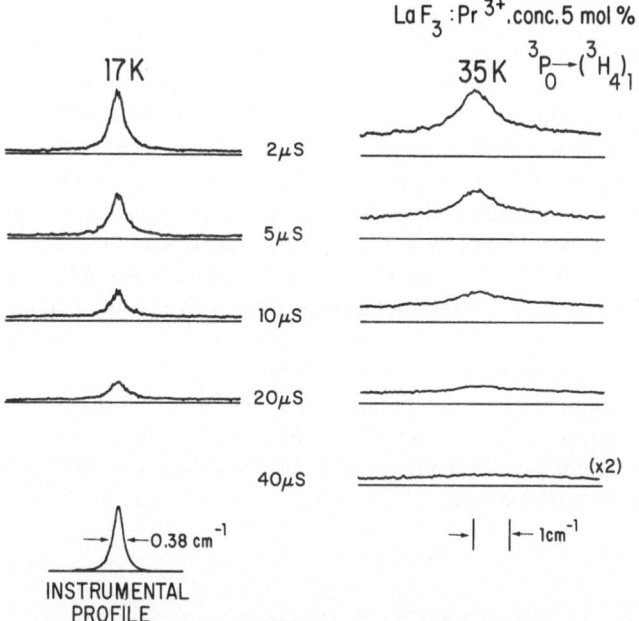

Fig. 3.4. TRFLN studies of the resonant $^3P_0 - (^3H_4)_1$ transition in a 5 at. % $LaF_3:Pr^{3+}$ at 17 and 35 K. The laser beam excited the center of the $(^3H_4)_1 - {}^3P_0$ inhomogeneous absorption profile and the signal was analyzed with a Fabry-Perot interferometer [3.21]

subsequent fluorescence was photoelectrically detected and spectrally analyzed with a Fabry-Perot (F–P) interferometer with a free spectral range of ~ 1.3 cm^{-1}. Time-resolved spectra were obtained by electronically gating the photodetection system at different delays after the pulsed laser excitation. Typical results are presented in Fig. 3.4. The width of the observed FLN signal at 17 K was limited by the resolution of the detection system. The time-dependent background signal at both temperatures was proposed to be due to fluorescence from the inhomogeneous line which appeared in overlapping orders of the F–P spectrum. This was due to spectral transfer of energy from the initially excited ions. It can be seen that the FLN signal decayed more rapidly, while the background fluorescence increased at a greater rate at higher temperatures, indicating a dependence of the spectral transfer rate on temperature. The problem of overlapping F–P orders was eliminated by *Selzer* et al. [3.27] who used a spectrometer to filter the overlapping orders. They monitored the nonresonant $^3P_0-(^3H_6)_1$ fluorescence which had a broader FLN component (~ 6 cm^{-1}) than the resonant emission due to "accidental degeneracy" effects of the crystalline field. While the broader FLN signal was due to the terminal state of the transition, the energy transfer dynamics were characteristic of the initial state and so yielded the same information as that obtained by studying the resonant transition. By narrowing various parts of the inhomogeneous profile they concluded that the transfer rate was asymmetric, being faster when the high energy side of the line was excited. They also reported that the transfer rate increased rapidly with the sample temperature. This technique was improved upon by *Hamilton* et al. [3.28] and *Huber* et al. [3.24], who obtained the trace with the high signal-to-noise ratio presented in Fig. 3.2. In this figure the full inhomogeneous lineshape immediately appeared as a background and rose uniformly, thus the mechanism responsible for the transfer of energy was independent of the energy mismatch, ΔE_{12}, in the transfer step. The linewidth of the FLN component did not broaden noticeably as a function of time. Both of these characteristics have also been observed for $3d$ ion doped crystals [3.29] as well as for $4f$ ions in glasses [3.30] where the inhomogeneous linewidth can be in excess of $100\,\mathrm{cm}^{-1}$ [3.31].

In order to extract quantitative information from TRFLN experiments it is best to measure the temporal evolution of the narrowed fluorescence, I_N, normalized to the total fluorescence from the transition $(I_N + I_B)$. This has been described by the quantity $R(t)$ where

$$R(t) = \frac{I_N(t)}{I_N(t) + I_B(t)}, \tag{3.2}$$

and essentially represents the possibility that an initially excited ion will remain excited at a time t in the absence of radiative decay. This approximation requires that the radiative decay rate γ_R be constant across the inhomogeneous emission profile, which is generally true for dilute rare earth concentrations where the dopants are relatively isolated. At higher concentrations, features in the spectra which are due to the effects of pairs can appear with different radiative lifetimes [3.32]. The presence of these artifacts can produce "pseudo-diffusion" effects which result in misleading conclusions [3.33]. The large variation in local fields experienced by optically active ions in glasses results in a

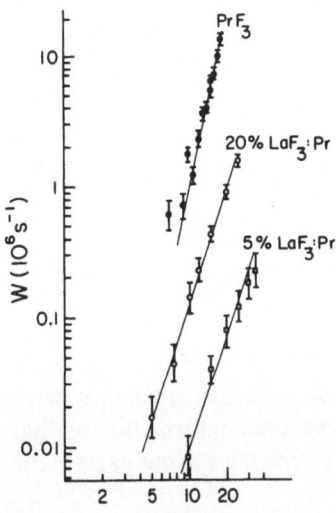

Fig. 3.5. Comparison of the spectral transfer rates within the inhomogeneous $^3P_0$ state of $LaF_3:Pr^{3+}$. For the 5 and 20 at % samples the transfer rate increases as T^3, whereas for PrF_3 the appropriate dependence is $T^{4.3}$ [3.28]

large variation in radiative lifetimes across the emission profile, thus a modification of the data analysis would be required as noted by *Motegi* and *Shionoya* [3.34].

It should be clear that the macroscopic quantity $R(t)$ represents an ensemble average of ions with various configurations of neighboring dopant ions and is typically a nonexponential function of time. A detailed discussion of the averaging procedure will be deferred to Sect. 3.3. We merely state at the moment that the energy transfer rate W was initially obtained from data by assuming that the model of *Inokuti* and *Hirayama* [3.17] was valid for $R(t)$ [3.21, 27]. Later on W was defined as the time taken for $R(t)$ to reach $1/e$ of its initial value [3.28], an assumption which was later confirmed as valid. The transfer rates extracted from $R(t)$ for three samples with different Pr^{3+} concentrations are plotted in Fig. 3.5. For the lower dopant concentrations (5 at.% and 20 at.%) the intraline ion-ion transfer rate is seen to increase at T^3 while for PrF_3 the rate varies with $T^{4.3}$.

3.2.2 Microscopic Analysis of TRFLN

The properties of TRFLN signals which we have discussed cannot be explained satisfactorily by the early energy transfer theories [3.3, 4]. This transport problem was reformulated by *Holstein* et al. [3.35] who considered the total system Hamiltonian to consist of the host phonon bath in addition to the two-center Hamiltonian of *Förster* and *Dexter* [3.3, 4] and also included a coupling term to represent the interaction between the vibrating lattice and the ions. Thus, the energy eigenvalues of the system are

$$|i\rangle = |1^*, 2; n_{s,q}\rangle, \tag{3.3}$$

while the Hamiltonian is expressed as

$$H = \{H_e^{(1)} + H_e^{(2)} + H_{int}\} + V_p + H_p. \tag{3.4}$$

The term in braces is the Förster-Dexter two-body Hamiltonian, V_p describes the vibrating lattice, and H_p represents the ion-phonon interaction term. The dynamics of the system are described by calculating ion–ion energy transfer probabilities in increasing orders of perturbation. A detailed account of this procedure is presented by *Holstein* et al. [3.25].

Parametric dependencies of W on certain quantities, e.g., T and ΔE_{12}, can be extracted from the calculation by making various assumptions. In $LaF_3:Pr^{3+}$, TRFLN studies have indicated that $^3P_0$ intraline transfer processes are dominated by a two-site nonresonant interaction (1 phonon, 2nd order process). This process is characterized by a T^3 dependence on temperature which is observed for the 5 at.% and 20 at.% samples. (PrF_3 will be discussed presently.) The transfer rate is predicted to be independent of the ion–ion energy mismatch which is observed experimentally. The entire inhomogeneous fluorescence profile is detected at all times after the initial narrowband laser excitation pulse. The asymmetry in transfer rates from an FLN signal on the high energy side compared to that on the low energy side of the inhomogeneous profile is simply due to the thermal population effects of the phonons which mediate the transfer process.

For the above-mentioned process the transfer rate is given by

$$
W_{12} = \frac{2\pi J^2}{\hbar} (\Delta f)^4 \sum_{s,s',q,q'} \hbar^{-2} (\omega_{s,q}\, \omega_{s',q'})^{-2}
$$

$$
\times (2Mv_s^2)^{-2} (n_{s',q'}+1)(n_{s,q}) \cdot h_{\text{II}}
$$

$$
\times \delta(\hbar\omega_{s',q'} - \hbar\omega_{s,q} - \Delta E_{12}). \tag{3.5}
$$

Here M and v_s are the crystal mass and average velocity of sound in the material, $n_{s,q}$ represents the thermal population of phonons with energy $\hbar\omega_{s,q}$, and ΔE_{12} represents the energy mismatch between the two ions; J is the ion–ion coupling term described by the Förster-Dexter theory and $\Delta f = (f - g)$, where f and g are the ion–phonon coupling strengths for the ground and excited states of an ion at a single site. The symbol h_{II} represents a phonon phase or coherence factor

$$
h_{\text{II}} = |\exp(i\mathbf{q} \cdot \mathbf{r}_{12}) + \exp(i\mathbf{q}' \cdot \mathbf{r}_{12}) - 1 - \exp[i(\mathbf{q}-\mathbf{q}') \cdot \mathbf{r}_{12}]|^2, \tag{3.6}
$$

and plays an important role in determining both temperature and range dependencies for the interaction. The phonon lattice sum in W_{12} can be replaced by an appropriate Debye-like integral. For dilute Pr^{3+} dopant concentrations, $qr \gg 1$ and so one obtains

$$
W_{12} \sim T^3. \tag{3.7}
$$

For PrF_3 where the separation of the optically active ions is much less on average than for lower dopant concentrations, the approximation $qr \gg 1$ is no longer valid so that coherence effects due to h_{II} can be significant. *Hamilton* [3.28] considered the case of $qr = \frac{\pi}{2}$ and other intermediate values for the specific case of PrF_3 and deduced a $T^{6.1}$ temperature dependence for W_{12}. This somewhat larger value than the observed $T^{4.3}$ dependence is thought to be due to the sensitivity of the lattice sum to the phonon density of states in the neighborhood of the interacting ions. The range dependence of h_{II} can enhance the distance over which transfer can effectively take place, as is illustrated in Fig. 3.6 where the phase factor is plotted as a function of inter–ion separation [3.36]. This predicted enhancement in interaction range has been observed by *Hegarty* et al. [3.37].

The ion–ion coupling strength J may be calculated using intermediate coupling eigenstates and follows a formalism developed by *Kushida* [3.38]. For the $^3P_0$ state of Pr^{3+}, intraline transfer occurs via the electric dipole–electric dipole interaction. This introduces a r^{-6} range dependence for W_{12} which in turn translates into a quadratic dependence on dopant ion concentration, c_d,

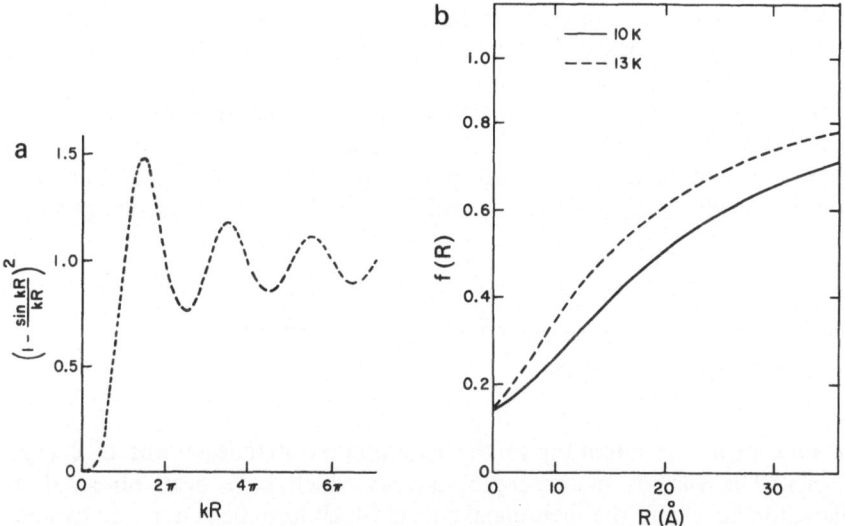

Fig. 3.6. (**a**) The phonon phase factor h_{II} which appears in the "one-phonon second order" process and which produces interference effects in the microscopic transfer rates. (**b**) Enhancement factor in the transfer rate as a function of distance [3.36, 37]

for $qr \gg 1$. The data in Fig. 3.5 yield a $c_d^{1.9}$ dependence for the 5 at.% and 20 at.% samples which is consistent with the electric dipole–electric dipole interaction. This type of scaling, however, does have limited applicability, especially at higher concentrations, due to the range enhancement effects of h_{II}.

TRFLN experiments have not been limited to $LaF_3 : Pr^{3+}$, though this system has undoubtedly received more attention than any other. Intraline spectral dynamics have also been observed for the $({}^4F_{3/2})_1 - ({}^4I_{9/2})_1$ transition in $Nd_x La_{(1-x)} P_5 O_{14}$ at low temperatures by *Broer* et al. [3.39]. The transfer in an $x = 0.2$ sample has been identified as a one-site resonant process with an electric dipole–electric dipole ion pair interaction for $17\,K < T < 24\,K$. In this case, the transfer rate exhibits an exponential temperature dependence

$$W_{12} \sim \exp(-\Delta/kT), \tag{3.8}$$

which is characteristic of an energy activation process. The reported value of $\Delta = 94\,cm^{-1}$ corresponds to the $({}^4I_{9/2})_2 - ({}^4I_{9/2})_1$ energy separation. Spectral energy migration within an inhomogeneously broadened line has also been reported in ruby by *Selzer* et al. [3.40]. Line-narrowed fluorescence from the R_1 line of samples with Cr^{3+} concentrations in the range 0.025–0.9 at.% was observed at low temperatures (5–50 K). The nonresonant donor–donor transfer rate in this case increased linearly with temperature, indicating a direct one-phonon process, and was considered due to the dominant effects of weakly exchange coupled ion pairs which were folded into the main R_1 line. A similar direct one-phonon process has also been reported by *Brundage* and *Yen* [3.41] for Yb^{3+} in a silicate glass, where the TRFLN technique was applied to the $(^2F_{5/2})_1-(^2F_{7/2})_1$ transition from 4.2 to 51 K. The observed spectra were consistent with a one-phonon-assisted process with what was believed to be an electric quadrupole–electric quadrupole interaction between the Yb^{3+} ions. This interaction shows an R^{-13} dependence on site separation (R^{-10} is characteristic of quadrupole–quadrupole) and so it is speculated that an even shorter ranged exchange interaction might contribute to the transfer process.

3.3 Macroscopic Aspects of Energy Transfer: Intraline Dynamics

We now turn our attention to the macroscopic manifestations of energy transfer. The various macroscopic features which have been observed in different systems and the theoretical efforts which have been made to explain these features in terms of the microscopic parameters will be discussed. We will initially consider energy transfer within an inhomogeneously broadened line, and then expand our discussion to include transfer processes among unlike ions, or groups of ions, which includes sensitization, trapping and quenching of fluorescence.

The technique of TRFLN has allowed the transfer of energy within the same state of similar ions to be probed. The inability of conventional spectroscopy to probe this interaction resulted in gross assumptions on the dynamics of like–ion transfer being made, which has led to questionable conclusions based on these models.

The theoretical analysis required to make the necessary transition from microscopic interactions to macroscopic behavior, such as fluorescence lifetime and the quantity $R(t)$, entails various averaging procedures over the microscopic parameters and in general can pose a formidable task. Consider, for example, a dilute system. While the host lattice may be ordered, the optically active centers, both donors and acceptors, may be distributed at random on the lattice sites. For transfer within an inhomogeneously broadened line, the exact nature of the broadening mechanism must be understood.

Current successful theoretical developments in this area are based on a set of coupled rate equations of the form

$$\frac{dP_n(t)}{dt} = -\left(\gamma_r + X_n + \sum_{n' \neq n} W_{nn'}\right)P_n(t) + \sum_{n' \neq n} W_{n'n}P_{n'}(t). \tag{3.9}$$

Here $P_n(t)$ is the probability that an optically active ion, at site n, is in an excited state at a time t; γ_r is the reciprocal of the lifetime of the excited state; $W_{nn'}$ is the energy transfer rate from an excited ion at lattice site n to an unexcited ion at n'; and $W_{n'n}$ represents the back transfer rate to n from n'. The parameter X_n denotes the total rate at which energy is removed from the level of interest to other species which do not allow any back transfer to occur. This trapping or quenching of the fluorescence is important. However, for the moment it will be ignored by setting $X_n = 0$. The solution to these coupled rate equations has been reviewed by *Huber* [3.42], so only the important results will be mentioned here.

If $R(t)$ represents the probability that an ion initially excited by the laser pulse will still be excited at time t, it is relatively easy to see that the initial ($t \simeq 0$) decay of $R(t)$ is given by

$$\frac{dR(t)}{dt} \simeq \frac{dP_0(t)}{dt} = -c \sum_n W_{0n}, \tag{3.10}$$

where W_{0n} is the transfer rate from an initially excited ion at site 0 to an ion at site n. [The summation over n in (3.10) and all subsequent equations excludes the case of $n = 0$.] The dopant concentration is c and represents the probability that an optically active dopant ion is present at site n. Thus, the initial behavior of $R(t)$ follows a simple exponential decay [3.43]. Various approximations have been made to calculate $R(t)$, with two distinct temporal regimes considered: short time behavior, $t \sim 0.1/W_0$ (W_0 is the nearest neighbor transfer rate) and the long time asymptotic behavior, $t \sim \infty$. It has been argued that only the short time behavior is necessary to describe the observed dynamics for most $4f$ systems since the radiative lifetime is comparable to all transfer times. Thus, the emission intensity will be weak when the long time asymptotic transfer characteristics become dominant.

The most useful short time approximation for $R(t)$ which is valid for all donor concentrations c is given by

$$R(t) = \prod_n [1 - c + c \exp(-W_{0n}t)f(W_{0n}t)]. \tag{3.11}$$

An exact lattice sum can be carried out if the nature of the ion–ion interaction is assumed. For multipolar interactions,

$$W_{0n} = W_0(R_0/R_{0n})^s = \beta/(R_{0n})^s, \tag{3.12}$$

R_0 representing the nearest-neighbor separation and R_{0n} being the distance from lattice site "0" to "n". The form of the function $f(W_{0n}t)$ depends on the donor concentration c. For $c > 0.5$ a reasonable fit can be found to $R(t) \sim 0.05$ for

$$f(W_{0n}t) = 1 + \tfrac{1}{2}(W_{0n}t)^2, \tag{3.13}$$

while for $c < 0.2$ a good fit over the same range of $R(t)$ is obtained with

$$f(W_{0n}t) = \cosh(W_{0n}t). \tag{3.14}$$

In the absence of any back transfer of energy, an exact expression for $R(t)$ is obtained with

$$f(W_{0n}t) = 1. \tag{3.15}$$

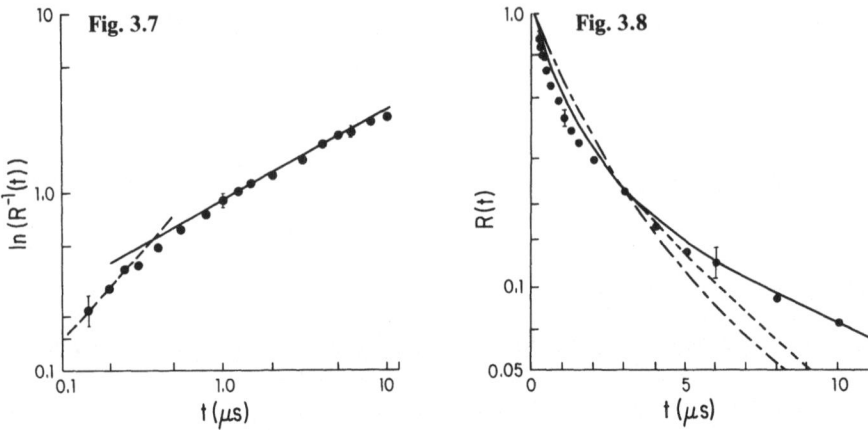

Fig. 3.7. A plot of $\ln[R(t)^{-1}]$ as a function of t, for the results shown in Fig. 3.2. The broken line represents the initial pure exponential decay and the solid line exhibits a $t^{3/6.5}$ dependence, indicative of an electric dipole-electric dipole interaction [3.24]

Fig. 3.8. Comparison of experimental and theoretical modeling results. The best fit is obtained for $f(t) = \cosh(W_{0n}t)$ ($-$), which treats back transfer to all orders. The dot-dash line represents the Inokuti-Hirayama results for $f(t) = 1$, which does not consider back transfer. The dashed line represents $f(t) = 1 + (1/2)(W_{0n}t)^2$ which assumes only one exchange in the excitation [3.24]

This is a generalized result of an expression first obtained by *Inokuti* and *Hirayama* [3.17]. A further simplification of $R(t)$ can be obtained for $c \ll 1$ by assuming a multipolar interaction

$$R(t) = \exp[(-4\pi/3)n(\beta t)^{3/s}\Gamma(1 - 3/s)2^{(3/s - 1)}].\tag{3.16}$$

In order to extract a meaningful transfer rate W_0 from the experimental data, the correct multipolar ion–ion interaction must first be determined. This is achieved by plotting $\ln R(t)$ versus $\ln t$. Equation (3.16) predicts that the data will lie on a straight line with slope $3/s$. This is confirmed for the data in Fig. 3.7 [3.24], where $s = 6.5$ points to an electric dipole–electric dipole interaction. It can be seen that for $t \sim 0$, $R(t)$ varies exponentially with t as predicted by (3.10). The transfer rate can now be obtained by fitting (3.11) to the data, for the various forms of $f(W_{0n}t)$, with $s = 6$ and W_0 an adjustable parameter. It can be seen from Fig. 3.8 that the best fit is obtained for $f(W_{0n}t) = \cosh(W_{0n}t)$ with $W_0 = 0.37 \times 10^6 \text{ s}^{-1}$ [3.24]. This parameter is for data obtained at $T = 14$ K and is found to have a T^3 dependence. An analysis similar to the above has been carried out on the $^3P_0$ fluorescence from PrF_3 and yields a best fit for $f(W_{0n}t) = 1$ with $W_0 = (0.4 \pm 0.05) \times 10^6 \text{ s}^{-1}$ at $T = 14$ K [3.28], indicating that the probability of the excitation returning to the initially excited ion is smaller in the concentrated system.

The fit to the data, in Fig. 3.8, at early times is not exact as $R(t)$ decays faster than predicted by the model. This is thought to indicate the presence of a short-range interaction, e.g., quadrupole–quadrupole or superexchange, whose influence would be most significant at early times [3.24]. The effects of such short-ranged interactions on fluorescence quenching will be discussed in more detail in Sect. 3.4.2.

Once the values of W_0 and s have been decided, it then becomes possible to model the original TRFLN data as indicated in Fig. 3.2. In this model, it is assumed that the nature of the inhomogeneous broadening is microscopic. Thus, the distribution of optically active ions surrounding any given one is truly random. The disagreement between the fit and the data in the valley region adjacent to the narrowed feature is taken as evidence of a certain degree of ordering or clustering in the crystal [3.24]. The effects of clustering will again be discussed in more detail with reference to fluorescence quenching in Sect. 3.4.2.

3.4 Interline Spectral Transfer

In this section energy transfer between two distinct optically active centers will be considered. Various aspects of this process, which instigated the study of fluorescence dynamics, have been reexamined in recent years and successfully modeled as a result of the information on donor-donor and acceptor-acceptor dynamics which has been obtained by TRFLN techniques.

This form of energy transfer can manifest itself in various ways. When the donors and acceptors are composed of different ionic species, then phenomena such as trapping of the donor fluorescence or sensitization of the acceptor emission can occur. Energy transfer which occurs between different levels of similar ions can lead to cross relaxation or concentration quenching as indicated in Fig. 3.1(iii). In these situations the transfer of energy is character-ized by a large energy mismatch ΔE between the levels, so that back transfer is rarely significant. The relatively sharp optical features of rare earth ions has made them ideal candidates for the study of interline transfer processes. In addition, materials codoped with $3d$ and $4f$ ions have been characterized where the broad absorption bands of the transition metal ions could provide effective sensitization of the rare earth emissions.

Many of these studies were performed before the development of TRFLN techniques, so that data analysis was based on phenomenological models with inherent limiting assumptions in which the only known parameters were the dopant concentrations [3.11, 44]. These theories have generally considered the temporal evolution of the donor fluorescence following broadband, pulsed excitation. Theoretical models have been developed which deal with two distinct regimes. In both cases, the initial decay is nonexponential but asymptotically approaches an exponential decay rate for $t \sim \infty$. When the donor–donor transfer is relatively slow (compared to the donor–acceptor transfer), the asymptotic donor fluorescence decay constant is believed to be expressible in terms of a donor–donor diffusion constant. This treatment was developed by *Yokota* and *Tanimoto* [3.18]. For relatively fast donor–donor transfer, the hopping model, proposed by *Burshtein* [3.19], has been used to predict the donor fluorescence decay. The asymptotic decay rate has been expressed in terms of a hopping time, τ, which represents the average time that the excitation remains on a donor ion. The validity of these models is restricted, as discussed by *Huber* [3.42].

Recently, the problem has been reexamined by *Huber* [3.42, 45] in terms of the series of coupled rate equations, (3.9). The decay of the donor fluorescence following pulsed excitation is of the form

$$n_d(t) = n_d(0) \exp(-\gamma_r t) f(t), \tag{3.17}$$

where n_d represents the number of excited donors at time t, γ_r is the radiative decay rate and $f(t)$ is the fraction of donors which would remain in the initial excited state at time t in the absence of fluorescence decay.

The exact form of $f(t)$ depends on the relative magnitudes of the donor–donor and donor–acceptor transfer rates [$W_{nn'}$ and X_n in (3.9)], and has a simple representation in the two limiting cases. For no donor–donor transfer, i.e., $W_{nn'} = 0$, $f(t)$ is of the form

$$f(t) = \prod_n [1 - c_A + c_A \exp(-X_{0n} t)], \tag{3.18}$$

where c_A is the fractional acceptor concentration and X_{0n} denotes the transfer rate from an excited donor at site "0" to an acceptor at site "n". If it is assumed that a multipolar interaction governs the donor–acceptor transfer mechanism then X_{0n} can be expressed as

$$X_{0n} = X_0 (R_0/R_{0n})^s = \alpha/(R_{0n})^s, \tag{3.19}$$

where X_0 is the nearest-neighbor transfer rate and R_0 is the separation of the nearest neighbors. Equation (3.18) is a generalization to all values of c_A of the result first obtained by *Inokuti* and *Hirayama* [3.17] for the case of $c_A \ll 1$. In the rapid donor–donor limit, $f(t)$ can be expressed as a single exponential

$$f(t) = \exp\left(-c_A \sum_n X_{0n} t\right), \tag{3.20}$$

which reflects the fact that the donor–donor transfer is so rapid that all the donors have an equal probability of being excited.

It is in the intermediate region between these two limits that various approximations are necessary and where the phenomenological models have been applied. *Huber* [3.45] has concentrated on the analysis of the Laplace transform of $f(t)$

$$\hat{f}(s) = \int_0^\infty dt \exp(-st) f(t), \tag{3.21}$$

based on the average T-matrix approximation (ATA). The solutions obtained with this approach yield the diffusion and hopping models in the proper circumstances and have served to define the situations in which these models are valid.

In the following sections we will discuss three distinct cases in which the theories described above have been applied: the trapping of optical energy in an ordered system, the quenching of fluorescence in a random system, and energy transfer to two distinct disordered acceptors.

3.4.1 Trapping of Optical Excitation in a Concentrated System

In this section the results of a study by *Hegarty* et al. [3.37] into the trapping of $^3P_0$ excitations in the ordered system PrF_3 doped with small amounts of Nd^{3+} ions will be reviewed. The lifetime of the $^3P_0$ state decreased rapidly and changed its nature as Nd^{3+} was introduced in increasing concentrations into the host crystal as shown in Fig. 3.9. These results were obtained using "conventional" spectroscopic techniques where the whole inhomogeneous line was excited with a broadband pulsed laser source and broadband emission, again from the total inhomogeneous profile, was monitored. Thus, emissions from all possible sites which could participate in the energy transfer process were sampled. The variation in the $^3P_0$ temporal dynamics with Nd^{3+} concentration was taken as evidence of transfer between the two ionic species. It was not clear exactly which states were involved in this nonresonant process, however, the absence of any detectable fluorescence corresponding to the Nd^{3+} $^4F_{3/2}$ metastable state was taken as suggesting a cross-relaxation process. The

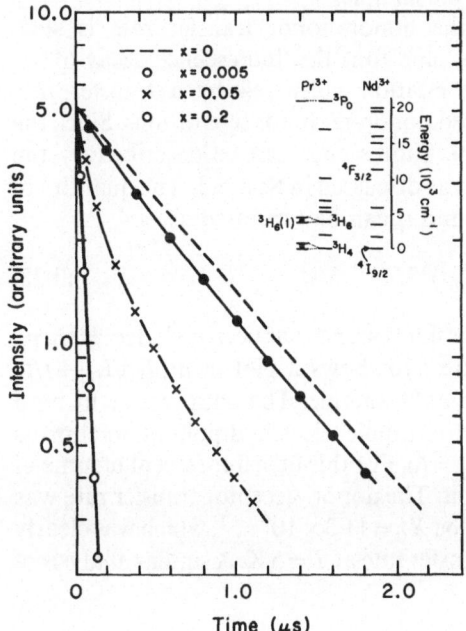

Fig. 3.9. Total fluorescence decays of the $^3P_0$ state of $Nd_xPr_{1-x}F_3$. Traces were taken at 5 K and show the change in the decay characteristics from exponential ($x=0$) to nonexponential and returning to exponential ($x=0.2$). The inset denotes the various Pr^{3+} and Nd^{3+} levels of interest [3.37]

donor–acceptor transfer rate has been found to be only weakly temperature dependent at low temperatures (5–40 K) suggesting that one or more high energy phonons are involved in the transfer process [3.15].

The energy transfer dynamics within the donor system were measured using TRFLN of the $^3P_0$ fluorescence from the Pr^{3+} ions. A good fit to the observed data at all temperatures was obtained by assuming (i) that the host was completely ordered, i.e., c_D (donor concentration) ~ 1 and (ii) that so few Pr^{3+} ions were initially excited that the effects of back transfer could be ignored. Thus, the effects of donor–donor transfer on the fluorescence decay could be described by (3.11) with $f(W_{0n}t)=1$, yielding a dependence of the form $\exp(-W_{DD}t)$, where

$$W_{DD}=c_D \sum_n W_{0n}=c_D W_0 \sum_n (R_0/R_{0n})^6 . \tag{3.22}$$

It was assumed that the donor–donor transfer was mediated by an electric dipole–electric dipole interaction as found in PrF_3. At $T=14$ K a best fit to the data was obtained with $W_{DD}=5.46 \times 10^6$ s^{-1} for the 5 at.% Nd^{3+} sample. A simple lattice summation, such as that in (3.22) yielded for PrF_3 $W_0=4.7 \times 10^5$ s^{-1}, a near-neighbor transfer rate which agreed well with the value found in pure PrF_3, where the rate varies with temperature as $T^{4.3}$. Thus,

$$W_{DD}=64.39 \, T^{4.3}, \quad \text{and} \tag{3.23}$$

$$\beta(T)=\beta(0) \, T^{4.3} . \tag{3.24}$$

Therefore, $PrF_3 : Nd^{3+}$ was considered to be a useful test system for various energy transfer models, as the magnitude of β relative to α could be varied by simply changing the temperature of the sample.

In the low temperature limit, the donor–donor transfer rate is slow compared to the donor–acceptor rate, and thus the fluorescence decay of the $^3P_0$ level following a broadband excitation pulse reflects a transfer rate averaged over the randomly distributed donor–acceptor separations. Since the acceptor concentration c_A is low, the fluorescence can be described by the model of *Inokuti* and *Hirayama* [3.17] as discussed in Sect. 3.3. The quantity of interest here is $I(t)$ rather than $R(t)$, and it can be expressed as

$$\ln[I(t)/I_0]+\gamma_R t = -(4\pi/3)n_A R_0^3 (X_0 t)^{3/s} \Gamma(1-3/s), \tag{3.25}$$

where the intrinsic lifetime of the donor fluorescence has been included. Here n_A is the density of acceptor ions. Figure 3.10 shows a plot of $\ln[I(t)/I_0]+t/t_0$ (where $t_0=1/\gamma_R$) versus $t^{1/2}$ for the 5 at.% sample. The data can be fit by a straight line indicating that the electric dipole–electric dipole interaction is dominant in the transfer process (i.e., $s=6$). For this fit, a fluorescent lifetime of $t_0=900$ ns gave the best straight-line fit. The donor–acceptor transfer rate was calculated from the slope of the line to be $X_0=11.3 \times 10^6$ s^{-1}, which was clearly much faster than the donor–donor transfer rate at $T=5$ K. A similar analysis of

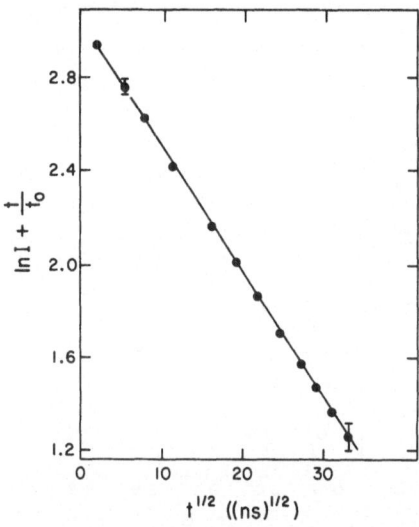

Fig. 3.10. Decay of the $^3P_0$ state in the 5 at.% Nd in PrF$_3$ sample shown in Fig. 3.9, according to the Inokuti-Hirayama model. The fit given by this model yields $s=6$ and identifies the Pr$-$Nd trapping as arising from an electric dipole-electric dipole interaction [3.37]

the $^3P_0$ fluorescence decay for samples with higher Nd^{3+} concentrations yielded the same value for X_0, with $s=6$, provided the intrinsic decay time of the $^3P_0$ level was increased (up to 1200 ns for 20 at.% Nd^{3+}). This was due to a reduction in $^3P_0$ self-quenching by cross relaxation between Pr^{3+} ion pairs as the Pr^{3+} concentration was reduced. (This subject will be discussed in detail in the following section.)

The $^3P_0$ decay rate was found to increase with temperature and to approach a single exponential at $T \sim 40$ K, as illustrated in Fig. 3.11, at which point the decay rate became temperature independent. In this regime the migration of energy among the donors was believed to be rapid so that each donor had the same probability of transferring energy to the acceptors, which can be expressed as

$$X_{DA} = c_A \sum_n X_{0n} = c_A X_0 \sum_n (R_0/R_{0n})^6 . \tag{3.26}$$

Here c_A allows for the probability that site "n" contains an acceptor ion. Thus the fluorescence should evolve as

$$I(t) = I_0 \exp[-(t_0^{-1} + X_{DA})t] , \tag{3.27}$$

for each concentration c_A. The intrinsic lifetime t_0 should again be concentration dependent and should equal the low temperature value obtained from the Inokuti-Hirayama fit. Figure 3.12 shows a plot of $\log X_{DA}$ versus $\log c_A$ from which a linear dependence of X_{DA} on c_A can be deduced using (3.26). For the 5 at.% Nd^{3+} sample, with $t_0 = 900$ ns, the nearest-neighbor transfer rate X_0 was found to be 11.65×10^6 s^{-1}, in close agreement with the low temperature data. For $T = 40$ K, β was calculated to be 1.7×10^{-37} cm^6 s^{-1} whereas $\alpha = 5.46 \times 10^{-38}$ cm^6 s^{-1}, which indicated that the condition for rapid migration, $\beta \gg \alpha$, was indeed valid.

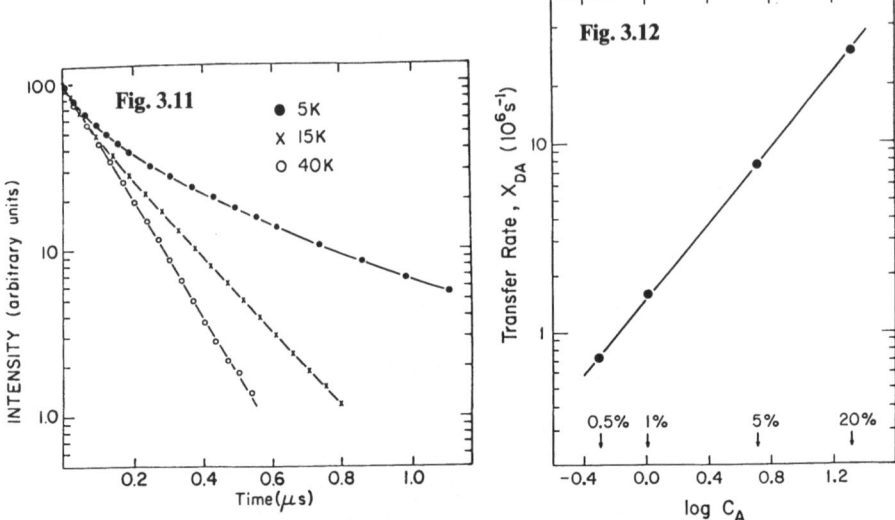

Fig. 3.11. Temperature dependence of the decay of the $^3P_0$ state of PrF_3 doped with 5 at.% Nd^{3+}. The 40 K trace yields the saturated rates in the fast diffusion limit from which the trapping rate X_{DA} can be calculated [3.37]

Fig. 3.12. Plot of the saturated values of the transfer rate X_{DA} inferred from the fast diffusion donor limit plotted against Nd^{3+} concentration in PrF_3. This plot allows the extraction of a near-neighbor trapping rate X_{01} for the $^3P_0$ excitation which is consistent with the value obtained through the use of Fig. 3.10 [3.37]

In the intermediate temperature regime, $5\,K < T < 40\,K$, the $^3P_0$ decay was more complex since the two energy transfer processes competed. The two phenomenological models, mentioned previously, should apply in this region. For $\beta \ll \alpha$, the diffusion model of *Yokota* and *Tanimoto* [3.18] should be applicable, whereas when $\beta \geq \alpha$, at higher temperatures, the hopping model of *Burshtein* [3.19] should yield satisfactory results.

Both models predict that the fluorescence decay should be nonexponential at early times but should approach a single exponential asymptote in the long time limit. Such behavior can be observed in the $T = 40\,K$ decay in Fig. 3.11.

The diffusion model of Yokota and Tanimoto predicts that the asymptotic decay rate c_{YT}, after subtraction of the intrinsic decay rate, is given by

$$c_{YT} = 8.6 n_A \alpha^{1/4} D^{3/4}, \tag{3.28}$$

where D is the diffusion constant for energy transfer among the donors. *Huber* [3.45] has derived an approximation for D for a cubic lattice:

$$D = \tfrac{1}{6} \sum_n W_{0n} R_{0n}^2 = (\beta/6) \sum_n (1/R_{0n})^4, \tag{3.29}$$

assuming an electric dipole–electric dipole interaction. Thus

$$c_{YT} \propto \beta^{3/4} \propto W_{DD}^{3/4}. \tag{3.30}$$

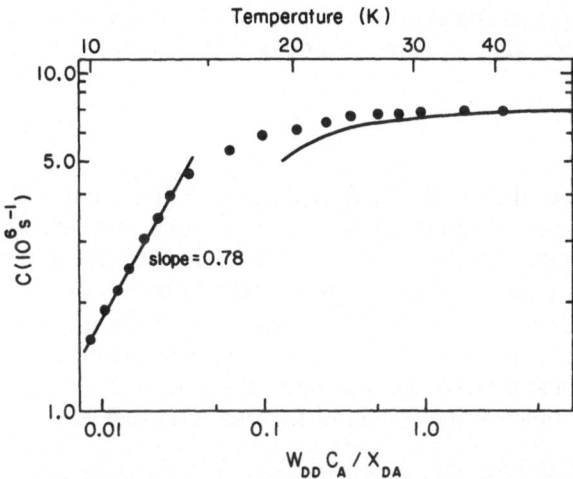

Fig. 3.13. Plot of the asymptotic slope, c, of the $^3P_0$ fluorescence in PrF_3 : 5 at.% Nd^{3+} as a function of the donor-donor transfer rate, W_{DD}, normalized to the rapid-transfer decay rate X_{DA}/c_A. Temperature is indicated on the upper scale. The straight line indicates the region over which the diffusion model applies and the upper line is the behavior predicted by the hopping model. The region between 15 and 23 K represents the transition region $\alpha \gtrsim \beta$, where both models fail [3.37]

From (3.22) and (3.26) it can be seen that

$$\beta/\alpha = W_{DD}c_A/X_{DA}. \tag{3.31}$$

Thus, the plot in Fig. 3.13, $\log c$ versus $\log(W_{DD}c_A/X_{DA})$ is equivalent to plotting $\log c$ versus $\log(\beta/\alpha)$, which is independent of sample concentration. For $10\,\text{K} < T < 40\,\text{K}$, i.e., $0.01 < \beta/\alpha < 0.05$, a straight-line relationship was observed with a slope of 0.78 ± 0.8, in good agreement with the expected value. A value for D was calculated from the data, which yielded at $T = 12\,\text{K}$

$$D = 6.81 \times 10^{-9}\,\text{cm}^2\,\text{s}^{-1}. \tag{3.32}$$

A theoretical value of D was calculated numerically for noncubic lattices. For PrF_3 this yielded a value 5 times smaller than the experimental value and was ascribed to the interference effects of the phonon phase factors discussed previously in Sect. 3.2.2. When such effects were taken into account, a value

$$D_{th} = 5.54 \times 10^{-9}\,\text{cm}^2\,\text{s}^{-1}, \tag{3.33}$$

was obtained in close agreement with the measured diffusion constant.

At higher temperatures, where $\beta \gtrsim \alpha$, the hopping model of Burshtein was found to be valid. The asymptotic exponential decay rate c_B, can be described by the equation

$$c_B = c_A \sum_n X_{0n}/(1 + X_{0n}\tau), \tag{3.34}$$

where c_A is the acceptor concentration as before. Here τ is the "hopping" time or the average time that the excitation remains on any donor ion and has been identified [3.37] as

$$\tau = W_{DD}^{-1}. \tag{3.35}$$

Using the experimental values already obtained for W_{DD} and α, the variation of c_B with temperature was calculated and is represented by the upper solid line in Fig. 3.13. The agreement with experimental data is good in the applicable region, i.e., $\beta/\alpha \geq 1$, but the rapid migration regime is seen to take over very quickly at $\beta/\alpha > 2$.

This study permitted all the models to be applied to a single system for the first time. It has provided an appreciation of the limits of applicability of each and has illustrated the dynamics of the system in the transition region.

3.4.2 Quenching of Fluorescence by Cross Relaxation

In the previous section it was noted that the intrinsic fluorescence decay rate decreased significantly with decreasing ion concentration. This is just one manifestation of quenching of fluorescence in crystals with high concentrations of optically active ions [3.46, 47]. We now extend our discussion to include the phenomenon where the acceptor ions are pairs of donor ions. A possible cross-relaxation process for LaF_3 doped with Pr^{3+} is illustrated in the insert to Fig. 3.14. In this case, the excited Pr^{3+} ion shares its excitation with an

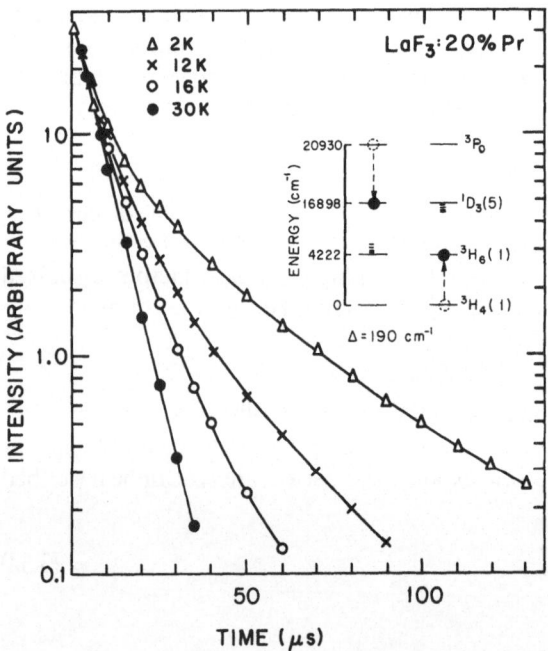

Fig. 3.14. Plot of the decay of the $^3P_0$ state in a 20 at.% sample of $LaF_3:Pr^{3+}$ as a function of temperature, showing the effects of increased donor-donor transfer on fluorescence quenching by cross relaxation. A schematic of the cross-relaxation channel is shown in the inset [3.48]

unexcited neighboring Pr^{3+} ion, so that energy is lost from the donor system. This situation is somewhat more complicated than that considered in the previous section since the lower Pr^{3+} dopant concentrations imply that the donor system, while correlated with the acceptors, is itself disordered.

A study of the temperature dependence of quenching of the $^3P_0$ fluorescence from LaF_3 doped with 20 at.% Pr^{3+} has been reported by *Hegarty* et al. [3.48]. The analysis is similar to that performed on $PrF_3 : Nd^{3+}$ [3.37] and is based on the relative magnitudes of the competing processes described by β and α where α now represents the cross-relaxation interaction.

At very low and very high dopant concentrations the decay of the $^3P_0$ fluorescence after broadband pulsed excitation was exponential at low temperatures. The decay rate constant decreased from 51 μs for dilute Pr^{3+} concentrations to 700 ns for PrF_3. The evolution of the fluorescence for a 20 at.% sample is illustrated in Fig. 3.14. At 2 K the decay was nonexponential, however, as the temperature was increased the decay became faster and approached a single exponential with a saturated decay rate at $T \simeq 30$ K. This saturation effect was taken to indicate the onset of rapid donor–donor migration where $\beta \gg \alpha$.

The nearest-neighbor donor–donor transfer rate, W_0, was calculated to be approximately $1000\,s^{-1}$ for $T = 2$ K from previous TRFLN studies [3.24] on samples with similar dopant concentrations. This is considerably slower than the observed fluorescence dynamics. Thus, it was assumed that donor–donor transfer could be ignored at 2 K and that the temporal evolution of the $^3P_0$ fluorescence could be described by the model of *Inokuti* and *Hirayama* [3.17] (3.25) if a multipolar cross-relaxation interaction was assumed. The fluorescence data were fitted with this model in a manner similar to that shown previously in Fig. 3.10 for $PrF_3 : Nd^{3+}$ and yielded a best fit for $s = 6$, indicating an electric dipole–electric dipole interaction, with

$$X_0 = 8.9 \times 10^4\,s^{-1}. \tag{3.36}$$

It was assumed that this value could then be used to predict the temporal behavior of the low temperature fluorescence at all other dopant concentrations. In particular, for PrF_3 a value of 830 ns for the lifetime was calculated, when suitable allowance for the shrinkage of the lattice due to the lanthanide contraction was made. This value agreed well with the observed lifetime of 760 ns.

At temperatures intermediate between 2 K and ~32 K the relative magnitudes of β and α changed drastically, as discussed above, so that the previous phenomenological models should once again be applicable. For a 20 at.% dopant concentration it was found that W_0 varies as T^3 [3.24], as will β. An extrapolation of the calculated values of β for the temperature range 8 K $< T <$ 32 K indicated that $\beta \gg \alpha$ and so the hopping model of *Burshtein* [3.19] was used to describe the asymptotic exponential behavior of the $^3P_0$ fluorescence. A modified version of (3.34), assuming an electric dipole–electric dipole interaction, was used to fit the data [3.42], giving

$$W_{AS}=(2\pi^2/3)n_A(\alpha/\tau_0)^{1/2}[1-(2/\pi)\tan^{-1}(R_c^6/\alpha\tau_0)^{1/2}],\qquad(3.37)$$

where W_{AS} represents the asymptotic decay rate, τ_0 is the hopping time and R_c is a critical radius which is chosen so that W_{AS} yields the correct decay rate for rapid migration at higher temperatures. The hopping time is related to the TRFLN quantity $R(t)$ by [3.45]

$$\tau_0=\int_0^\infty R(t)dt,\qquad(3.38)$$

and was completely determined from independent data, based on the TRFLN models described in Sect. 3.3. It was estimated that

$$\tau_0=1.389/W_0.\qquad(3.39)$$

Figure 3.15 shows the observed variation of W_{AS} with temperature, the solid line was calculated using (3.37). The agreement is quite good, indicating that the hopping model provides a consistent description of the dynamics of cross relaxation in the temperature regime under consideration.

When the temperature was increased further above 32 K the exponential fluorescence decay rate increased dramatically. This was attributed to the onset of an additional thermally activated cross-relaxation process of the form

$$W=A\exp(-\Delta/kT),\qquad(3.40)$$

where $\Delta=190\text{ cm}^{-1}$. An examination of the cross-relaxation process involving the $(^1D_2)_5$ and $(^3H_6)_1$ intermediate levels as indicated in Fig. 3.14 yields an energy deficit of 190 cm^{-1} which could be provided by a phonon of this energy. Such a process would indeed exhibit an activation behavior and points to the importance of this cross-relaxation channel.

The $^3P_0$ fluorescence decay was observed to contain a long-lived exponential tail, with a lifetime of 51 μs. This is characteristic of the decay rate of isolated

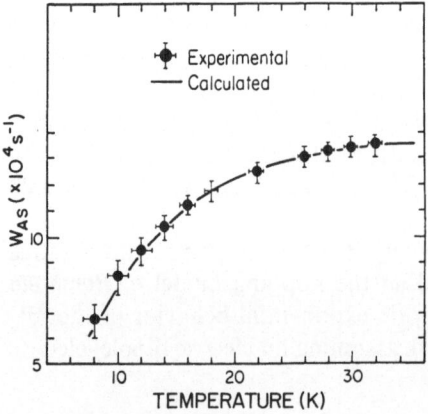

Fig. 3.15. Plot of the asymptotic decay rate of the adjusted fluorescence of the $^3P_0$ state in a 20 at.% sample of Pr^{3+} in LaF$_3$ as a function of temperature. The solid line is the behavior predicted by the Burshtein model [3.48]

Pr^{3+} ions in dilute samples, and thus was regarded by *Hegarty* [3.48] as an emission from isolated ions which did not interact with the main body of cross-relaxing Pr^{3+} pairs. It was estimated that 18% of the Pr^{3+} ions were in domains of low dopant concentration. This long-lived fluorescence persisted at higher temperatures although the lifetime shortened, indicating that some energy transfer process was active. The measured transfer rate had a $T^{3.5}$ temperature dependence, comparable to that of the donor–donor dynamics. The proposal that domains of isolated Pr^{3+} ions existed was considered correct as it was necessary to deconvolute the long-lived exponential component from the fluorescence data, at all temperatures, in order to obtain the self-consistent energy transfer model outlined above.

An alternative explanation of the observed temporal features was offered by *Vial* and *Buisson* [3.32] based on a short range cross relaxation caused by a superexchange interaction. They studied the decay of luminescence following pulsed laser excitation of satellite lines in the vicinity of the sharp $^3P_0-(^3H_4)_1$ emission line in LaF_3 with dilute (0.1%) amounts of Pr^{3+} (Fig. 3.16). These features had already been identified as $^3P_0$ emission from Pr^{3+} ions perturbed by Pr^{3+} near neighbors [3.49]. At low temperatures, the emission from each line was characterized by a single exponential decay with a lifetime τ which was shorter than that of the main line, τ_0. *Vial* and *Buisson* [3.32] argued that the observed difference was due to a cross relaxation of the perturbed Pr^{3+} ion with its unexcited Pr^{3+} near neighbor, the cross-relaxation rate W being given by

$$W = \tau^{-1} - \tau_0^{-1}. \tag{3.41}$$

The feature labeled "f" in Fig. 3.16 was found to have the largest cross-relaxation rate and was assigned to a Pr^{3+} ion with another Pr^{3+} ion at its

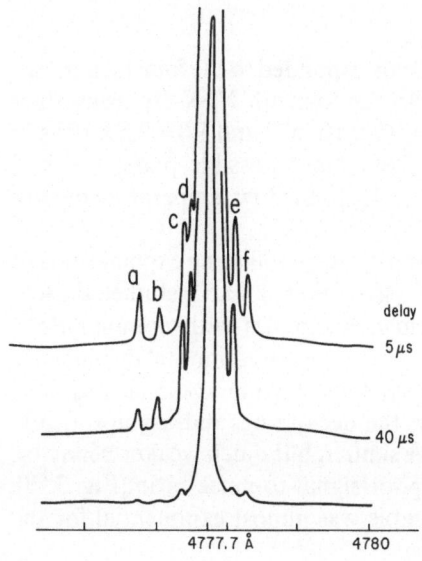

Fig. 3.16. The lower curve shows the absorption spectrum of $LaF_3 : 0.1\% \ Pr^{3+}$ in the region of the $(^3H_4)_1 - {}^3P_0$ transition. The two other curves show time-resolved excitation spectra of the $^3P_0$ fluorescence with two different delays after the laser pulse. (Note that the gain for the 40 μs spectrum is twice that for the 5 μs spectrum.) [3.32]

nearest-neighbor (NN) site. Feature "*a*" had the next largest W and was associated with a Pr^{3+} ion with a neighbor at a slightly more distant site, and so on. Clearly, the transfer rates should scale as R^{-s} for a multipolar interaction. However, it was found that W decreased faster than predicted by any reasonable value of s, so a short range superexchange interaction was proposed.

The temporal evolution of the fluorescence can be modeled exactly by (3.18). (Note that *Vial* and *Buisson* [3.32] used W instead of X to denote cross relaxation.) For a near-neighbor-only interaction the infinite product can be expressed by

$$I(t) = I_0 \exp(-\gamma_R t) \prod_i [1 - c + c \exp(-W_i t)]^{N_i}, \tag{3.42}$$

where c is the Pr^{3+} concentration and N_i represents the number of near neighbors of type "*i*". This equation predicts a single exponential with intrinsic lifetime γ_R^{-1} at long times, for small c:

$$I(t) = I_0 \exp(-\gamma_R t) \prod_i (1 - c)^{N_i}. \tag{3.43}$$

The initial fluorescence decay rate assumes a simple form for a short range interaction

$$\frac{d}{dt} \ln(I/I_0) = -\gamma_R - c \sum_i N_i W_i. \tag{3.44}$$

For a multipolar interaction the initial decay rate is given by

$$\frac{d}{dt} \ln(I/I_0) = -\gamma_R - c W_0 \sum_n (R_0/R_{0n})^s, \tag{3.45}$$

where W_0 is the NN cross-relaxation rate.

The authors [3.32] argued that site "*f*" corresponded to the four 1st and two 2nd NNs and site "*a*" was associated with the four 4th NNs. By using their microscopic cross-relaxation rates of $W_f = 9.7 \times 10^4 \text{ s}^{-1}$ and $W_f = 2.5 \times 10^4 \text{ s}^{-1}$ and ignoring the smaller contributions from other possible pairs, *Vial* and *Buisson* [3.32] obtained an excellent fit of (3.42) to the low temperature data of *Hegarty* et al. [3.48].

In an attempt to reconcile these two apparently conflicting explanations of the cross-relaxation process in $LaF_3:Pr^{3+}$, *Morgan* et al. [3.50] studied the low temperature $^3P_0$ fluorescence decay from five samples with nominal Pr^{3+} concentrations of 1, 5, 20, 50, and 80 at.%. At $T = 2$ K, the temporal dependence of the emission was nonexponential at early times for the 1%, 5%, and 20% samples as shown in Fig. 3.17. However, the decay rates did asymptotically approach that of the isolated ion (51 μs). A similar, but much weaker behavior, was observed for the 50% sample with poor signal-to-noise ratio (Fig. 3.18). The fluorescence from the 80% Pr^{3+} sample was almost exponential for the

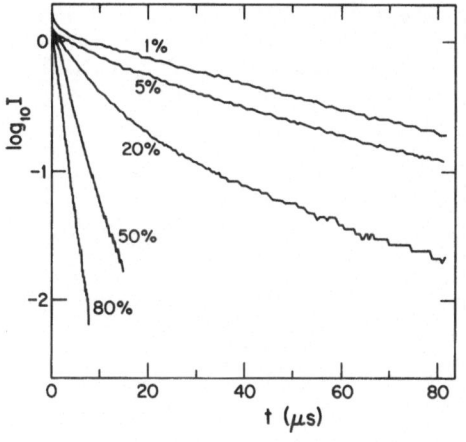

Fig. 3.17. $^3P_0$ fluorescence decay patterns observed from five samples of LaF$_3$ doped with 1, 5, 20, 50 and 80 at.% Pr^{3+} at 2 K [3.50]

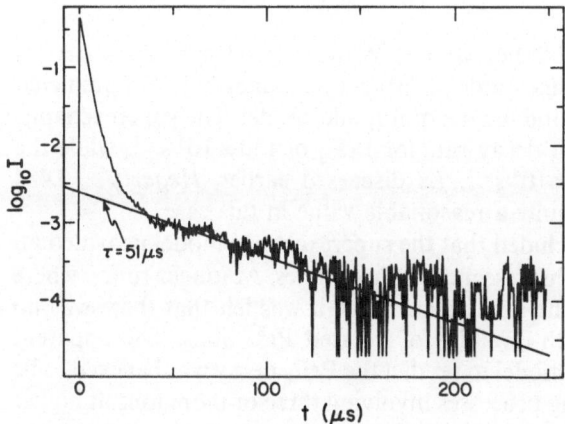

Fig. 3.18. Detail of the $^3P_0$ fluorescence decay from LaF$_3$:20 at.% Pr^{3+} at 2 K. An exponential component at 51 μs is indicated by the straight line [3.50]

first few microseconds, but the decay rate increased gradually at later times before the signal was submerged in the noise.

The data for all samples were fit by both the short range superexchange model of *Vial* and *Buisson* [3.32] and the Inokuti-Hirayama long range multipolar model used by *Hegarty* et al. [3.48]. The proposed 51 μs single-exponential contribution of the isolated ions was first deconvoluted from the raw data. The intensity of this "apparent" single ion exponential feature could be calculated from (3.43). A good agreement with the observed data was obtained for $N_a + N_f = 10$, as predicted [3.32], ranging from 86% for 1% Pr^{3+} to 0.12% for the 50% Pr^{3+} sample. The best fit of the multipolar model for each sample was for the electric dipole–electric dipole interaction and yielded a nearest-neighbor cross-relaxation rate $W_0 \sim 10^5$ s^{-1} for all samples. However, a consistently good fit of the short range model with the fitted values of W_a and W_f, comparable to the observed macroscopic rates, was also obtained (Fig. 3.19). The initial fluorescence decay rates were predicted by (3.43) and

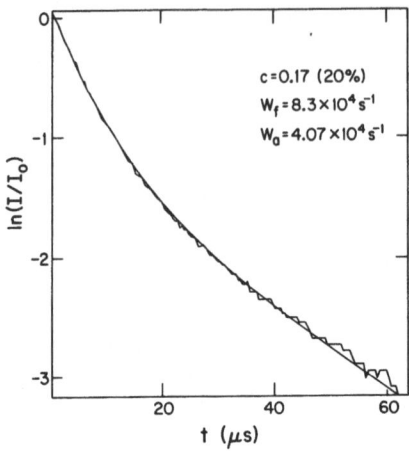

Fig. 3.19. Fit of the two-parameter short range interaction model [3.32] to the fluorescence decay from $LaF_3 : 20$ at.% Pr^{3+} at 2 K. The fit yields $W_f = 8.3 \times 10^4 \, s^{-1}$ and $W_a = 4.07 \times 10^4 \, s^{-1}$ [3.50]

(3.44) for the parameters obtained above. While a good consistent fit was obtained for the superexchange model, a large discrepancy between predicted and measured values was found for the multipolar model. The superexchange model did however predict a decay rate for PrF_3 of $130 \times 10^4 \, s^{-1}$, while the observed value is only $70 \times 10^4 \, s^{-1}$. As discussed earlier, *Hegarty's* [3.48] multipolar treatment gave quite a reasonable value in this case.

Morgan et al. [3.50] concluded that the superexchange model provided an excellent description of the fluorescence at short times. At longer times, where both models could predict the decay accurately, it was felt that there was no need to invoke a model with domains of isolated Pr^{3+} ions. The apparent failure of the superexchange model to predict the PrF_3 decay was believed to be due to the onset of quenching processes involving three or more ions at higher dopant concentrations. Although the results were not conclusive it appeared that the superexchange interaction contributed significantly to the quenching of $^3P_0$ fluorescence by cross relaxation.

3.4.3 Fluorescence Quenching in a Disordered System

In this section we will extend our discussion of optical energy transfer and trapping to include the situation where both the donors and acceptors are randomly distributed on the lattice sites of the host material.

Recently *Glynn* et al. [3.51] have observed the temporal evolution of the $^3P_0$ emission, excited by a pulsed laser source, in the material LaF_3 doped with 25 at.% Pr^{3+} and 3 at.% Nd^{3+}. Three channels for the transfer of energy exist in this system: donor–donor transfer among Pr^{3+} ions; energy transfer from the Pr^{3+} to Nd^{3+} ions; and a Pr^{3+}–Pr^{3+} cross-relaxation interaction. The interaction parameters for these three processes are labeled β, α_A and α_B, respectively, where α and β have the same meanings as in the previous sections. We summarize the results from Sects. 3.4.1, 2 in which *Hegarty* et al. [3.37, 48]

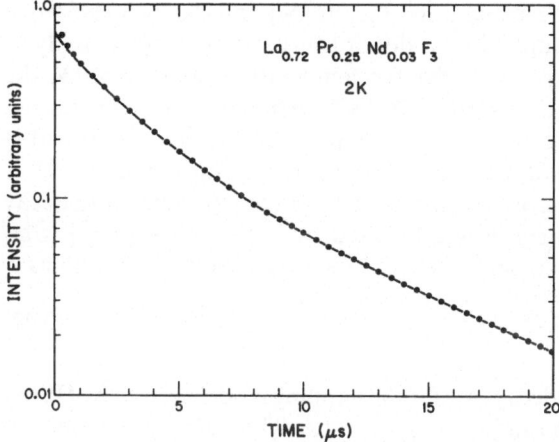

Fig. 3.20. Plot of the decay of $^3P_0$ state of Pr^{3+} in $La_{0.72}Pr_{0.25}Nd_{0.03}F_3$ at 2 K. The points represent the experimental data and the solid curve the best fit to (3.46) [3.51]

obtained $\alpha_A = 5.46 \times 10^{-38}$ cm^6 s^{-1}, and $\alpha_B = 3 \times 10^{-40}$ cm^6 s^{-1} below 30 K. In this temperature region β varied with temperature as T^3 and had a value of 18.8 $\times 10^{-40}$ cm^6 s^{-1} at 14 K.

The decay of the donor fluorescence following pulsed excitation at low temperatures, where donor–donor transfer can be ignored, can be described by (3.17), with a slightly modified form of (3.18) for $f(t)$, i.e.,

$$f(t) = \prod_n [1 - c_A - c_B + c_A \exp(-X_{0n}^A t) + c_B \exp(-X_{0n}^B t)], \qquad (3.46)$$

which merely reflects the two quenching channels for the fluorescence. In this case $c_A = 0.03$ and $c_B = 0.25$.

The observed nonexponential $^3P_0$ decay at 2 K was fitted with (3.46) rather than the usual Inokuti-Hirayama continuum approximation of (3.25). An electric dipole–electric dipole interaction was assumed responsible for both quenching processes and the lattice product in (3.46) was carried out over the first 300 neighboring sites. Regarding α_A and α_B (i.e., X_0^A and X_0^B) as adjustable parameters, the best fit, illustrated in Fig. 3.20, was obtained for $\alpha_A = 2.5 \times 10^{-38}$ cm^6 s^{-1} and $\alpha_B = 1.5 \times 10^{-40}$ cm^6 s^{-1}, which were approximately a factor of two smaller than those reported by *Hegarty* [3.37, 48]. The data were also fitted with (3.46) for $c_B = 0$ (i.e., no Pr–Pr cross relaxation) and yielded $\alpha_A = 5.8 \times 10^{-39}$ cm^6 s^{-1}, which was a factor of 10 smaller than Hegarty's value. A final fit was made to the data where the Pr–Pr cross relaxation was modeled on the short range superexchange interaction proposed by *Vial* and *Buisson* [3.32]. This resulted in a best fit for $\alpha_A = 8.1 \times 10^{-39}$ cm^6 s^{-1} with a 10% discrepancy between the fit and the data at early times. This result was also regarded as unsatisfactory by *Glynn* et al. [3.51].

An examination of the transfer parameters at higher temperatures indicates that the dynamics of the system are quite complicated. For example at $T = 14$ K, $\beta/\alpha_A = 0.08$ suggests that the diffusion model of *Yokota* and *Tanimoto*

[3.18] would be appropriate to describe the Pr–Nd transfer process, while $\beta/\alpha_B = 13$ indicates that the rapid donor–donor transfer model would apply to the Pr–Pr cross-relaxation process. For the temperature range 35–70 K the corresponding values are 1.17–9.4 and 176–1808, assuming a T^3 temperature dependence for β. Thus the hopping model proposed by *Burshtein* [3.19] should apply for the Pr–Nd transfer, and the rapid migration model is appropriate for the Pr pair cross relaxation. A fit of the latter model to decay data in this temperature range was attempted by *Glynn* et al. [3.51], who proposed that the asymptotic decay rate at long times could be written as

$$W_H = \gamma_R + W_{AS} + c_B \alpha_B \sum_n R_{0n}^{-6}, \tag{3.47}$$

where W_{AS} is the asymptotic decay rate for the hopping model derived by Huber (3.37) and γ_R is the intrinsic radiative decay rate. They also considered an alternative expression to describe the asymptotic form of the decay rate for the hopping model, proposed by *Sakun* and co-workers [3.52, 53]:

$$W_S = \gamma_R + \pi(2\pi/3)^{5/2} n_D n_A (\alpha_A \beta P_0)^{1/2} + c_B \alpha_B \sum_n R_{0n}^{-6}. \tag{3.48}$$

Here P_0 is the probability that during a random walk of the donor excitation in the lattice the excitation will never return to the starting point. For a simple three-dimensional random walk, $P_0 = 0.66$ [3.53]. At elevated temperatures it was noted that n_A, the density of Nd^{3+} ions in the ground $(^4I_{9/2})_1$ state, decreased fractionally from 0.92 at 30 K to 0.62 at 70 K due to thermal excitation of higher states, which might not contribute effectively to the energy transfer process. Also, *Hegarty* [3.48] reported that the Pr–Pr cross relaxation rate increased above 32 K exhibiting an energy activation behavior. The parametric temporal dependence of the cross-relaxation rate was modeled on the data in [3.48] and included in (3.47, 48) as

$$c_B \alpha_B \sum_n R_{0n}^{-6} = [0.93 + 400 \exp(-288/T)] \times 10^5 \, s^{-1}. \tag{3.49}$$

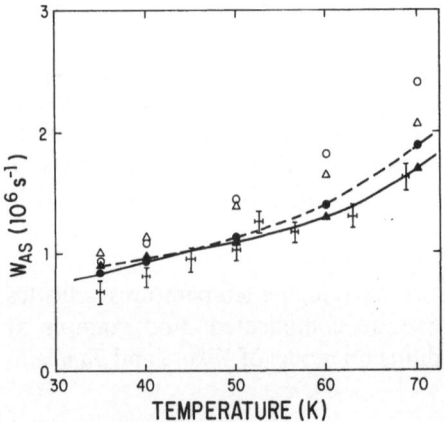

Fig. 3.21. Temperature dependence of the asymptotic decay rates in the 35–70 K region. + are the experimental data, Δ are calculated using (3.47) with n_A = constant. ▲ are calculated by using (3.47) with $n_A = n_{A0}(T)$. ○ are calculated by using (3.48) with n_A constant. ● are calculated using (3.48) with $n_A = n_{A0}(T)$ [3.51]

The asymptotic fluorescence decay rate at long times was calculated for both models represented by (3.47, 48), using the transfer parameters measured at low temperature, for both n_A constant and n_A dependent on temperature [i.e., $n_A = n_{A0}(T)$]. The results are reproduced in Fig. 3.21. While no definitive conclusions could be drawn it appeared that the general assumptions of the model were correct and that the thermal depopulation of the Nd^{3+} ground state did influence the Pr–Nd transfer efficiency.

In the temperature region $18\,K < T < 30\,K$, where $0.16 < \beta/\alpha_A < 0.74$ and $12 < \beta/\alpha_B < 719$ the rapid diffusion model should be appropriate for the Pr–Pr cross relaxation (which is independent of temperature) [3.48]. However, the Pr–Nd system is in a regime intermediate between the hopping and diffusion models where as yet no model is known to work.

3.4.4 Macroscopic Spectral Dynamics in Other Systems

It has already been noted that quite a large body of literature exists relating to energy transfer in rare-earth systems. Much of it, however, preceded the recent experimental and theoretical advances which have been discussed in earlier sections. Therefore, many of the conclusions reached can be considered suspect. In the space available in this chapter we cannot adequately survey all the reports in this area, but will concentrate on a system other than Pr^+ in which the results of a correct laser spectroscopic study have led to a better insight into the energy transfer processes active in the system.

Various Nd^{3+} doped materials have received much attention in the past, due mainly to their potential importance as infrared laser sources. Many of these studies have been reviewed [3.13, 44], but it must be recognized that the treatments have suffered from the limitations of traditional spectroscopic methods.

TRFLN techniques have been applied to a number of Nd doped materials recently and energy transfer between Nd^{3+} ions in dissimilar crystallographic sites has been reported by *Powell* and collaborators [3.54–56]. Most of this work, however, has involved a nonresonant excitation process to populate the metastable $^4F_{3/2}$ level, due to the lack of a tunable IR coherent source at $\sim 1\,\mu m$. Unfortunately, the effects of accidental coincidence [3.27] due to the nonresonant excitation processes complicated the fluorescence lineshapes so that no conclusive analysis of the donor–donor dynamics could be made.

The recent introduction of H_2 Raman down converters and F-center lasers [3.57, 58] has allowed resonant TRFLN techniques to be extended well into the near IR spectral region. Direct FLN measurements of Nd^{3+} $^4F_{3/2}$ properties in various glass hosts were reported by *Pellegrino* et al. [3.59]. *Broer* et al. [3.39, 60] observed donor–donor transfer in the $^4F_{3/2}$ state of a Nd-doped pentaphosphate system.

This latter system, $Nd_xLa_{(1-x)}P_5O_{14}$, neodynium lanthanum pentaphosphate (LNPP), is an example of a stoichiometric compound which exhibits unusual characteristics and in which considerable interest has been shown since its initial inception [3.61, 62]. The quenching of the $^4F_{3/2}$ fluorescence varies linearly with Nd^{3+} concentration so that even for NdP_5O_{14}, i.e., $x=1$, only weak quenching is observed [3.63]. This is in contrast to $Nd:YAG$ where the dependence of fluorescence quenching on dopant concentration is seen to be quadratic [3.64] and limits the maximum useful Nd^{3+} concentration in YAG lasers. The decay of the $^4F_{3/2}$ fluorescence, following pulsed excitation, is exponential over the period of measurement for all Nd concentrations. Similar effects have also been observed in related Pr-doped compounds [3.65, 66].

Various mechanisms have been proposed to explain the optical properties of these systems [3.62, 65]. The absence of significant concentration quenching is attributed to a weak cross-relaxation interaction caused by large energy mismatches between the intermediate states involved in the process. A fast donor–donor interaction was proposed to explain the single exponential decay of the $^4F_{3/2}$ fluorescence whereby the system was believed to be in the rapid diffusion regime. Finally, the linear dependence of concentration quenching was thought due to the presence of unknown trapping centers.

Flaherty and *Powell* [3.56] looked for spectral dynamics involving the $^4F_{3/2}$ state which was excited nonresonantly via the $^2G_{5/2, 7/2}$ states with a visible dye laser. No transfer either between the regular Nd^{3+} ions or to Nd^{3+} ions in distinct centers was observed and it was concluded that no spectral migration of energy existed and that a rapid resonant donor–donor transfer was the cause of the exponential fluorescence decay. It was also proposed that quenching of the fluorescence was produced by surface effects.

More recently, *Broer* et al. [3.39, 60] have succeeded in narrowing the $^4F_{3/2}$ emission from LNPP at low temperatures using resonant excitation. They reported observing an onset of spectral energy transfer within the inhomogeneous emission profile above 20 K, for $x=0.2$ and 0.75. Some of their results are presented in Fig. 3.22. Below $T=20$ K no such transfer was observed, however, the narrowed fluorescence emission did not broaden with time, which indicated that no incoherent resonant transfer occurred. It was proposed that if no donor–donor transfer occurred, (3.18) could be used to explain fluorescence quenching if it were due to cross relaxation between Nd pairs. Because of the weak trapping rate, (3.19) was approximated by

$$f(t)=\exp\left(-tx\sum_n X_{0n}\right), \tag{3.50}$$

where x is now the Nd concentration. The observed decay rate was written in the form

$$\tau^{-1}=\gamma_R+x\sum_n X_{0n}, \tag{3.51}$$

which predicted single exponential fluorescence decay behavior. The system therefore would not develop beyond the early time behavior, because the $^4F_{3/2}$ excited state would be depopulated radiatively in a time t much less than X_{01}^{-1}. This is in total agreement with the observed behavior, as it can be seen from

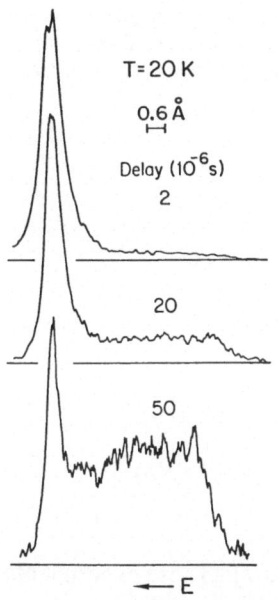

T=20 K

0.6 Å

Delay (10^{-6}s)

2

20

50

← E

(3.51) that the quenching rate varies linearly with Nd concentration. Furthermore, by assuming an electric dipole–electric dipole cross-relaxation interaction *Broer* [3.39] carried out the lattice sum in (3.51) using the room temperature value, $X_{01} = 1.1 \times 10^3 \, s^{-1}$ with $\gamma_R = 2.9 \times 10^{-3} \, s^{-1}$, reported by *Lenth* et al. [3.63] and obtained excellent agreement with the measured quenching rates for the $x = 0.2$, 0.75 and 1 samples.

An analysis of the $T > 20$ K TRFLN data, for the $x = 0.2$ sample, following the treatment outlined in Sect. 3.3 indicated that the donor–donor transfer occurred via an electric dipole–electric dipole interaction. A modified form of (3.11) and (3.14), i.e.

$$R(t) = \prod_n \left\{ 1 - x + x \exp\left[-(X_{0n} + W_{0n})t \right] \cosh(W_{0n}t) \right\}, \tag{3.52}$$

was used to fit the data with $X_{01} = 1.1 \times 10^3 \, s^{-1}$; W_{01} was treated as a variable parameter and the lattice product was calculated explicitly for each value. The values of W_{01} which yielded the best fits at different temperatures are plotted in Fig. 3.23, and have an exponential temperature dependence with an activation energy of 94 cm^{-1}. The TRFLN data suggested that the donor–donor transfer rate was dependent on the energy mismatch ΔE, so it was concluded that the one-site resonant process was responsible for the spectral energy transfer. A room temperature value for W_{01} was extrapolated from the results in Fig. 3.23 from which a diffusion constant D describing the spatial migration of the donor excitation was obtained [3.42]. This method yielded a value for D which was only 20 times smaller than one obtained using four-wave-mixing techniques

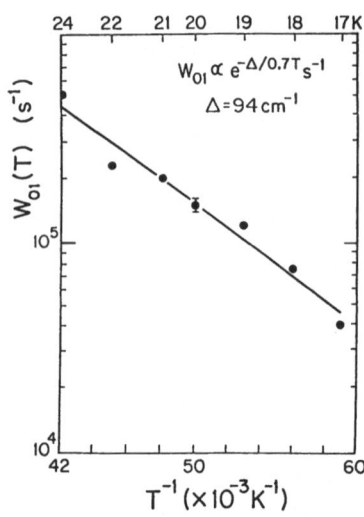

Fig. 3.23. Nearest-neighbor donor-donor transfer rates $W_{01}(T)$ showing an exponential temperature dependence with an activation energy $\Delta = 94\,\mathrm{cm}^{-1}$ [3.39]

which measure spatial diffusion of energy directly [3.67]. (These experiments will be discussed in the next section.)

The results of the TRFLN study explain in a consistent manner all the features of the Nd-doped pentaphosphate system. It should also be apparent that the direct measurement of the donor–donor interaction was a key element to unraveling the problem correctly.

3.5 Resonant Energy Transfer and Spatial Migration

Our discussion of energy migration in optically active systems has so far concentrated on the nonresonant process of spectral migration, within an inhomogeneously broadened feature and among dissimilar centers. The models have not incorporated spatial aspects of transfer, except for calculating the dependence of the transfer rate on interionic separation. No predictions of the characteristics of spatial migration have been made due principally to the fundamental limitations of TRFLN which does not probe any spatial features of the dynamic optical system. The technique has been most effective in providing quantitative estimates of phonon assisted nonresonant donor–donor processes but has only provided indirect indications on the existence of any resonant energy transfer.

In this section, various experiments will be reviewed which have attempted to elucidate the nature of the spatial migration of energy and the extent to which resonant energy transfer occurs within an ensemble of optically active ions.

3.5.1 Spatial Transfer of Energy

Of the several techniques which have been developed in recent years to measure the range of spatial migration of energy [3.68], the "transient grating" technique has proved to be the most popular. This method, based on a four-wave-mixing (FWM) process, has been applied to diverse situations where various transport phenomena have been studied [3.69]. When two coherent laser beams with the same frequency cross in a material at an angle θ, a standing wave in the electric field in a plane perpendicular to the bisector of the two beams is created. If the laser frequencies are in resonance with an optical transition in the material, a spatially modulated excited state population can be formed with a period

$$\Lambda = \lambda/2 \sin(\theta/2). \tag{3.53}$$

A third beam (which may have a different frequency from the other two) can then be Bragg scattered from the grating. In degenerate four-wave mixing this probe beam, which does have the same frequency as two writing beams, will satisfy the scattering condition if it counterpropagates along one of the writing beams. The scattered coherent signal is then found to counterpropagate along the other writing beam. A typical experimental setup to write and read the grating is presented in Fig. 3.24.

The intensity of the scattered beam I_s depends on the depth of the grating $\Delta n (= n_{peak} - n_{valley})$ where n represents the excited state population

$$I_s \propto I_p (\Delta n)^2, \tag{3.54}$$

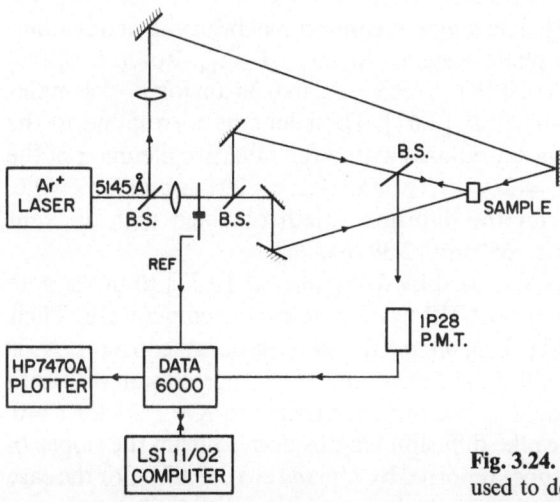

DETECTION OF FOUR-WAVE MIXING

Fig. 3.24. The experimental arrangement used to detect and analyze the decay of a laser-induced grating in $LaP_5O_{14} : Nd^{3+}$. (B.S. represents a beamsplitter) [3.77]

where I_p represents the probe beam intensity. Thus, if one of the writing beams is amplitude modulated, then the temporal behavior of the grating is probed by the scattered beam. In the absence of any spatial dynamics, the grating decays due to the depopulation of the excited state, with a rate characteristic of the fluorescence decay process. The presence of energy migration tends to increase the rate of decay as the grating contrast (Δn) decreases as energy diffuses from the peaks to the valleys. For diffusive motion along the grating dimension x, the excited state population $n(x, t)$ obeys the one-dimensional diffusion equation

$$\frac{\partial n(x, t)}{\partial t} = D \frac{\partial^2 n(x, t)}{\partial t^2} - \frac{n(x, t)}{\tau}. \tag{3.55}$$

The last term in (3.55) includes the effects of relaxation, and D represents the energy diffusion constant. When one of the writing beams is turned off the grating decays. Assuming an initial one-dimensional sinusoidal modulation of the excited state population, (3.54) and (3.45) predict that I_p will decay exponentially with a decay rate given by

$$K = 2\tau^{-1} + (32\pi^2 D/\Lambda^2)$$
$$= 2\tau^{-1} + (32\pi^2 D/\lambda^2) \sin^2(\theta/2). \tag{3.56}$$

Thus, the scattered beam decay rate can be changed by varying the grating period by an appropriate adjustment of the crossing angle of the writing beams, θ. A plot of K versus $\sin^2(\theta/2)$ should be linear and the diffusion constant D can be calculated.

The technique of FWM is extremely useful for studying energy diffusion in organic systems where diffusion coefficients are large [3.70, 71]. Several unsuccessful attempts have been made to observe similar effects in inorganic crystalline systems [3.72–75]. The single exception has been the neodynium-doped lanthanum pentaphosphate system, $Nd_xLa_{1-x}P_5O_{14}$. Room temperature spectral diffusion lengths of 0.18 μm for $x = 0.2$ to 0.36 μm for $x = 1$ samples have been reported by *Lawson* et al. [3.67]. These lengths correspond to the average distance l_a that the energy diffuses within the radiative lifetime τ of the Nd $^4F_{3/2}$ metastable levels, $l_a = (2D\tau)^{1/2}$. (For $x = 0.2$, $\tau \sim 294$ μs and for $x = 1.0$, $\tau \sim 124$ μs.) Other estimates for the diffusion length obtained with different techniques vary from 32 nm to 650 nm [3.39, 61, 76].

The FWM technique was also used by *Morgan* et al. [3.77] to investigate spatial energy migration in the LNPP system at room temperature. Their results are summarized in Fig. 3.25, where the grating decay constant, K, is plotted as a function of $\sin^2(\theta/2)$. The predictions of (3.56) agreed well with data for $5 \times 10^3 < \sin^2(\theta/2)$ and could be seen to extrapolate to $K = 2\tau^{-1}$ for $\theta = 0$. The diffusion rates, and hence the diffusion lengths derived from the slopes of the straight lines, agree with those reported by *Lawson* et al. [3.67]. For the case of $\sin^2(\theta/2) > 15 \times 10^{-3}$, the temporal decay of the scattered beam was exponential, as expected. However, as shown in Fig. 3.25, K unexpectedly

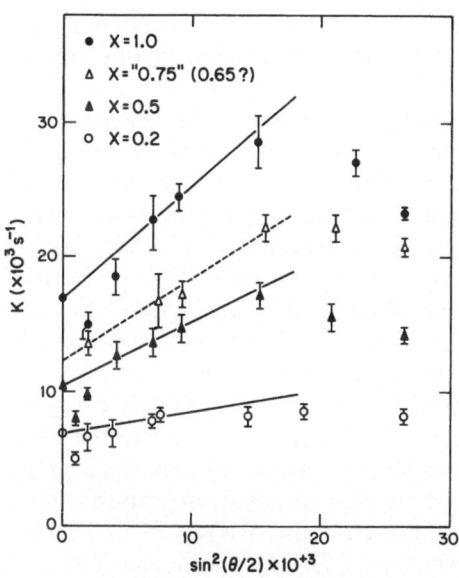

Fig. 3.25. Plot of the observed scattered probe beam decay rate K for various crossing angles θ of the writing beams, in $Nd_xLa_{(1-x)}P_5O_{14}$ at room temperature. The values for twice the fluorescence decay rates of the $4F_{3/2}$ state are plotted on the K axis. [3.50]

GRATING DECAY RATE (K) vs
CROSSING ANGLE OF WRITING BEAMS (θ)

decreased with increasing θ (i.e., with decreasing grating period Λ). *Morgan* et al. [3.77] argued that this was due to a short range trapping process whose effects became more significant when smaller dimensions were probed by the grating. At smaller crossing angles of the writing beams, i.e., $\sin^2(\theta/2) < 5 \times 10^{-3}$, the scattered beam decay consisted of two exponential components. The rate of one component was close to twice the fluorescence decay rate, and the faster component decreased rapidly with decreasing θ. The latter decay rate was plotted in Fig. 3.25 and the effect was ascribed to the presence of some other type of grating in the material. The most likely possibility, i.e., a thermal grating, was discounted, but the ferroelastic nature of this birefringent material, it was believed, would allow the formation of gratings from other nonlinear optical interactions.

Tyminski et al. [3.78] studied degenerate four-wave mixing in the same pentaphosphate system at various temperatures between 10 K and room temperature. They reported that an oscillatory pattern exists in the decay of the scattered beam at low temperatures ($T < 100$ K) and low pump powers ($P < 30$ mW) and analyzed the results along the lines developed by *Wong* and *Kenkre* [3.79]. The values extracted from the diffusion constant D from this treatment increased with decreasing temperature, which they took to indicate the presence of partially coherent excitonic migration in the system. This result would appear to be at odds with the TRFLN observations of *Broer* et al. [3.39, 60] as discussed in Sect. 3.4.4. This latter group succeeded in narrowing the $^4F_{3/2}$ inhomogeneous emission at low temperatures and saw no evidence for an incoherent resonant or rapid nonresonant transfer of energy among the donor Nd^{3+} ions. It is evident that some work is required in this area to reconcile these two models.

The technique of transient grating spectroscopy suffers from some major limitations when applied to the study of energy migration in inorganic materials. The physical dimensions over which transfer can be detected are quite coarse and are limited to the smallest grating period obtainable, namely $\Lambda = \lambda/2$. This corresponds to the case of counterpropagating writing beams where $\theta = 180°$. Clearly, a blue or UV laser would prove to be a finer probe than a red or IR source. The present range of lasers thus sets a lower limit of 200–300 nm on the smallest grating size. For migration to have a detectable effect on the grating decay, therefore, the optical excitation must migrate over ~ 1000 lattice sites (along the grating) within the lifetime of the excited state. That this criterion is satisfied in $Nd_x La_{1-x} P_5 O_{14}$ is possibly due to the weak effects of fluorescence quenching in concentrated samples where the average Nd–Nd separation is small enough to allow effective energy migration among the optically active ions. On the basis of this criterion, $Al_2 O_3 : Cr^{3+}$, i.e., ruby, was the subject of various FWM experiments [3.72–74]. The lifetime is long at low temperatures ($\tau \sim 4$ ms) and it was believed that a rapid migration of energy occurred among the ^2E metastable state of the Cr^{3+} ions [3.80]. No noticeable spatial effects of this migration, however, were detected and an upper limit of 30 nm was placed on the diffusion lengths in samples with Cr^{3+} concentrations between 0.05 and 1.55 at.% at 10 K.

These latter results will be discussed in the next section which deals with the topic of resonant energy transfer in solids.

3.5.2 Resonant Energy Transfer

Resonant energy transfer processes are difficult to probe directly and in many cases controversy exists as to the extent of resonant energy transfer. By resonant transfer we mean the migration of optical energy among ions which have excited states within a homogeneous and possibly line-narrowed profile. The TRFLN technique may be used to probe for resonant energy transfer only in some limited circumstances, particularly where the process is incoherent. The transfer mechanism would depend on the spectral overlap of almost resonant ions and as such would be analogous to that originally proposed by *Forster* and *Dexter* [3.3, 4] as represented by (3.1). The FLN signal would broaden (as energy was transferred to ions with partially overlapping spectral profiles) and would eventually fill the inhomogeneously broadened feature. One such observation has been reported by *Harig* et al. [3.81] in an organic system, however, no analogous process has been found to exist in 4f systems.

When the interionic coupling strength J is large and comparable to the energy shifts ($\Delta E \sim \Delta u_{inh}$) the single ion description is not considered appropriate. The optical excitation can no longer be regarded as residing on a specific ion but is shared coherently by the whole group of equivalent ions. When the optically active ions form an ordered array, the momentum of the delocalized exciton can be described by a wave vector k which must be conserved. The homogeneous linewidths of spectral features can be large due to the effects of dispersion on the excited states. In general, the interionic coupling is weak in $4f$ materials so that the model of excitation localized on a single ion generally provides an adequate description of transfer even in concentrated systems. There are some exceptions to this observation, however, which have been discussed by *Cone* and *Meltzer* [3.82].

One approach to directly probe for resonant energy transfer in the absence of spectral transfer has been developed by *Strauss* et al. [3.83] using a two-photon technique. One low-powered narrowband laser beam was used to line-narrow the $^3P_0-(^3H_4)_1$ fluorescence profile in $LaF_3:Pr^{3+}$. A second narrow, but more powerful laser beam was tuned to the $^3P_0-(^3H_6)_1$ transition and burned a hole in the broad emission which was caused by the effects of "accidental degeneracy" [3.27]. The process is illustrated in Fig. 3.26, which also shows some of the data obtained in the study. Resonant energy transfer would manifest itself by the refilling of the hole. No such dynamics were observed in samples with 80 at.% Pr^{3+} concentrations. PrF_3 could not be studied in this manner due to the lower threshold for laser action as observed by *Hegarty* and *Yen* [3.84]. *Strauss* [3.83] concluded that no resonant transfer occurred in this material due to the low concentration c_{eff} of resonant ions within the line-narrowed $^3P_0-(^3H_4)_1$ feature, given by

$$c_{eff} = c(\Delta v_l / \Delta v_{inh}). \tag{3.57}$$

Δv_l represents the laser linewidth in this case. This conclusion is in general agreement with the concept of microscopic broadening and confirms conclusions drawn from nonresonant studies in this system [3.27].

Recently, *Dahl* and *Schaack* [3.85] have investigated the $^3H_4$ multiplet in PrF_3 using Raman and infrared reflection and absorption spectroscopy with polarized light at temperatures between 1.5 and 300 K and magnetic fields up to 8T. An interionic interaction between Pr^{3+} ions mediated by optical phonons was reported to result in a Davydov splitting of the electronic excitations. The authors argued that this lattice-mediated Pr–Pr interaction is only valid for the ground state multiplet component and is similar to observations on $Tb(OH)_3$ reported by *Cone* and *Meltzer* [3.86].

The transition from a single ion model to one in which an excitation is shared by a collection of ions is of fundamental importance and of considerable interest. *Anderson* [3.87] has shown that when the transfer interaction potential falls off faster than the inverse third power of the separation between the sites, an abrupt transition from energy localization to rapid transfer occurs. In this model, energy transfer will not occur between optically active ions in regions of low dopant concentration in a sample where the average ion-pair separation is large. In more concentrated areas the shorter interionic distances give rise to a stronger interaction leading to a delocalization of the optical excitation.

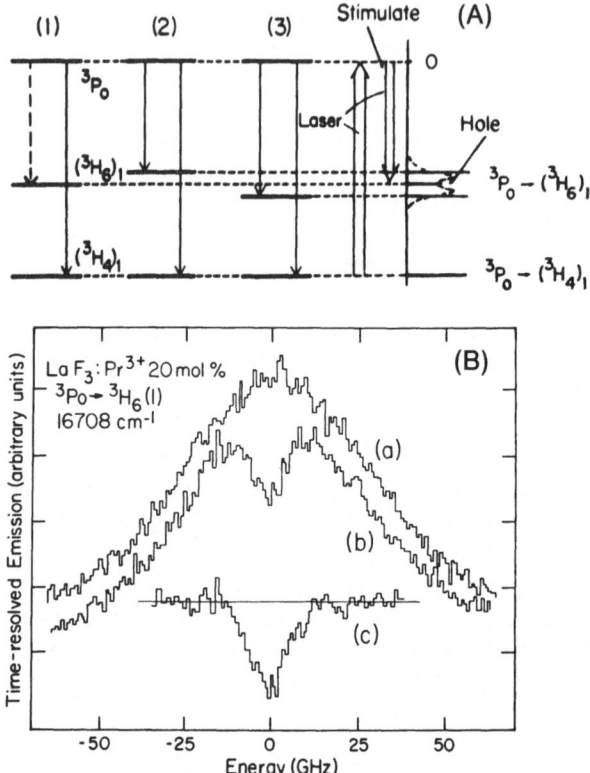

Fig. 3.26. (**A**) Illustration of the "accidental degeneracy" effect in FLN and the use of a two-photon method to burn a hole in the intermediate fluorescence. The dynamics of the hole are expected to yield evidence of resonant transfer in the upper state. (**B**) Optical hole burned into the $^3P_0$ population. The FLN pump laser had a width of 2 GHz; the stimulating laser was tuned to the center of the $^3P_0 - (^3H_6)_1$ degenerate fluorescence and its width was deliberately increased to 10 GHz. Traces were taken with a delay time of 4 µs and a gate of 3 µs. (*a*) Spectrum with no stimulation; (*b*) spectrum with stimulation; (*c*) hole calculated from (*a*) and (*b*) [3.83]

Proof of the existence of this "Anderson localization" phenomenon has been looked for in ruby (α-Al$_2$O$_3$:Cr^{3+}) for many years. *Imbusch* [3.80] first proposed a rapid migration of energy among single Cr^{3+} ions to explain the single exponential decay of the R_1 fluorescence at low temperatures following nonresonant broadband pulsed excitation. This system is known to transfer energy to Cr pair centers and thus enhance their N-line emission [3.88, 89]. The fluorescence of the N-lines was studied by *Imbusch* [3.80] who observed a temporal behavior consisting of an initial fast intrinsic decay followed by a slower exponential component with the same lifetime as that of the R_1 line. The slow component was attributed to excitation transfer from the single Cr^{3+} ions. *Selzer* et al. [3.29, 40] repeated this experiment using narrowband laser excitation to excite single Cr^{3+} ions directly to the metastable ^2E level. Their observations concurred with the picture of a strong resonant, nonradiative migration among the main body of single ions. They did, however, suggest that macroscopic strain broadening existed whereby resonant transfer could only occur between Cr^{3+} ions in particular regions of their samples. This model agreed with the FWM results reported by various groups [3.72–74] who set an upper limit of 30 nm on the spatial diffusion length in ruby.

The observation of Anderson localization in ruby was reported by *Koo* et al. [3.90], who monitored the normalized N_2 trap fluorescence as a narrowband laser beam was scanned through the low energy side of the R_1 absorption profile. A sudden decrease in the trap fluorescence intensity when the sample was excited in the wing of the inhomogeneously broadened R_1 line was believed to indicate a localization of the single ion excitation due to the low fractional concentration of ions with that energy. The experiment was repeated by *Chu* et al. [3.91] who used a narrow-band dye laser which could be tuned quickly across the entire R_1 profile. They concluded that their N_2/R_1 data provided no evidence for a sharp mobility edge.

It was pointed out earlier that the FWM technique which monitors spatial energy migration directly was really too coarse to allow definitive conclusions to be drawn in cases involving inorganic crystalline materials. An ingenious experiment which provided a much finer probe was first carried out by *Chu* et al. [3.92]. This experiment made use of the inequivalence of two types of cation sites in α-Al_2O_3, where an internal electric field acts in opposite directions on two sites. If an external electric field is applied along the optic axis it increases the net electric field acting on one site, while reducing the net field at the other site. If Cr^{3+} ions are located at both types of sites, labeled A and B, their energy levels will be Stark shifted in opposite directions and hence the R_1 lines of the two subgroups will be split. This linear pseudo-Stark splitting, which was first discovered by *Kaiser* et al. [3.93], can be quite large in ruby where the R_1 lines of the two subgroups can be separated by about 1 cm^{-1}, which is sufficient to move the two ionic groups out of resonance with each other.

The experiment of *Chu* et al. [3.92] can be explained schematically with the aid of Fig. 3.27 [3.94]. In (a) no electric field is applied so that the R_1 lines of groups A and B overlap. In (b) an external electric field is turned on so that the R_1 lines are separated. A short pulse from a narrowband dye laser then excites a subgroup of the A ions, which fluoresce immediately. This is represented by the shaded curve. In (c) the electric field is pulsed off so that the A and B ions are brought into resonance. If resonant energy migration exists in the material the optical excitation can be transferred to those B ions which are in resonance with the excited A ions. In (d) the field is turned on again so that the A and B ions are once again out of resonance. A FLN signal from the B ions is sought which is indicated by the additional shaded feature. The strength of the resonant interaction can be inferred from the intensity of this additional line-narrowed feature.

The actual spectra observed by *Chu* et al. [3.92] were complicated due to the ground state splitting of the Cr^{3+} ion. Nevertheless, they concluded that while a

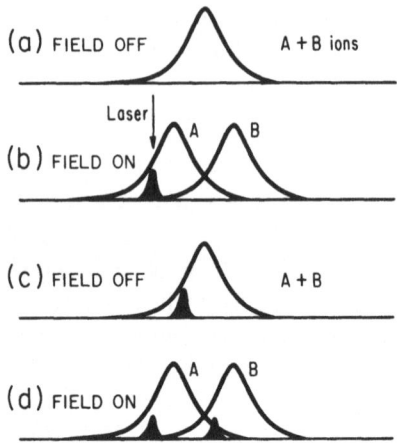

(a) FIELD OFF   A + B ions

Laser

(b) FIELD ON   A   B

(c) FIELD OFF   A + B

(d) FIELD ON   A   B

Fig. 3.27. Schematic representation of the various processes involved in the transient electric field experiments of *Chu* et al. [3.92]. The details are discussed in the text. (Redrawn after [Ref. 3.94, Fig. 12])

nonradiative resonant transfer occurred it was much slower than had been postulated. *Jessop* and *Szabo* [3.95] carried out a similar experiment on ruby and reached the same conclusion that resonant energy transfer was slow. *Gibbs* et al. [3.96] calculated the A to B transfer rate based on an electric dipole–electric dipole interaction with known dipole matrix elements and obtained a value close to that measured experimentally by the transient electric field techniques.

It is not possible from the results of these experiments to comment on the extent of resonant transfer within either sublattice A or B. *Duval* and *Monteil* [3.97] reported the results of an experiment in which a uniaxial stress applied in a specific direction on a sample of ruby differentiated two sublattices which were not equivalent to the A and B sublattices of the Stark-switching experiments. They concluded that nonradiative resonant energy transfer was rapid among ions of either the A or B subgroups, respectively, but was slow between ions of different groups. *Monteil* et al. [3.98] recently observed spectral energy transfer in the R_1 line of ruby in the presence of an electric field applied along the optical axis. They also reported that the transfer was faster between ions of the same group (A or B, respectively) than between different groups. Such a difference, they concluded, could be due to selective sublattice occupation in microscopic domains. This selective sublattice occupation would also account for a rapid resonant migration within the same sublattice and a slow migration of energy between the dissimilar sublattices.

No single experiment has thus far conclusively demonstrated the existence or absence of rapid resonant migration or Anderson localization in ruby. However, the generally held view at the moment is that a rapid transfer of energy among the single Cr^{3+} ions in ruby does not exist.

Although the search for Anderson localization in ruby has proven to be inconclusive, important insights into the dynamics of energy transfer have been obtained. Our understanding of the processes have extended to disordered, amorphous materials. As an example, a recent study of energy transfer in the mixed oxide system α-Al$_{2x}$Ga$_{2(1-x)}$O$_3$: Cr^{3+} has been reported by *Wasiela* et al. [3.99]. This material can be regarded as a disordered form of ruby where some of the Al^{3+} cations have been randomly replaced by optically inert Ga^{3+} ions [3.100]. Spectroscopic measurements have shown no evidence of a mobility edge, however, additional insights into the influences of disorder on the various energy transfer mechanisms have been obtained.

3.6 Concluding Remarks

We have attempted in this chapter to highlight some of the recent experimental work which has improved our understanding of energy transfer processes in inorganic insulating materials. These advances have unquestionably been provoked by the development of narrow-band tunable sources of laser radiation.

The technique of time-resolved fluorescence line narrowing as applied to rare-earth-doped systems, and in particular LaF$_3$ doped with Pr^{3+}, has permitted the direct measurement of donor–donor energy transfer. These results have encouraged the development of more sophisticated theoretical models which have provided unified microscopic and macroscopic descriptions of the dynamics of disordered systems. This success is partly due to the smaller number of unknown parametric quantities necessary to describe the energy transfer processes, and is reflected by the general agreement with earlier empirical models under appropriate conditions.

Our discussion has also highlighted some of the shortcomings of our present understanding of energy transfer. In particular, the area of resonant transfer has generated a considerable degree of unresolved controversy. Even here, TRFLN has played a key role in some of the more ingenious experiments. However, the observation and study of the transition regime between localized and collective behavior of energy transport phenomena awaits the development of more novel applications of laser spectroscopy and perhaps some more suitable inorganic materials.

Acknowledgements. One of us (GPM) would like to thank Pat Hoade for technical assistance in the production of this chapter.

References

3.1   P. Pringsheim: *Fluorescence and Phosphorescence* (Interscience, New York 1949)
3.2   H.W. Leverenz: *An Introduction to Luminescence in Solids* (Wiley, New York 1950)
3.3   T. Förster: Naturwissenschaften **33**, 166 (1946); Ann. Phys. **2**, 55 (1948)

3.4   D.L. Dexter: J. Chem. Phys. **21**, 836 (1953)
3.5   M.J. Weber: In *Handbook on the Physics and Chemistry of Rare Earths*, Vol. 4, ed. by
       K.A. Gschneider, L.E. Young (North-Holland, New York 1979) Chap. 35
3.6   R. Tomaschek: Ann. Phys. **75**, 561 (1924)
3.7   P.J. Bolden, F.A. Kroger: Physica **15**, 747 (1949)
3.8   L.G. Van Uitert: J. Electrochem. **107**, 803 (1960);
       L.G. Van Uitert, R.C. Linares, R.R. Soden, A.A. Ballman: J. Chem. Phys. **36**, 702 (1962);
       L.G. Van Uitert, R.R. Soden: J. Chem. Phys. **36**, 1797 (1962);
       L.G. Van Uitert, R.R. Soden, R.C. Linares: J. Chem. Phys. **36**, 1793 (1962);
       L.G. Van Uitert: In *Luminescence in Inorganic Solids*, ed. by P. Goldberg (Academic,
       New York 1966) Chap. 9
3.9   G.H. Dieke: *Spectra and Energy Levels of Rare Earth Ions in Crystals* (Wiley-
       Interscience, New York 1968)
3.10   S. Hüfner: *Optical Spectra in Transparent Rare Earth Compounds* (Academic, New York
       1978)
3.11   J.C. Wright: In *Radiationless Processes in Molecules and Condensed Phases*, ed. by F.K.
       Fong, Topics Appl. Phys., Vol. 15 (Springer, Berlin, Heidelberg 1976) Chap. 4
3.12   R. Reisfeld: Struct. Bonding **30**, 65 (1976)
3.13   R. Reisfeld, C.K. Jørgensen: In *Lasers and Excited States of Rare Earths*, Inorganic
       Chemistry Concepts, Vol. 1 (Springer, Berlin, Heidelberg 1977) Chap. 4
3.14   R. Orbach, M. Tachiki: Phys. Rev. **158**, 524 (1967)
3.15   T. Miyakawa, D.L. Dexter: Phys. Rev. B **1**, 2961 (1970)
3.16   J.C.W. Grant: Phys. Rev. B **4**, 648 (1971)
3.17   M. Inokuti, F. Hirayama: J. Chem. Phys. **43**, 1978 (1965)
3.18   M. Yokota, I. Tanimoto: J. Phys. Soc. Jpn. **22**, 779 (1967)
3.19   A.I. Burshtein: Zh. Eksp. Teor. Fiz. **62**, 1695 (1972) [English transl. Sov. Phys. – JETP
       46, 347 (1972)]
3.20   A. Szabo: Phys. Rev. Lett. **25**, 924 (1970)
3.21   R. Flach, D.S. Hamilton, P.M. Selzer, W.M. Yen: Phys. Rev. Lett. **35**, 1034 (1975)
3.22   A.L. Schawlow: In *Advances in Quantum Electronics*, ed. by J.R. Singer (Columbia
       University Press, New York 1961)
3.23   W.M. Yen, W.C. Scott, A.L. Schawlow: Phys. Rev. **137**, A271 (1964)
3.24   D.L. Huber, D.S. Hamilton, B.B. Barnett: Phys. Rev. B **16**, 4642 (1977)
3.25   T. Holstein, S.K. Lyo, R. Orbach: In *Laser Spectroscopy of Solids*, ed. by W.M. Yen,
       P.M. Selzer, Topics Appl. Phys., Vol. 49 (Springer, Berlin, Heidelberg 1981) Chap. 2
3.26   W.T. Carnall, H. Crosswhite, H.M. Crosswhite: "Energy Level Structure and Tran-
       sition Probabilities of the Trivalent Lanthanides in LaF$_3$", Argonne National
       Laboratory Report No. 60439, Argonne, IL (1977)
3.27   P.M. Selzer, D.S. Hamilton, R. Flach, W.M. Yen: J. Lumin. **12/13**, 737 (1976)
3.28   D.S. Hamilton, P.M. Selzer, W.M. Yen: Phys. Rev. B **16**, 1858 (1977)
3.29   P.M. Selzer, D.S. Hamilton, W.M. Yen: Phys. Rev. Lett. **38**, 858 (1977)
3.30   M.J. Weber, J.A. Paisner, S.S. Sussman, W.M. Yen, L.A. Riseberg, C. Brecher: J. Lumin.
       12/13, 729 (1976)
3.31   M.J. Weber: In *Laser Spectroscopy of Solids*, ed. by W.M. Yen, P.M. Selzer, Topics
       Appl. Phys., Vol. 49 (Springer, Berlin, Heidelberg 1981) Chap. 6
3.32   J.C. Vial, R. Buisson: J. de Phys., Lett. **43**, L-745 (1982)
3.33   R. Flach, D.S. Hamilton, P.M. Selzer, W.M. Yen: Phys. Rev. B **15**, 1248 (1977)
3.34   N. Motegi, S. Shionoya: J. Lumin. **8**, 1 (1973)
3.35   T. Holstein, S.K. Lyo, R. Orbach: Phys. Rev. Lett. **36**, 891 (1976)
3.36   D.S. Hamilton: Ph. D. Thesis, University of Wisconsin-Madison (1976)
3.37   J. Hegarty, D.L. Huber, W.M. Yen: Phys. Rev. B **23**, 6271 (1981)
3.38   T. Kushida: J. Phys. Soc. Jpn. **34**, 1318 (1973)
3.39   M.M. Broer, D.L. Huber, W.M. Yen, W.K. Zwicker: Phys. Rev. B **29**, 2382 (1984)
3.40   P.M. Selzer, D.L. Huber, B.B. Barnett, W.M. Yen: Phys. Rev. B **17**, 4979 (1978)
3.41   R.T. Brundage, W.M. Yen: Phys. Rev. B **34**, 8810 (1976)

3.42 D.L. Huber: In *Laser Spectroscopy of Solids*, ed. by W.M. Yen, P.M. Selzer, Topics Appl. Phys., Vol. 49 (Springer, Berlin, Heidelberg 1981) Chap. 3
3.43 R.K. Watts: In *Optical Properties of Ions in Solids*, ed. by B. DiBartolo (Plenum, New York 1974) p. 307
3.44 G. Blasse: In *Luminescence of Inorganic Solids*, ed. by B. DiBartolo (Plenum, New York 1978) p. 457
3.45 D.L. Huber: Phys. Rev. B **20**, 2307 (1979)
3.46 M.R. Brown, J.S.S. Whiting, W.A. Shand: J. Chem. Phys. **43**, 1 (1965)
3.47 L.G. Van Uitert, S. Iida: J. Chem. Phys. **37**, 986 (1962)
3.48 J. Hegarty, D.L. Huber, W.M. Yen: Phys. Rev. B **25**, 5638 (1982)
3.49 R. Buisson, J.C. Vial: J. de Phys., Lett. **42**, L-115 (1981)
3.50 G.P. Morgan, D.L. Huber, W.M. Yen: J. Lumin. **35**, 277 (1986)
3.51 T.J. Glynn, I. Laulicht, L.-R. Lou, A.J. Silversmith, W.M. Yen: Phys. Rev. B **29**, 4852 (1984)
3.52 V.P. Sakun: Fiz. Tverd. Tela (Leningrad) **21**, 662 (1979) [Sov. Phys. – Solid State **21**, 390 (1979)]
3.53 I.A. Bondar, A.I. Burshtein, A.V. Krutikov, L.P. Mezentseva, V.V. Sakun, V.A. Smirnov, I.A. Scherbakov: Zh. Eksp. Teor. Fiz. **81**, 96 (1981) [Sov. Phys. – JETP **54**, 45 (1981)]
3.54 L.D. Merkel, R.C. Powell: Phys. Rev. B **20**, 75 (1979)
3.55 M. Zokai, R.C. Powell, G.F. Imbusch, B. DiBartolo: J. Appl. Phys. **50**, 5930 (1979)
3.56 J.M. Flaherty, R.C. Powell: Phys. Rev. B **19**, 32 (1979)
3.57 L.F. Mollenauer: In *Laser Handbook*, Vol. 4, ed. by M.L. Stitch, M. Bass (North-Holland, Amsterdam 1985) Chap. 2
3.58 R.T. Brundage, W.M. Yen: Appl. Opt. **24**, 3687 (1985)
3.59 J.M. Pellegrino, W.M. Yen, M.J. Weber: J. Appl. Phys. **51**, 6332 (1980)
3.60 M.M. Broer, D.L. Huber, W.M. Yen, W.K. Zwicker: Phys. Rev. Lett. **49**, 394 (1982)
3.61 M. Blätte, H.G. Danielmeyer, R. Ulrich: Appl. Phys. **1**, 275 (1973)
3.62 H.G. Danielmeyer: In *Festkörperprobleme XV*, ed. by J.H. Queisser (Pergamon, Braunschweig 1975) p. 253
3.63 W. Lenth, G. Huber, D. Fay: Phys. Rev. B **23**, 3877 (1981)
3.64 H.G. Danielmeyer, M. Blätte, P. Balmer: Appl. Phys. **1**, 269 (1973)
3.65 B.C. McCollum, A. Lempicki: Mater. Res. Bull. **13**, 833 (1978)
3.66 H. Dornauf, J. Heber: J. Lumin. **22**, 1 (1980)
3.67 C.M. Lawson, R.C. Powell, W.K. Zwicker: Phys. Rev. Lett. **46**, 1020 (1981); Phys. Rev. B **26**, 4836 (1982)
3.68 H.P. Weber, P.F. Liao: J. Opt. Soc. Am. **64**, 1337 (1974)
3.69 H.J. Eichler: Opt. Acta **24**, 631 (1977); IEEE J. QE-**22**, 1194 (1986)
3.70 J.R. Salcedo, A.E. Siegman, D.D. Dlott, M.D. Fayer: Phys. Rev. Lett. **41**, 131 (1978)
3.71 M.D. Fayer: In *Spectroscopy and Exciton Dynamics of Condensed Molecular Systems*, ed. by V.M. Agranovich, R.M. Hochstrasser (North-Holland, Amsterdam 1983)
3.72 P.F. Liao, D.M. Bloom: Opt. Lett. **3**, 4 (1978); P.F. Liao, L.M. Humphrey, D.M. Bloom, S. Geshwind: Phys. Rev. B **20**, 4145 (1979)
3.73 H.J. Eichler, J. Eichler, J. Knof, C.H. Noak: Phys. Status Solidi **52**, 481 (1979)
3.74 D.S. Hamilton, D. Heiman, J. Feinberg, R.W. Hellwarth: Opt. Lett. **4**, 124 (1979)
3.75 R.C. Powell: J. de Phys. C **7**, 403 (1985)
3.76 P.F. Liao, H.P. Weber, B.C. Tofield: Solid State Commun. **16**, 881 (1975)
3.77 G.P. Morgan, S.-Z. Chen, W.M. Yen: IEEE J. QE-**22**, 1360 (1986)
3.78 J.K. Tyminski, R.C. Powell, W.K. Zwicker: Phys. Rev. B **29**, 6074 (1984)
3.79 Y.M. Wong, V.M. Kenkre: Phys. Rev. B **22**, 3072 (1980)
3.80 G.F. Imbusch: Phys. Rev. **153**, 326 (1967)
3.81 M. Harig, R. Charneau, H. Dubost: J. Lumin. **24/25**, 643 (1981)
3.82 R.L. Cone, R.S. Meltzer: In *Spectroscopy of Rare Earth Ions in Crystals*, ed. by A.A. Kaplyanskii, R.M. Macfarlane (North-Holland, Amsterdam 1987) Chap. 2
3.83 E. Strauss, W.J. Miniscalco, J. Hegarty, W.M. Yen: J. Phys. C **14**, 2229 (1981)

3.84   J. Hegarty, W.M. Yen: J. Appl. Phys. **51**, 3345 (1980)
3.85   M. Dahl, G. Schaack: Phys. Rev. Lett. **56**, 232 (1986)
3.86   R.L. Cone, R.S. Meltzer: J. Chem. Phys. **62**, 3573 (1975)
3.87   P.W. Anderson: Phys. Rev. **109**, 1492 (1958)
3.88   A.L. Schawlow, D.L. Wood, A.M. Clogston: Phys. Rev. Lett. **3**, 271 (1959)
3.89   F. Varsanyi, D.L. Wood, A.L. Schawlow: Phys. Rev. Lett. **3**, 544 (1979)
3.90   J. Koo, L.R. Walker, S. Geschwind: Phys. Rev. Lett. **35**, 1669 (1975)
3.91   S. Chu, H.M. Gibbs, A. Passner, S. Geschwind: Bull. Am. Phys. Soc. **24**, 894 (1979);
       Phys. Rev. B **24**, 7162 (1981)
3.92   S. Chu, H.M. Gibbs, S.L. McCall, A. Pasner: Phys. Rev. Lett. **45**, 1715 (1980)
3.93   W. Kaiser, S. Sugano, D.L. Wood: Phys. Rev. Lett. **6**, 605 (1961)
3.94   G.F. Imbusch: In *Energy Transfer Processes in Condensed Matter*, ed. by B. DiBartolo
       (Plenum, New York 1984) p. 471
3.95   P.E. Jessop, A. Szabo: Phys. Rev. Lett. **45**, 1712 (1980)
3.96   H.M. Gibbs, S. Chu, S.L. McCall, A. Passner: In *Coherence and Energy Transfer in
       Glasses*, ed. by P.A. Fleury, B. Golding (Plenum, New York 1984) p. 373
3.97   E. Duval, A. Monteil: In *Energy Transfer Processes in Condensed Matter*, ed. by B.
       DiBartolo (Plenum, New York 1984) p. 643
3.98   A. Monteil, E. Duval, A. Attar, G. Viliani: J. Phys. C **18**, 685 (1985)
3.99   A. Wasiela, Y. Merle D'Aubigné, D. Block: J. Lumin. **36**, 11, 23 (1986)
3.100 G.P. Morgan, T.J. Glynn, G.F. Imbusch, J.P. Remeika: J. Chem. Phys. **69**, 4859 (1978)

4. Laser Spectroscopy of Color Centers

Brian Henderson and Kevin P. O'Donnell

With 28 Figures

Color center physics has played an important role in solid state science for most of the present century. However, not until the 1950s did there appear a fundamental appreciation of the electronic structure of even the simplest of such centers [4.1a]. Then the application of electron spin resonance (ESR) and electron nuclear double resonance (ENDOR) spectroscopy led to a precise knowledge of the basic building blocks of a wide range of color centers in ionic crystals [4.2]. The primary concern of this review is the optical physics of *intrinsic lattice defects* and some defect–impurity complexes in which the impurity is optically inert. However, optically active impurities, i.e., *extrinsic lattice defects* are not discussed.

The present understanding of the electronic structure of color centers has evolved from many years of research which reached their peak in the mid-1960s. These studies revealed optical absorption and emission spectra across most of the visible and near-infrared regions; bandshapes representative of a very wide range of electron-photon coupling strengths were reported. It was evident that many new phenomena could be investigated through laser-based spectroscopic techniques. Some of the earliest applications were reviewed by H. Mahr in [4.1b]. In recent years renewed dynamism in color center physics has been generated by the potential applications of color center lasers. Such lasers may be optically pumped, broadly tuned and used to produce ultrashort (10^{-14} s) pulses. More specifically they work at longer wavelengths (700–4000 nm) than commercial dye laser systems, spanning the technologically important wavelength range of 1300–1550 nm. Many of the optical transitions of color centers are allowed electric dipole transitions, involving rather novel ionic relaxation processes in the excited electronic states. Relaxation of the defect environment by phonon creation is particularly suitable for investigation by laser techniques. In the optical spectroscopies prevalent in the 1960s it was difficult to obtain quantitative information about the dynamics of the optical pumping cycle. However, with the advent of single mode ring dye lasers and of tunable ultrashort laser pulses it became possible to carry out very high resolution spectroscopy in both the frequency and time domains.

In this chapter we begin by surveying the electronic structure of defects in the alkali halides (Sect. 4.1). The spectroscopy of such centers is characterized by the wide range of Huang-Rhys factors, $S \sim 4\text{--}40$, and a wide variety of transition shapes. The large Stokes' shifts between absorption and emission bands makes color center–containing crystals potentially suitable as the gain media for broadly tunable solid state lasers (Sect. 4.2) [4.3]. The nature of the phonons which *homogeneously* broaden color center spectra may be probed directly using Raman scattering (Sect. 4.3.1), whereas the mechanisms of *inhomogeneous broadening* of color center zero-phonon lines may be investigated by optical hole burning techniques (Sect. 4.3.2). Other phenomena that we discuss include pump-and-probe measurements of phonon lifetimes (Sect. 4.3.3), polarization effects in excited state absorption and two-photon absorption transitions (Sect. 4.3.4).

4.1 Models and Electronic Structures of Simple Color Centers

4.1.1 The F Center

The simple optical center illustrated in Fig. 4.1 consist of electrons or holes trapped at defect sites in alkali halide crystals. Such defect models are easily extended to analogous centers in more complex ionic solids [4.4, 5]. The F center is arguably the best characterized experimentally and best understood theoretically of all these defects: it consists of a single electron trapped by the Coulomb potential of an isolated anion vacancy (Fig. 4.1). In group theoretical language the electronic ground state is described by the representation $^2A_{1g}$ (*s*-like) of the octahedral group. Studies by ESR and ENDOR showed that some 70–80% of the electronic charge density is concentrated within the vacancy. The observation of hyperfine structure from up to sixteen shells of neighboring ions centered on the vacancy shows that significant electron charge density lies outside the vacancy. The lowest-lying excited states, labeled 2^2A_{1g} (2*s*-like) and 2^2T_{1u} (2*p*-like) were the subject of some controversy. The

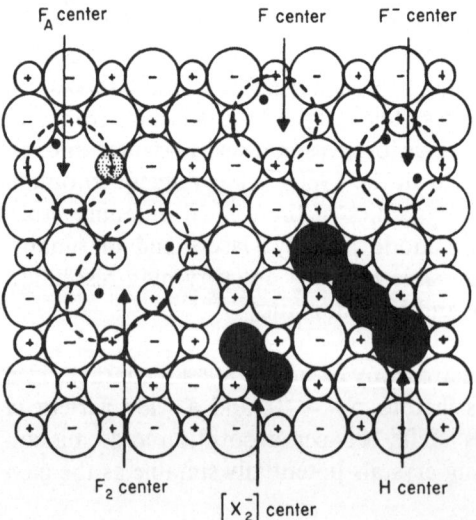

Fig. 4.1. Models of some common color centers in the alkali halides

first absorption transition, $A_{1g} \to T_{1u}$, occurs with the large oscillator strength ($f \sim 1$), expected of an allowed electric dipole transition. However, in emission the oscillator strength is much reduced ($f \sim 10^{-2}$), as is evidenced by the relatively long radiative lifetime ($\tau_R \sim 10^{-6}$ s). Many different experiments were directed at the nature of the first excited state of the F center, including measurements of the effects of uniaxial stress and electric field on optical absorption and emission bands, and subsequently ESR and ENDOR in excited electronic states. The absorption and emission bands of F centers are broadened by strong electron–phonon coupling to breathing mode vibrations; coupling to tetragonal and trigonal modes of vibration is weaker but still significant. The dichroism of the F center absorption band under a uniaxial stress indicates that the state reached in absorption, i.e., the *unrelaxed excited state*, is a fairly compact 2^2T_{1u} state. Stark effect measurements on the F-absorption band show that this 2^2P-like state lies lower than the nearby 2^2S-like state by about 0.1 eV. Similar perturbation measurements on the broad emission band of the F center show that after lattice relaxation the *relaxed excited state* is a spatially diffuse admixture of *s*- and *p*-like components, with the *s*-like component predominant. This accounts for both the small emission oscillator strength and the relatively long radiative lifetime.

4.1.2 The F_A Center

The simplest modification of the F center (Fig. 4.1) involves the replacement of one nearest-neighbor cation by an alkali impurity ion. This so-called F_A *center* has tetragonal C_{4v} symmetry in the electronic ground state. Actually there are *two types* of F_A center, decided by the size of the impurity cation relative to that of the host cation that it replaces [4.6]. The F_A(Na) center in KCl is an example of an $F_A(I)$ *center*, in which the relatively large Na$^+$ impurity ion produces only a small elastic mismatch on replacing the host cation. These centers are optically very similar to the F center (Table 4.1). Two linearly polarized absorption bands (F_{A_1} and F_{A_2}) result from the partial removal of excited state orbital degeneracy by the tetragonal crystal field. Although spatially compact in the ground state and unrelaxed excited state, the F_A center wavefunction is spatially diffuse in the relaxed excited state, just as it is in the F center. The diffuseness of the excited state leads to a reduced oscillator strength in emission, a relatively long radiative lifetime and a single unpolarized emission band. The diffuse excited state is rather insensitive to the perturbation introduced by the cation impurity. Table 4.1 also shows that the optical emission properties of $F_A(II)$ *centers* are quite different from those of the F_A(I) center and the F center. In these centers the impurity cation is a small alkali ion (e.g., Li) relative to the host cation. The F_A(Li) center in KCl is an example of the F_A(II) center. The Stokes' shift between the absorption and emission band is very large, the radiative lifetime is reduced by almost an order of magnitude relative to the F center and the single emission band is strongly linearly polarized. The relatively

Table 4.1. Absorption and emission properties of F_A centers in alkali halides. (Adapted from [4.6])

System	F_A type	Absorption peak [eV]		Emission peak [eV]	τ_R [10^{-8} s]
		F_{A_1}	F_{A_2}		
KCl					
F		2.31		1.24	58
F_A(Na)	I	2.35	2.12	1.12	53
F_A(Li)	II	2.25	1.98	0.46	8.5
KBr					
F		2.06		0.92	111
F_A(Na)	I	2.07	1.90	0.84	100
F_A(Li)	II	2.00	1.82	0.75	10
RbCl					
F		2.05		1.09	60
F_A(Na)	I	2.09	1.85	0.93	60
F_A(Li)	II	1.95	1.72	0.45	9

short radiative lifetime arises because the wavefunction remains compact in the relaxed excited state, whereas the luminescence polarization arises because the emitting state is the saddle-point configuration (Fig. 4.2) in the tunneling oscillations of one of the anion neighbors of the impurity midway between the two extreme locations of the anion vacancy.

Gellermann et al. [4.8] have reported the optical spectra of electrons trapped at anion vacancy–Tl^+ ion pairs: such complexes are referred to as $F_A(Tl)$ or $Tl^0(1)$ centers [4.8–10]. In the crystal field model of Tl^0 the ground configuration ($6p^1$) is split by spin-orbit interaction into a $^2P_{1/2}$ ground state and $^2P_{3/2}$ excited states. The anion vacancy produces an odd-parity crystal field term which further mixes the $6p$ terms, giving enhanced splittings and significant oscillator strength for all transitions. In order of increasing energy the crystal field states are labeled as $\Phi_\pm, \chi_\pm, \Psi_\pm$, and $\Sigma_\pm$, the first three of which are derived mainly from the $6p$ manifold, whereas the $\Sigma_\pm$ state is mainly derived from the atomic $7s$-state. In KCl the $\Phi_\pm \to \chi_\pm$ and $\Phi_\pm \to \Psi_\pm$ transitions occur at peak wavelengths 1040 nm and 720 nm respectively. The $\Phi_\pm \leftrightarrow \chi_\pm$ absorption transition covers the range 950–1200 nm and the emission transition spans the range 1380–1720 nm. The $\Phi_\pm \to \Sigma_\pm$ transition has a much larger oscillator strength than either of the near-infrared bands: it is observed at a peak wavelength of 550 nm, very close to the F band in this crystal, hence the identification as an F_A-like center [4.8]. Subsequent ESR, polarized absorption-luminescence and spin-dependent magnetic circular dichroism [4.9–12] confirmed the crystal field nature of the spectra and demonstrated that

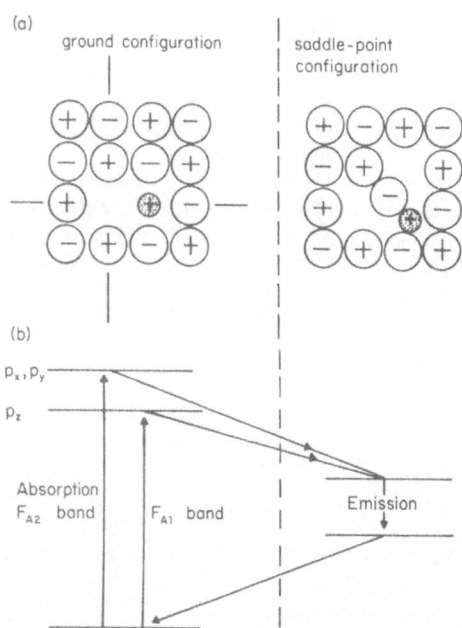

(a)

ground configuration

saddle-point configuration

(b)

p_x, p_y

p_z

Absorption
F_{A2} band

F_{A1} band

Emission

Fig. 4.2. (a) The configuration of an $F_A^{(II)}$ center in the ground and relaxed excited states. **(b)** Absorption and emission transition associated with these states. (From [4.7])

in KCl and KBr the trapped electron is strongly localized on the Tl site. Very recent optically detected magnetic resonance studies of these centers, together with detailed theoretical analysis, has shown that in the excited states the center of charge moves towards the vacancy, especially in the Na^+ halides [4.12].

4.1.3 F Center Aggregates – Divacancy Centers

Figure 4.3 elaborates on a number of possible variations of the F_2 aggregate: two geometrical arrangements are possible (Figs. 4.3a, b) depending upon whether the second anion vacancy is along a $\langle 110 \rangle$ or $\langle 112 \rangle$ direction. This discussion concentrates upon the first of these possibilities. Other modifications result from the different numbers of electrons which may be trapped by the divacancy. A single electron is trapped in the F_2^+ *center*, two electrons in the F_2 *center*, and three in the F_2^- *center*. The F_2^+ centers have been important in the development of tunable color center lasers. However, these centers are rather unstable against optically induced dissociation during laser pumping. Figures 4.3c and d show two possible mechanisms of enhancing the stability of the divacancy structure. Whether stabilization is achieved by the cation vacancy or the O^2-impurity the electronic structure is essentially unchanged.

Aegerter and *Lüty* [4.13] first reported systematic studies of the formation kinetics and optical properties of F_2^+ centers: these defects resemble the H_2^+ molecular ion. The energy levels may be accurately predicted using a model which treats the F_2^+ center as an H_2^+ ion embedded in a medium with dielectric

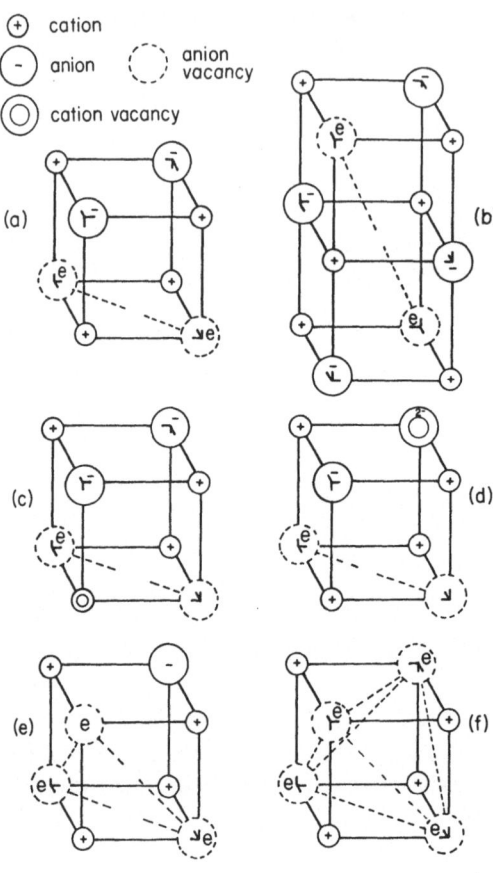

Fig. 4.3a–f. Schematic ionic configurations for F_2, F_3 and F_4 centers in alkali halides. (a) and (b) show two different configurations of F_2 centres, whereas (c) and (d) show possible structures of stabilized F_2^+ centers, F_2^+ cation-vacancy and $F_2^+ O_2^-$ complexes respectively. The F_3 and F_4 aggregate centres are shown in (e) and (f) respectively

constant κ [4.14]. The energy levels and distances then scale according to

$$E(F_2^+)=E(H_2^+, R_{12})/\kappa^2 \quad \text{and} \quad R_{12}=r_{12}/\kappa, \tag{4.1}$$

in which R_{12} is the vacancy–vacancy separation and r_{12} is the interproton spacing in the molecular ion. The energy levels calculated in this way are shown in Fig. 4.4. For F_2^+ centers having orthorhombic symmetry and a $\langle 110 \rangle$ symmetry axis, the point group symmetry is D_{2h}, and the electronic states transform as irreducible representations of this group. The lowest-lying levels are labeled $^2A_{1g}$, $^2B_{1u}$, $^2B_{2u}$, and $^2B_{3u}$, in order of increasing energy. In general, there are three strongly allowed absorption transitions out of the ground state: $^2A_{1g} \rightarrow ^2B_{1u}$, $^2A_{1g} \rightarrow ^2B_{2u}$, and $^2A_{1g} \rightarrow ^2B_{3u}$, the latter two of which are very close in energy. At high temperature, $T > 77$ K, one observes only a single luminescence transition $^2B_{1u} \rightarrow ^2A_{1g}$, the higher-energy emission band associated with $^2B_{2u}$, $^2B_{3u} \rightarrow ^2A_{1g}$ transitions being overwhelmed by nonradiative decay to the $^2B_{1u}$ level. At lower temperature, $T < 50$ K, both emission transitions may be observed.

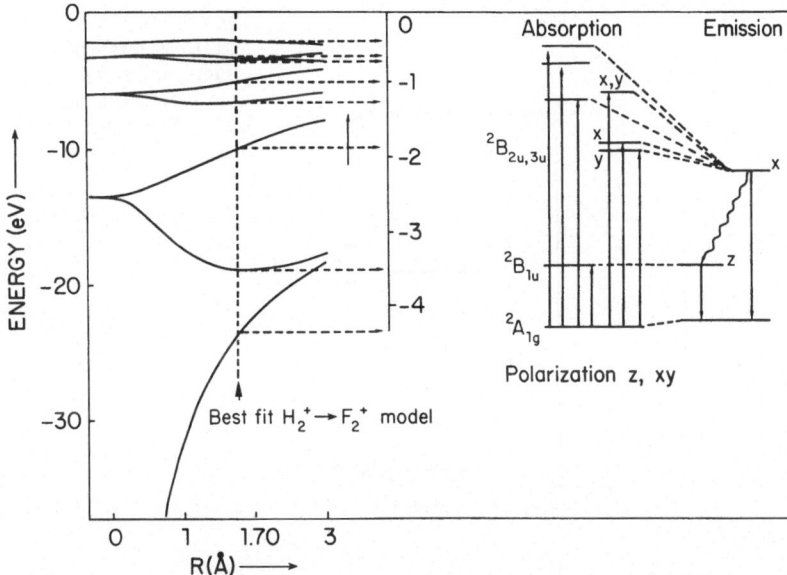

Fig. 4.4. The lowest-energy molecular orbital levels of F_2^+ centers calculated using the dielectric continuum model and the optical transition which are possible between these levels [4.15]

The ground state of neutral F_2 centers in D_{2h} symmetry is the spin singlet $^1A_{1g}$, both electrons occupying $1s$-like states associated with the individual F centers. A schematic energy level diagram for F_2 centers showing possible optical transitions is shown in Fig. 4.5. Absorption transitions out of the $^1A_{1g}$ ground state are to excited states which are products of a $1s$-like state on one site and $2p$-like state on the second site. Because p-states are of odd parity the resulting product state is also of odd parity. There are three such excited states: $^1B_{1u}$, $^1B_{2u}$, and $^1B_{3u}$, and the three allowed electric dipole transitions $^1A_{1g} \rightarrow {}^1B_{1u}$, $^1B_{2u}$, and $^1B_{3u}$ are linearly polarized parallel to the z, y, and x axes of the F_2 center, respectively. In general, the B_{1u}–B_{2u} energy separation is much larger than the B_{2u}–B_{3u} energy separation and the $^1A_{1g} \rightarrow {}^1B_{2u}$, $^1B_{3u}$ transitions are almost degenerate and underlie the F-band. As the optical absorption spectrum of F-aggregate centers in KCl shows (Fig. 4.6) the $^1A_{1g} \rightarrow {}^1B_{1u}$ [$F_2(1)$ band] is well separated from the F-band. Since the F_2 center is a two-electron center, spin triplet states are also expected. Usually on pumping into the $^1A_{1g} \rightarrow {}^1B_{1u}$, $^1B_{2u}$, $^1B_{3u}$ bands, deexcitation of the excited states results in $^1B_{1u} \rightarrow {}^1A_{1g}$ emission *or* nonradiative decay to the lowest-lying triplet state, i.e., $^3A_{1g}$ (Fig. 4.5). Because the $^3A_{1g} \rightarrow {}^1A_{1g}$ transition is spin and parity forbidden the $^3A_{1g}$ state is long-lived (~ 50 s in KCl), ideal for the study of excited state absorptions into other spin triplet states [4.1]. The $^3A_{1g}$ is relatively low-lying, typically < 1 eV above the $^1A_{1g}$ ground state, it decays slowly via multiphonon nonradiative decay. By comparison the $^1B_{1u} \rightarrow {}^1A_{1g}$ emission process, which is

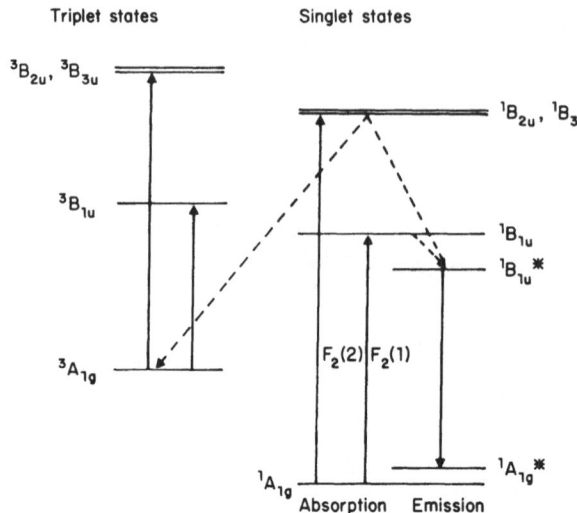

Triplet states   Singlet states

Fig. 4.5. The lowest singlet and triplet levels of F_2 centers, and some transitions between them. The broken lines indicate nonradiative transitions

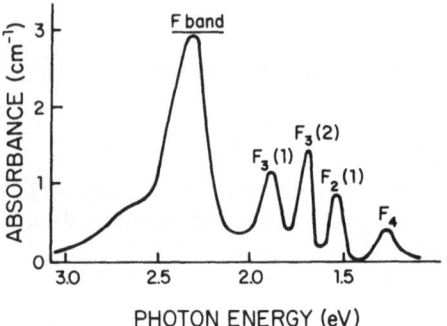

Fig. 4.6. Optical absorption spectrum of KCl after x-irradiation at room temperature. Measurement temperature $T = 95$ K. (After [4.15])

observed irrespective of whether the B_{1u}, B_{2u} or B_{3u} state is optically excited, is very efficient with radiative lifetime of order 10–30 ns.

4.1.4 F Center Aggregates – Trivacancy Centers

As with divacancy centers there are several possible ionic configurations for trivacancy centers in an alkali halide crystal. The simplest, most symmetrical arrangement is of three anion vacancies in nearest-neighbor sites on a {111} plane of the crystal (Fig. 4.3). Such a vacancy aggregate provides a deep trapping potential for electrons. Although F_3^+ centers (2 electrons), F_3 centers (3 electrons) and F_3^- centers (4 electrons) have been observed, there has been no report of F_3^{2+} centers (1 electron). Since the different charge states of these trivacancy aggregates are produced by photoionization reactions it is pertinent to comment on their respective electronic structures. According to the molecular orbital method [4.5] the lowest one-electron orbitals are an orbital

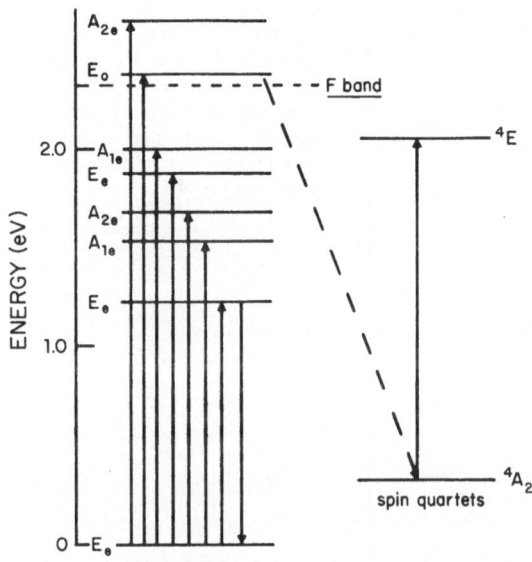

Fig. 4.7. Proposed energy level diagram for F_3 centers in KCl. The subscripts e and o on the various states indicate that they are even and odd parity respectively for reflections in the plane of the triangle of vacancies. (After [4.16])

singlet (a_1) and an orbital doublet (e) some 1–2 eV higher in energy. For the F_3^+ center the $(a_1)^2$ configuration leads to the $^1A_1$ ground state, whereas the excited configuration $(a_1)^1(e)^1$ leads to a ^1E state. Thus the fundamental optical absorption transition is $^1A_1 \rightarrow {}^1E$. The ground configuration for the F_3 center is $(a_1)^2(e)$ giving the ^2E state lowest. Possible excited configurations of the F_3 center are $(a_1)^1(e)^2$ and $(e)^3$, which produce states ^2E, $^2A_1$, $^2A_2$, and ^2E in order of increasing energy: the $(e)^3$ configuration leads to a low-lying $^4A_2$ state. Allowed electric dipole transitions can be made only between spin doublet states, as is illustrated in Fig. 4.7. The electronic states of the F_3^- center are derived from the $(a_1)^2(e)^2$ configuration which is split by electron–electron interaction into the three states $^1A_1$, ^1E, and $^3A_2$. Hund's rules would then favor $^3A_2$ as the ground state. The excited states derived from the $(a_1)^1(e)^3$ configuration are ^1E or ^3E, so that one might anticipate $^1A_1 \rightarrow {}^1E$ or $^3A_2 \rightarrow {}^3E$ transitions. However, no ESR signals have been observed for F_3^- centers; apparently, the ground state is the spin singlet $^1A_1$. Piezospectroscopy confirms that $A \rightarrow E$ absorption transitions are involved at centers with trigonal symmetry.

The optical absorption spectrum shown in Fig. 4.6 shows bands due to F_2, F_3, and F_4 centers. Other transitions of both F_2 and F_3 centers underlie the F-band. Evidence of these assignments comes from measurements on the kinetics of formation, polarized absorption and luminescence, perturbation spectroscopy and optically induced electron spin resonance [4.1–6]. At low temperatures the optical spectra of the F aggregate centers show structure typical of a wide range of electron-phonon coupling strengths. For example, in LiF, where the F centre has a Huang-Rhys factor S of about 40, the F aggregate

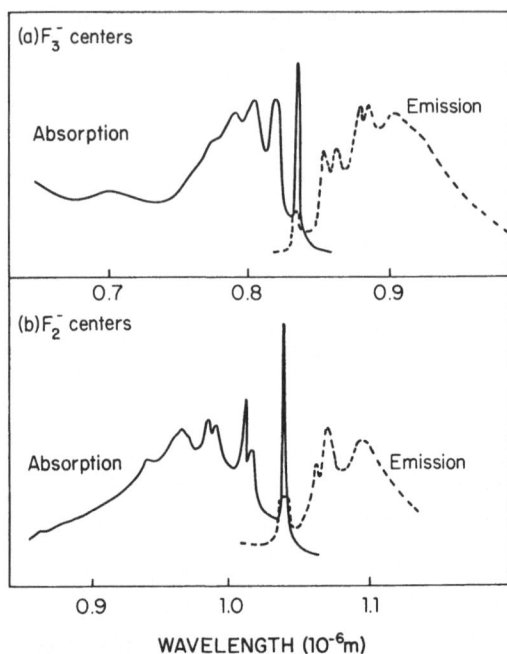

Fig. 4.8. Absorption and emission spectra of F_2^- and F_3^- centers in LiF measured at $T=4\,K$. Sharp structure in each spectrum is strongly contrasted with the broadband absorption spectra shown in Fig. 4.6

centers have $S \sim 5.5$ for F_2^+ and F_3^+ centers, $S \sim 2.5$ for F_2^- and F_3^--centers, with $S \sim 18$ and 3.6 for F_2 and F_3 centers respectively. In general, with $S < 7$ optical transitions show a resolved structure zero-phonon transition and more-or-less resolved vibronic peaks on the side of a broad band. Examples of structure in the absorption and emission spectra of F_2^- centers and F_3^- centers in LiF are shown in Fig. 4.8. In both cases the zero-phonon lines occur at identical wavelengths whereas the vibronic side bands are shifted to shorter wavelengths in absorption and longer wavelengths in emission. Note the general similarities in the shapes of the absorption and emission sidebands for each center. However, the absorption spectra are hardly mirror images of the emission spectra. That these emission spectra were excited using a pulsed ruby laser at 693 nm shows the importance of higher-lying states and of nonradiative decay to the lowest lying excited state. It is this resolved structure which makes high-resolution techniques (hole burning, fluorescence line narrowing) so powerful in addressing dynamic processes in optical spectroscopy.

When zero-phonon lines are observed, perturbation spectroscopy (stress, electric field, magnetic field) may be used to characterize the symmetry properties of the color centers [4.5]. Such measurements reveal that the F_3 center have trigonal symmetry with electronic states that transform as the irreducible representations A_1, A_2, and E of the C_{3v} group. For centers with an even number (2 or 4) of electrons, the electronic states may be either spin singlets $(S=0)$ or triplets $(S=1)$: only transitions associated with spin singlets have been reported. For neutral F_3 centers, where the total spin may be $S=\frac{1}{2}$ or $S=\frac{3}{2}$, the principal absorption and emission spectra involve the spin doublet states. These absorption transitions are identified in the energy level diagram shown in Fig. 4.7. As in Fig. 4.6 shows, the $F_3(1)$ and $F_3(2)$ bands are clearly seen: other transitions coincide with transitions of both F centers and other F aggregate centers. However, vibrational relaxation in the excited states after absorption in any of the broad bands may lead to intersystem crossing to a low-lying spin quartet state. Since this state is very long lived (in KCl 14.5 s) it forms an ideal medium from which to study excited state absorptions involving spin quartet states [4.1].

4.1.5 Photoionization of Alkali Halides Containing F Centers

Illuminating alkali halide crystals containing F centers at 77–250 K with F-band light photoionizes F centers with the concomitant formation of F^+ centers (empty anion vacancies) and F^- centers. The photoionization reaction can be written as

$$F \xrightarrow{\ h\nu\ } F^+ + e^-$$

$$F \xrightarrow{\ h\nu\ } F^- + h^+$$

to indicate that an electron is released from the F center with full quantum efficiency, leaving an empty anion vacancy. The released electron migrates through the crystal and is trapped at another F center forming the F^- center. If the optical bleaching is carried out at low temperature (77–100 K) no aggregation of centers takes place. Warming the crystal to some higher temperature (270–300 K) enables the F^+ center to migrate through the crystal until it is trapped by an impurity or defect. Obviously, if the trapping site is an F^- center, then one produces an F_2 center. Should the trapping site be an alkali impurity ion then an F_A center is formed, because at this higher temperature the F^- center are thermally unstable, releasing an electron to be trapped at an F_A^+ center [4.5]. The release of electrons may be enhanced by infrared light. These processes, photoionization and anion vacancy mobility, are at the heart of mechanisms for the formation of complex centers involving defects and impurities (e.g. F_A, F_2^+, F_{2A}^+...) [4.1, 5, 6]. As such they are important in preparing color-centre-containing crystals as the gain media for tunable lasers and as the media for optical storage devices.

One interesting feature is the use of F band light to produce photoionization of centers by thermal excitation from a relaxed excited state. In general, F bands in the alkali halides are quite broad (~ 0.25 eV), and aggregate centers (F_A, F_2, F_3) have optical transitions within the F band range of energies. Not surprisingly, therefore, irradiation with F band light also results in the indiscriminate photoionization of F aggregate centers. This may or may not be desirable. The formation of F_A centers is in many ways a model case of defect aggregation. As we have already noted the F_A center has two principal optical absorption bands, and these may be used to determine the nature of the state from which photoionization occurs. Once created, F_A centers undergo the same

. type of photoionization as F centers. At low temperatures the lifetime of the relaxed excited state of the F_A center is determined only by radiative decay. As temperature is increased then there is a thermally activated probability that the electrons will be excited into the conduction band. The surprising feature is that irrespective of whether the F_A centers are excited in the F_{A_1} or F_{A_2} bands (Fig. 4.2) the energy barrier against thermal activation out of the relaxed excited state is the same; of the order 0.1 eV. This demonstrates, as did the emission studies, the existence of a diffuse relaxed excited state slightly below the conduction band, similar to the relaxed excited state of the F center.

Anion vacancy mobility is also important in the formation of F_2^+ centers. To stabilise large concentrations of these charged defects requires the presence of deep electron traps (e.g. Mn^{2+}, Ni^{2+}, Pd^{2+}) in the crystal [4.17]. Then F_2^+ centers may be created by additive coloration and suitable optical treatment. Alternatively irradiation at 120–180 K with fast electrons (1–2 MeV) produces F centers and F^+ centers: in addition, the divalent cations, M^{2+}, trap electrons to become monovalent, M^+. The irradiated crystals are warmed to 300 K for 10–15 min to enable the F^+ centers to migrate and aggregate with F centers and so produce F_2^+ centers. The crystal must then be cooled to 77 K, otherwise the F_2^+ centers dissociate. Unfortunately F_2^+ centers are not especially stable. As we discuss below, several different schemes for stabilizing F_2^+-type centers were developed to permit long-term stable laser action over extended tuning ranges. A two-step photoionization process was used by *Mollenauer* and *Bloom* [4.17] to achieve almost complete $F_2 \rightarrow F_2^+$ conversion in electron-irradiated NaF and KF crystals containing divalent transition metal ions added to the crystals to act as electron traps. The same procedure was used by *Foster* and *Schneider* [4.18] to produce high concentrations of F_{2A}^+ centers in additively colored KI : Li crystals. After additive coloration, F centers were converted to F_A, F_2, and F_{2A} centres by exposing the crystal to blue-green light at 240–250 K. The F_{2A} concentration was further enhanced by exposing the crystal to the same light for different times at 200 K and 150 K. The crystal is then exposed to radiation at 1.06 μm from a *Q*-switched Nd : YAG laser. This procedure, which ionizes F_{2A} centers by sequential absorption of two photons by the F_{2A} centers, is a highly selective way of producing photoionization since the energy required to pump the lowest energy transition of the F_{2A} center is not sufficient to photoionize other centers (F, F_A, etc.). The electrons released by the F_{2A} center are trapped by F and F_A centers acting as intrinsic electron traps. It seems likely that other highly selective multiple-photon transitions may be used for photoconversion of a range of aggregate centers in these crystals.

4.1.6 Trapped Hole Centers

The products of UV x-ray and particle irradiation of alkali halides include trapped electron centers and trapped hole centers, the structures of which depend upon crystal purity, irradiation type and temperature. The principle intrinsic hole centers are now referred to as $[X_2^-]$ centers and H centers (Fig. 4.1). The $[X_2^-]$ ion is a molecular ion, $[F_2^-]$, $[Cl_2^-]$, etc., depending upon whether the ion is in a fluoride, chloride, or other halide lattice. The H center is an interstitial $[X_2^-]$ center stabilized in a split-interstitial configuration centered on a single halide ion vacancy. Each of these defects may in its own way be associated with an impurity. The centers may be produced by ionizing radiation at a sufficiently low temperature that hole migration is quenched. Furthermore the crystal must contain a sufficient density of electron traps to prevent hole annihilation by recombination. These traps may be either impurity ions or other defects (e.g., F^+ centers).

Molecular ions such as F_2^-, Cl_2^-, Br_2^- are stable in free space. In an alkali halide crystal such an ion is referred to as a self-trapped hole since no other defects or impurities are involved. The self-trapping is a spontaneous consequence of the hole being localized in a covalent bond between two adjacent halide ions. In consequence the molecular axis is a crystal $\langle 110 \rangle$ direction. The electronic wavefunction of the trapped hole is strongly localized on the two anions. Since the model involves a positive hole in a p-shell, the center has a single unpaired electron spin. It is, therefore, paramagnetic and easily identified by its ESR spectrum [4.19–20]. The nature of the electronic states of the self-trapped hole may be constructed from a halide ion (X^-), ground term $^1S_0$, and a halide atom (X), ground term $^2P_{3/2, 1/2}$ with the $^2P_{3/2}$ level lying lower in energy. At large separations the states of $[X_2^-]$ are those of a free halide ion and a halogen atom in its lowest $^2P_{3/2}$ or $^2P_{1/2}$ state. As Fig. 4.9a shows, this gives two states at large separations, which split into two bonding orbitals and two antibonding orbitals. The degeneracy of states with twofold orbital degeneracy is raised by the combined effects of spin-orbit coupling and crystal field interaction. The point symmetry of the $[X_2^-]$ centers, Fig. 4.1, is D_{2h}; such symmetry lifts all orbital degeneracy and the electronic states are orbital singlets which transform as irreducible representations of the D_{2h} point group. The highest energy transition takes place between the $^2B_{1u}$ ground state and the $^2B_{1g}$ excited state. The two other transitions, $^2B_{1u}$ to $^2B_{2g}$ and $^2B_{3g}$, occur in the infrared; because of their large widths they are not always resolved (Fig. 4.9b), especially in the lighter halide crystals. The polarization properties of the optical absorption transitions of the self-trapped hole centers are revealed by measurements of the dichroism of the absorption spectrum in certain circumstances (Fig. 4.9b and c). Such dichroic absorption, which arises from the reorientation of the centers during illumination with light in the ultraviolet band, facilitates the identification of the three polarized absorption bands which have very different strengths and half-widths. The following transition assignments follow from the dichroic absorption spectra in KBr:

$$^2B_{1u} \rightarrow {}^2B_{1g}; \; z\text{-polarized } [110]; \; 380 \text{ nm} \equiv 3.25 \text{ eV},$$

$$^2B_{1u} \rightarrow {}^2B_{2g}; \; y\text{-polarized } [1\bar{1}0]; \; 930 \text{ nm} = 1.33 \text{ eV},$$

$$^2B_{1u} \rightarrow {}^2B_{3g}; \; x\text{-polarized } [001]; \; 770 \text{ nm} = 1.00 \text{ eV}.$$

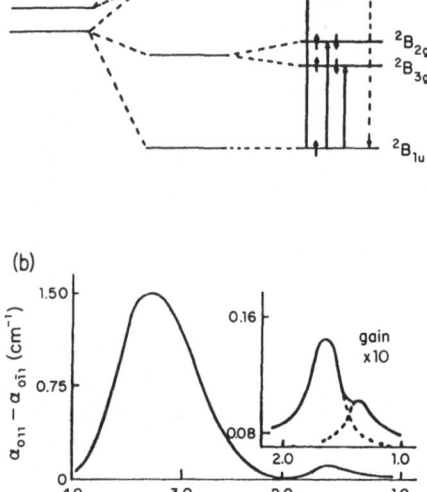

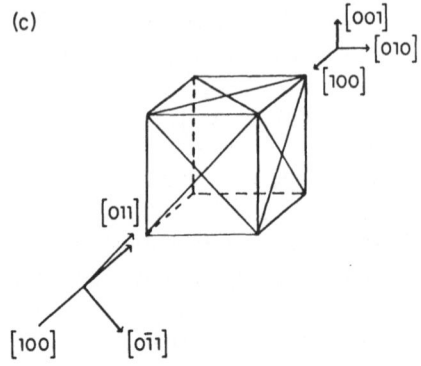

Fig. 4.9. (a) Schematic energy levels, and optical transitions of $[X_2^-]$ centers in alkali halides, (b) dichroic absorption spectrum of X_2^- centers in KBr and (c) the arrangement for observing dichroism in an absorption spectrum

(The first is an allowed electric dipole transition in which strong charge density oscillations occur along the molecular axis. Such behavior is confirmed in every case, $[F_2^-]$, $[Cl_2^-]$, $[Br_2^-]$, and $[I_2^-]$, by the experimental results.)

The splitting between the infrared bands (i.e., ~ 0.3 eV) is just the spin-orbit coupling energy. Similar behavior is observed for the $[I_2^-]$ centers in KI, where the spin-orbit energy is of order -0.5 eV. Only a single infrared absorption band has been observed for $[F_2^-]$ and $[Cl_2^-]$ centers, because the spin-orbit energy is rather small. However, in KCl, measurement of the difference in the optical density for light polarized along [100] and [011] axes produces a splitting of the infrared band which is much larger than the spin-orbit interaction. This splitting between the infrared bands of about 0.14 eV is the crystal field splitting.

Optically induced reorientation by 60° or 90° rotation of the defect axes leads to an enhancement of the occupation of one orientation of centers over other possible orientations. Parallel studies of the changes in the optical and ESR spectra following polarized bleaching have been used to tie the ESR spectrum to a particular optical band. They are an established feature of hole center folklore and constitute an unmatched aid to identification of centers [4.20, 21]. Reorientation of centers may also be thermally activated. By measuring ESR and optical spectra following thermal reorientation of a pre-aligned spectrum it has been shown that 90° jumps do not occur with significant probability in either KCl or KI. The measured activation energy for reorientation by 60° jumps is 0.54 eV in KCl and 0.27 eV in KI [4.21].

Although the $[X_2^-]$ center is the simplest trapped hole center in halide lattices, its formation in irradiated crystals is seldom so efficient that it is the predominant defect in pure crystals. Since it is charged relative to the host lattice, electron trapping impurities are required to avoid electron-hole recombination preventing the buildup of $[X_2^-]$ centers. Rather than these centers, interstitial halogen atoms or H centers are the predominant species produced by irradiation at low temperatures. As Fig. 4.1 shows, the H center is a split interstitial center, which consists of an $[X_2^-]$ ion centered on a single anion site. Obviously such a center produces a certain strain in the host lattice and on this account one might anticipate that the H center would have the greater space available to it in the $\langle 111 \rangle$ direction. At least in KCl and KBr, the energy compensation from weak covalent bonding to the two halide ions at either end of the molecular ion must offset the increased strain energy. In these crystals the H center might be loosely described as an $[X_4^{3-}]$ molecular ion. This structure is required by the observed ESR spectra. Conversely, the H center in LiF has been shown by ESR and ENDOR to be oriented along $\langle 111 \rangle$ directions, the release of strain energy due to this orientation offsetting the reduction in energy due to the additional weak bonding effects [4.20, 21]. The analysis of optical absorption transitions for the H center proceeds much as in the case of the $[X_2^-]$ centers, except that the H center transitions occur at higher photon energies due to the closer spacing of the molecular ion in the H center relative to the self-trapped hole center. As with the $[X_2^-]$ center, the optical transitions of the H center are strongly polarized since the molecular ion rotates during optical excitation. The dipole nature of the transition may be deduced from measurements of the induced dichroism at low temperature (< 10 K). However, if the temperature is raised to 10.9 K the dichroism disappears, because the centers reorient via a thermally activated process with an activation energy of 0.031 eV. The H center decays by diffusion parallel to its molecular axis above 40–45 K. The fate of such mobile H centers is that they annihilate at F centers and so restore a perfect lattice in this region. Alternatively they may be trapped at impurities.

In mixed alkali halide crystals a wide variety of impurity-related trapped hole spectra may be observed. For example, in crystals containing two different halide ions $[XY^-]$ centers can be produced by suitable irradiation and thermal treatment. These centers are stable at rather higher temperatures than the conventional $[X_2^-]$ and H centers. The intrinsic centers may also be stabilized by replacing one of the alkali ions neighboring the $[X_2^-]$ or H centers by an impurity alkali ion. Such centers are usually called $[X_2^-]_A$ centers and H_A centers. The $[X_2^-]_A$ center differs only slightly from the self-trapped hole center. However, there are numerous H_A centers, their behavior and properties being dependent on both the particular alkali halide and the impurity alkali halide [4.20].

4.2 Color Center Lasers

The simplest laser cavity consists of an active medium placed between a pair of parallel mirrors. Such an arrangement causes photons propagating parallel to the cavity axis to be reflected backwards and forwards through the medium so stimulating further photon emission by atoms in their excited states. In this situation the beam intensity parallel to the axis of the cavity, given by

$$I(v) = I_0(v) \exp[\gamma(v)l - L]$$ (4.2)

is amplified when $\gamma(v)l > L$. In this equation l is the optical thickness of the medium, $\gamma(v)$ is the *small gain coefficient* of the medium and L is included to represent the losses in the laser cavity. The small gain coefficient is related to the *population inversion* Δn between the participating states by

$$\gamma(v) = \frac{c^2}{8\pi n^2} \left(\frac{\eta}{\tau}\right) \left(\frac{g(v)}{v^2}\right) \Delta n,$$ (4.3)

in which v is the radiation wavelength, η is the quantum efficiency of the transition, τ is the emission lifetime and $g(v)$ is the shape function of the optical transition. The population inversion required to overcome the cavity losses, i.e., the threshold population (Δn_T), is then defined by

$$\frac{c^2}{8\pi n^2} \left(\frac{\eta}{\tau}\right) \left(\frac{g(v)}{v^2}\right) \Delta n_T \geq \frac{L}{l}.$$ (4.4)

This *threshold condition* must be met before a beam can be produced.

The tunability of the laser enters through the shape function of the transition, $g(v)$. The ruby laser is a single frequency laser because the shape function is essentially a very narrow delta function. Color center luminescence bands are, by contrast, broad and structureless as a consequence of homogeneous broadening. In the presence of strong electron–phonon coupling, the *single configurational coordinate model* (Fig. 4.10) implies a large coordinate offset between the harmonic oscillator potentials of ground and excited states. This indicates that the pump band is broad because the absorption transitions sample the vibrational coupling in the excited electronic state. In a similar way the luminescence spectrum reflects the electron–phonon coupling in the ground state. The total energy (photon plus phonons) released in the laser transition is fixed; tuning is determined by which of the different phonon levels is the final level of the optical transition. The optical pumping cycle shown in Fig. 4.10 is a highly simplified representation of the optical properties of color centers. Nevertheless, it shows clearly that color center lasers are *four-level systems:* the pump transition (1)→(2) populates the relaxed excited state (3) very efficiently by virtue of the phonon relaxation between vibronic levels (2) and (3). Because nonradiative transitions from level (4) to (1) in the ground state are also very

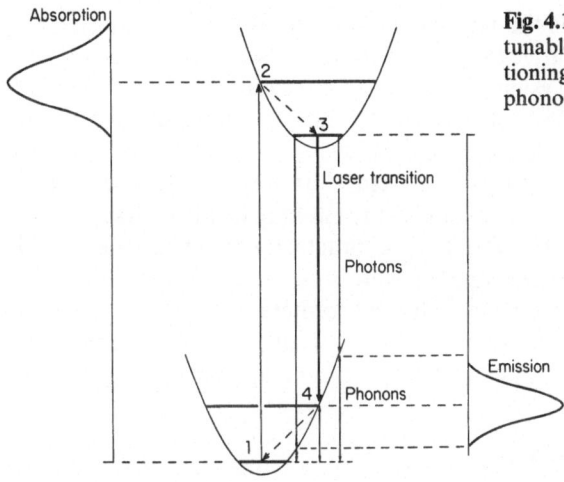

Absorption

Laser transition

Photons

Phonons

Emission

Fig. 4.10. The four-level nature of a tunable vibronic laser, and the partitioning of energy between emitted phonons and photons

fast, relative to the radiative lifetime, $\Delta n \sim n_3$ since $n_4 \sim 0$. A rate equation analysis gives [4.4]

$$\Delta n = \frac{P(1 - A_{34}/W_{41}^{nr})}{(W_{34} + A_{34})}, \tag{4.5}$$

where P is the pump rate, A_{34} is the Einstein coefficient for *spontaneous emission*, W_{34} is the *stimulated emission rate* and W_{41}^{nr} is the nonradiative decay rate between levels (4) and (1). Above the threshold pump rate the population inversion Δn_T remains constant because W_{34} increases very rapidly. Recalling that the radiation density $\varrho(v)$, Einstein B coefficient and intensity, $I(v)$ are related through $W_{34} = \varrho(v)B_{34} = I(v)B_{34}/c$, we see that the laser output

$$I(v) = \frac{c(1 - A_{34}/W_{41}^{nr})}{B_{34}\Delta n_T} P - \frac{cA_{34}}{B_{34}} \tag{4.6}$$

increases linearly with pump rate above threshold. Obviously unless $W_{41}^{nr} > A_{34}$ there can be no population inversion: in color center lasers $A_{34} \sim 10^8$ s^{-1} whereas $W_{41}^{nr} \sim 10^{13}$ s^{-1} because phonon-assisted nonradiative decay is very efficient. The pump power required to reach threshold is given by $P_T = \Delta n_T A_{34}$, Δn_T being defined by (4.4). For color center lasers, values of P_T may be relative modest.

According to (4.4) the gain variation with frequency reflects the shape of the broad luminescence band. For a Gaussian band shape the small gain coefficient $\gamma(v_0)$ at the peak of the band is

$$\gamma(v_0) \simeq \frac{c^2}{8\pi n^2} \left(\frac{\eta}{\tau}\right) \frac{n_3}{v_0^2 \delta v}, \tag{4.7}$$

where δv is the full width at the half maximum intensity points, and $\Delta n \simeq n_3$ because W_{41}^{nr} is very large relative to A_{34}. At first sight the F center appears to be an ideal tunable four-level system, because the nonradiative decay rates W_{23}^{nr} and W_{41}^{nr} are large ($\sim 10^{13}$ s^{-1}) relative to the spontaneous decay rate $A_{34} \simeq 10^7$ s^{-1}. Substituting in (4.7) the values $v_0 = 3 \times 10^{14}$ Hz, $\tau/n = 600 \times 10^{-9}$ s, $\delta v = 6.3 \times 10^{13}$ Hz and $n_3 \sim 10^{16}$ cm^{-3} appropriate to F centers in KCl gives $\gamma(v_0) \simeq 0.04$. Such a small gain coefficient may be rather less than the cavity losses, implying that F centers are impractical candidates for tunable solid state lasers. For the same excited state population inversion, $\gamma(v_0)$ is 4.2 cm^{-1} for $F_A(Li)$ centers, 3.5 cm^{-1} for F_2^+ centers and 2.7 cm^{-1} for Tl^0 centers, all in KCl. In consequence, there has in recent years been considerable development of tunable laser systems based on such aggregate color centers.

4.2.1 F_A and F_B Center Lasers

Although the first operation of an $F_A(Li)$ center laser in KCl was reported as long ago as 1965 [4.23], major development occurred only after 1974 when *Mollenhauer* and *Olson* [4.7] observed genuinely tunable CW operation of $F_A(II)$ centers in both KCl:Li and RbCl:Li (Fig. 4.11). In KCl, the $F_A(Li)$ center laser has pump bands which peak at 550 nm and 630 nm, so that it may be pumped with the line output from either Kr$^+$ or Ar$^+$ ion lasers. The luminescence peak is at 2700 nm, and the tuning range stretches from 2500 nm to 2900 nm. Typically for an input power of about 2.5 W the output power at the band peak is about 200–250 mW. F_B centers involve two impurity cations in the nearest neighbor shell of the anion vacancy. Such centers have very similar optical properties to the $F_A(II)$ centers, and have also been used as tunable lasers. F_A and F_B centers in potassium- and rubidium-based alkali halides

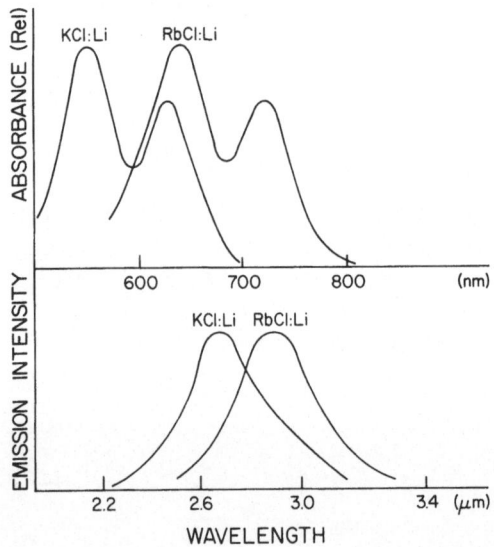

Fig. 4.11. Tuning curves for F_A centers in potassium- and rubidium-based alkali halides [4.7]

enable tunable laser operation from 2000–3400 nm. Unfortunately, the quantum efficiency of these centers is less than unity and decrease monotonically with increasing temperature. In consequence, the laser thresholds of F_A and F_B centers are much lower at lower temperatures. Generally, these lasers are very reliable because F_A and F_B centers are stable under optical pumping and under thermal cycling between 77 and 300 K.

As we have already noted, the optical properties of the $Tl^0(1)$ centers [or $F_A(Tl)$ centers] are quite different from those of the $F_A(I)$ and $F_A(II)$ centers. The low energy optical transitions of the $Tl^0(1)$ centers have a relatively small Stokes' shift and a quantum efficiency of one. $Tl^0(1)$ lasers in KCl and KBr may be efficiently pumped using the 1.06 μm line from the Nd:YAG laser, giving tunable laser operation over the wavelength range 1400–1700 nm in both crystals. Indeed $Tl^0(1)$ center lasers have been operated in NaCl, KF, KCl, KBr, and RbCl crystals all at wavelengths close to 1.5 μm. In cw operation, laser output power levels exceeding 1 W have been achieved with input powers of 4–5 W. The $Tl^0(1)$ center laser is an archetypal four-level system, with low threshold pump power and good slope efficiency [4.8, 10, 24].

4.2.2 F_2^+ Center Lasers

The F_2^+ center laser operates on the absorption and emission transitions between the ground ($^2A_{1g}$) and first excited ($^2B_{1u}$) states. The Stokes' shift is comparatively small so that the energy conversion efficiency is high ($\sim 80\%$). Since both absorption and emission cross sections are large ($\sim 10^{-16}$ cm^2) the single pass gain in 1–2 mm thick crystals is large. Furthermore, the quantum efficiency is large and independent of temperature. Clearly the vibronically broadened $^2A_{1g} \rightarrow {}^2B_{1u}$ transitions of the F_2^+ centers have almost ideal characteristics for tunable four-level laser systems. In Fig. 4.12 we summarize the absorption and emission properties of F_2^+ centers in nine different alkali halides. Absorption profiles range from 550 nm (LiF) to 1700 nm (KI), whereas

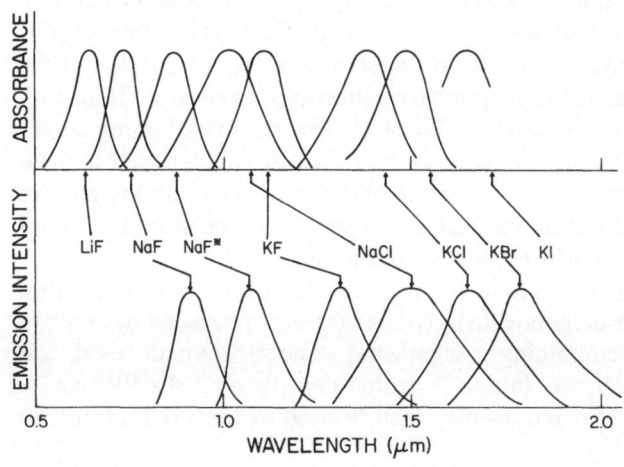

Fig. 4.12. Compilation of infrared absorption and emission spectra for F_2^+ centers in alkali halides. (Adapted from [4.3])

the laser emission spans the infrared region from 920 to 2400 nm [4.17]. Although the $^2A_{1g} \rightarrow {}^2B_{1u}$ transition is the most effective in pumping the laser, the $^2A_{1g} \rightarrow {}^2B_{2u}$, $^2B_{3u}$ transitions are important because they provide a means of reorientating the F_2^+ center axes to align all centers into a particular $\langle 110 \rangle$ orientation. All centers will then contribute to a fully polarized laser mode. Unfortunately, despite the very attractive properties of F_2^+ center lasers there are several operational difficulties. Two- or three-photon absorption of the pump beam leads to reorientation of the F_2^+ centers into the five equivalent orientations of the defect axes, which absorb and emit radiation less effectively than centers in the principal laser orientation. Hence there is a gradual degradation of laser performance. A further source of degradation is the bleaching of F_2^+ centers under intense laser pumping. Finally, once prepared the F_2^+ centers have a relatively short shelf-life at room temperature, even in the dark, as a consequence of thermally activated reorientation and dissociation. Any successful laser system must overcome these problems.

Mollenauer [4.17] reported that in heavily irradiated NaF: Mn^{2+} crystals the F_2^+ absorption and emission bands decay at room temperature to be replaced by other bands at longer wavelengths. These bands were attributed to $(F_2^+)^*$ centers, the * indicating the involvement of an unknown defect or impurity. Subsequent work has identified the stabilizing species as a nearest-neighbor cation vacancy (Fig. 4.3c). The $(F_2^+)^*$ center is much more stable than the isolated F_2^+ center. As an alternative to the M^{2+} ions as electron traps, *Lüty* and his associates [4.24] used controlled additions of OH^- or SH^- ions: the electron traps are created out of the debris of radiation damage of the OH^- or SH^- ion. In NaF: OH^- crystals the F_2^+ centers also decayed at 300 K to be replaced with $(F_2^+)^{**}$ centers, the emission band of which was displaced further into the infrared than that of the $(F_2^+)^*$ center. These new centers are F_2^+-like in every respect; nevertheless they are created by electron irradiation which also creates other defects that may be the cause of instability. The additive coloration technique was subsequently used to produce stable $(F_2^+)^{**}$ centers in NaCl [4.25] and in a range of other crystals (KCl, KBr, and RbCl) [4.26]. These recent studies establish that the stabilizing entity is the O^{2-} ion located in a near-neighbor substitutional anion site (Fig. 4.3d). Just which one of four possible ionic configurations has yet to be established. A comparison of the positions of the lowest energy absorption transition and the emission transition for these $F_2^+(O^{2-})$ centers is shown in Fig. 4.13. Typical power tuning curves and threshold characteristics for this laser in KCl are shown in Fig. 4.14. However, the most important characteristic of this laser is its stability relative to F_2^+ centers: once additively colored the crystals may be stored at room temperature for many months without degradation.

Schneider et al. [4.27] stabilized the F_2^+ centers with monovalent cation impurities in the nearest-neighbor shell. The defects were produced by additive coloration of crystals containing a significant concentration of alkali ion impurity (~ 200 ppm), to an initial F center density of $2\text{–}4 \times 10^{18}$ cm^{-3}. Selective photoionization reactions may then be used to convert F centers to

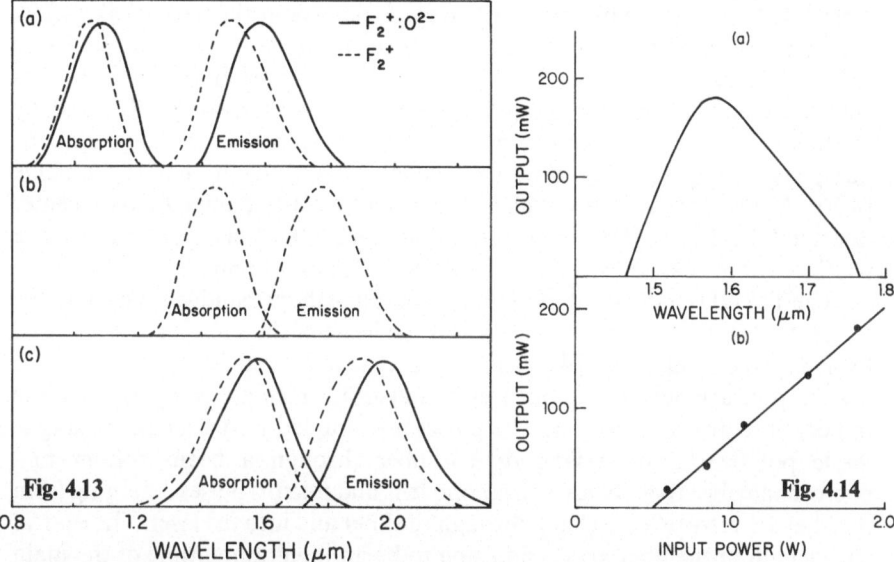

Fig. 4.13. Absorption and emission transitions of $F_2^+ : O^{2-}$ centers in (a) NaCl, (b) KCl, and (c) KBr. (After [4.26])

Fig. 4.14a, b. The cw operating characteristics of an $F_2^+ : O^{2-} : NaCl$ laser during pumping with an input power of 2 W from a 1.06 μm Nd:YAG laser. (a) The power output vs wavelength using a 5% output coupler; (b) output power versus input power at the band peak. (After [4.25])

(mainly) F_{2A}^+, F_A^-, and F^- centers [4.17, 18]. Using such crystals it has been possible to achieve broad band tuning of color centers out to ~ 4000 nm. For example, in KCl crystals double-doped with Li^+ and Na^+ it has been possible to tune continuously from 1670 to 2460 nm, using the emission from both $F_{2A}^+(Li)$ and $F_{2A}^+(Na)$ centers [4.27]. Furthermore, $F_{2A}^+(Li)$ center lasers in KI and RbI, respectively, have tuning ranges 2380–3990 nm and 2840–3680 nm. In the case of RbI the $F_{2A}^+(Li)$ center concentration was insufficient to attain the full tuning range of the laser [4.28].

4.2.3 Ultrashort Pulses and Soliton Effects

The generation of picosecond pulse trains by mode-locking has been a primary goal of many laser development programmes. Several reports of mode-locking of color center lasers have appeared. *Mollenauer* and *Bloom* [4.17] observed picosecond pulses when mode-locking the F_2^+ center in LiF laser by synchronously pumping with a mode-locked Kr^+ ion laser. Subsequently, they mode-locked the F_2^+ laser in KF and NaCl by synchronous pumping with the 1.06 μm Nd-YAG laser. In addition mode-locked pulses of 5 ps duration have been generated in a synchronously pumped $F_2^+ : O^{2-}$ center laser in NaCl tunable from 1.47 to 1.73 μm with average power of 450 mW at the band peak [4.25]. The generation of pulse trains at 2.6 μm by the F_A(Li) center in KCl was reported by *Isganitis* et al. [4.29], whose reported pulse durations exceeded 100 ps. Somewhat shorter mode-locked pulses ($\sim$40 ps) were obtained by mode-locking a synchronously pumped F_A(Li):KCl laser at 2.653 μm, the pump laser being a mode-locked Ar^+ ion laser operating on the 514.5 nm line [4.30]. More recently *Islam* et al. [4.31] synchronously pumped F_A(Li) center lasers in KCl:Li and RbCl:Li using a mode-locked Kr^+ laser, obtaining pulse widths as short as 7.6 ps tunable between 2.54 and 3.15 μm.

The Tl^0(1) laser does not mode-lock over all of its cw bandwidth as does the F_2^+ center laser [4.32]. Nevertheless, it is the heart of the soliton laser invented by *Mollenauer* and *Stolen* [4.33], in which a mode-locked Tl^0(1) center laser is used to produce pulses of about 1 ps bandwidth, which are compressed in the optical fibre arm of the laser. The pulses are coupled into a length of single-mode polarization-preserving optical fiber through a beam splitter and microscope objective. By use of a second lens and mirror, pulses emerging from the fiber are retroreflected back through the fiber and into the laser. The optical path length of the fiber arm is adjusted to be an integral multiple of the main cavity length so that pulses returned from the fiber are coincident with those present in the laser cavity. This forces the laser itself to produce narrower pulses, a process which rapidly builds up with successive round trips through the main laser cavity and feedback arm. Soliton pulses are generated in optical fibers as a result of the interaction between the negative group velocity dispersion and the nonlinear refractive index. For the fundamental soliton, the pulse broadening effects of the dispersion are exactly balanced by the pulse narrowing effect of the nonlinearity, so that it never changes its shape. Higher-order solitons, produced at higher input powers, undergo a periodic sequence of narrowings and splittings. At its widest (beginning and end of any period), the $N = 2$ soliton has a $sech^2$ shape, with peak power four times that of the fundamental soliton of the same width. The periodic narrowing is an effective means of producing subpicosecond pulses; the shortest pulse produced so far is of only 19 fs duration [4.34].

4.3 Laser Spectroscopy

Emission lines from gas lasers and early solid state lasers have been used as optical sources for spectroscopy for many years. Although many different discrete wavelengths are available, such lasers do not provide a continuous range of wavelengths. The impact of lasers in spectroscopy has been all the more profound since the development of tunable dye lasers. Organic dyes have efficient luminescence outputs with wavelegth ranges of 20–30 nm FWHM. Such tuned laser outputs may have the twin benefits, although not at the same instant, of extremely narrow bandwidth (50–100 kHz) and ultra-short pulses (10–100 fs). These complementary facilities enable an ever-widening range of spectroscopic experiments to be performed in both the time domain and in the frequency domain.

4.3.1 Light Scattering by Color Centers

The development of light scattering techniques predated the laser by approximately one century. Phenomena discovered many years ago by Rayleigh, Brillouin and Raman have found important applications in solid state physics. The classical experimental geometry for studying Rayleigh scattering is to observe the scattered radiation at 90° to the incident radiation. Rayleigh discovered that the intensity scattered per molecule decreased by a factor of order ten on condensation to the liquid phase, with a smaller decrease on solidification. Brillouin predicted that elastic waves in solids should give rise to fine structure in the Rayleigh scattered light at the Bragg coherence condition $\lambda_1 = 2\lambda_p \sin(\phi/2)$, where λ_1 is the wavelength of light, λ_p is the wavelength of those phonons responsible for scattering the light, and ϕ is the scattering angle. The motion of the scattering centers Doppler shifts the laser frequency by

$$\Delta v_1 = \pm v_p = \pm 2v_1(v/c)\sin(\phi/2), \tag{4.8}$$

where v_p is the phonon frequency and v is the velocity of sound in the medium. For visible radiation those phonons probed by Brillouin scattering have frequencies in the tens of gigahertz (10^{10} Hz) region. The Brillouin scattered components are completely polarized in the 90° scattering geometry. Although Rayleigh scattering has been applied to studies of exciton localization in semiconductor multiple quantum well structures [4.35], and Brillouin scattering to the observation of the phonon bottleneck in $MgO:Ni^{2+}$ [4.36], neither techniques has found use in color center physics.

In Raman scattering the spectrum of light scattered at 90° by gases, liquids and solids was observed to contain wavelengths different from that of the incident monochromatic light. The frequency displacements of the Raman scattered lines were observed to be independent of the frequency of the incident light, in contradistinction to both fluorescence excitation and Brillouin scattering. The frequency shifts in the Raman spectrum of solids are related to local atomic vibrations. Raman scattering occurs when a change in polarizability is involved during atomic vibrations. This usually means that Raman transitions occur between vibrational states of the same parity. Thus infrared and Raman spectra give complementary information about the vibrational spectra of optical centers. Observation of the Raman effect has been applied to various color center problems, to give information about the spectrum of vibrations coupled to the electronic states of the defects and about the structure of defect aggregation.

The electron-phonon interaction at color centers may be inferred from measurements of optical bandwidths and peak positions as a function of

Table 4.2. Absorption and emission bands of F centers in sodium and potassium alkali halides

Crystal	FWHM at 0K absorption Γ_0 [eV]	$\hbar\omega$ [eV]	S	Peak energy [eV]	
				Absorption	Emission
NaCl	0.270	0.0177	42	2.746	0.975
NaBr	0.388	0.0247	44	2.345	–
NaI	0.276	0.0176	44	2.063	–
KCl	0.195	0.0150	31	2.295	1.215
KBr	0.160	0.0095	51	1.059	0.916
KI	0.146	0.0084	54	1.874	0.827

temperature. The results are interpreted in terms of the configurational coordinate model, which assumes that the spectrum of coupled vibrational modes is *qualitatively* well represented by a single vibrational mode. Analysis of the experimental results yields values of the Huang-Rhys parameter S and the effective phonon energy, $\hbar\omega$. The data for F centers in the alkali halides (Table 4.2) show that the Huang-Rhys parameters lie between 28(NaF) and 61(LiCl), clearly precluding the observation of resolved vibronic structure at low temperatures. The effective phonon energies, when compared with the phonon spectrum for alkali halide crystals, suggest that the F band width is determined by coupling to acoustic band phonons for alkali halides with smaller atomic masses (LiF, NaF, NaCl) and to optic phonons for those with heavier masses (KI, KBr). The essential information which the band-shape analysis yields is the single average frequency representative of the modes interacting with the center. If the F center electron interacts primarily with modes well separated from the center, then those modes will be longitudinal optical modes of the lattice. On the other hand if localized modes predominate involving mainly the nearest-neighbor ions, much lower vibrational frequencies than those of typical lattice phonons are observed. Just which situation occurs can sometimes be deduced from Raman scattering experiments. In fact, Raman scattering measurements give very detailed information about lattice vibration frequencies. The earliest measurements were reported by *Worlock* and *Porto* [4.37], who observed a continuous spectrum of frequencies that contribute to the first order Raman scattering of F centers in NaCl and KCl, excited by laser lines in the low energy tails of the F bands. Their results for NaCl are shown in Fig. 4.15. In a perfect alkali halide crystal there is no first order Raman scattering, so that the lattice vibrations must affect the F center. The Raman scattering results from modulation of the transition energies and mixing of the nearly degenerate s-like and p-like excited states of the F centers. *Worlock* and *Porto* [4.37] showed that the major part of the scattering at the F centers in NaCl is from modes with frequencies near $\hbar\omega = 100$ cm^{-1} rather than at the longitudinal optical frequency of $\hbar\omega = 265$ cm^{-1}. Hence the vibrational interaction is characteristic of the lattice close to F centers, i.e., localized modes.

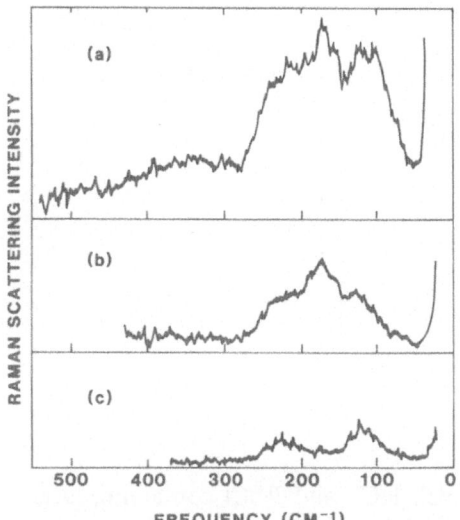

Fig. 4.15 a–c. Raman scattering in NaCl containing F centers. (a) Unpolarized light, (b) polarized parallel to laser light, and (c) polarized perpendicular to laser light. Spectra are measured using the 514.5 nm line from an Ar^+ laser. (After [4.37])

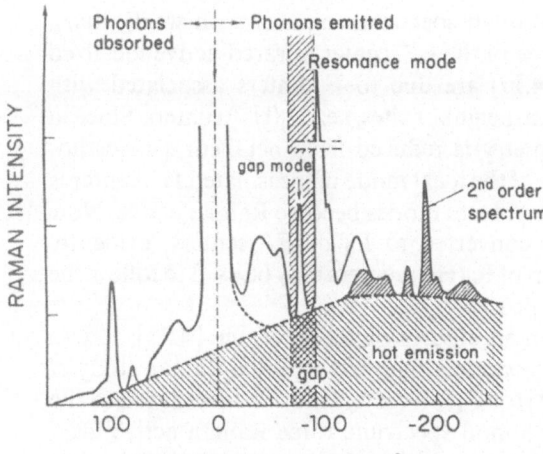

Fig. 4.16. The Raman spectrum of KI containing F centers. (After [4.38])

F centers can also give rise to a local mode, the frequency of which falls in the gap between the acoustic and optical branches of the frequency spectrum of the host crystal. Gap modes have been observed in the Raman spectra of F centers in NaBr, NaI, KBr, and KI. As Fig. 4.16 shows, in KI the gap between acoustic and optical branches extends from 70 to 96 cm^{-1}, and a Raman-active gap mode is observed at $\hbar\omega = 78.4\ cm^{-1}$ together with a resonance mode at 82.0 cm^{-1} [4.36]. Also evident is the second order spectrum and the anti-Stokes shifted components. High frequency localized modes have been observed in hydrogenated alkali halides. Substitutional H-atoms which trap an electron to form an H^- ion (or U center) are not Raman-active on account of the high

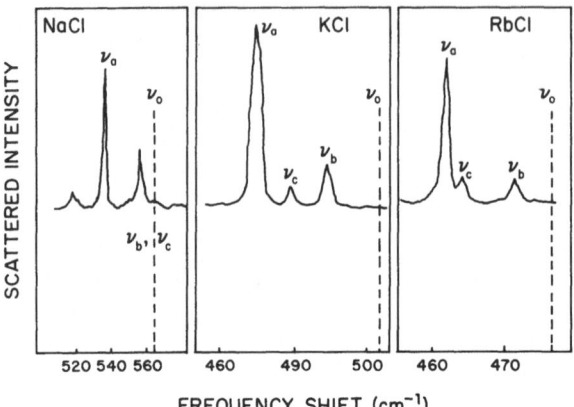

Fig. 4.17. Raman spectrum of $F_H(H^-)$ localized modes in NaCl, KCl and RbCl [4.39]

symmetry. However, in crystals of NaCl, KCl, and RbCl containing large concentrations of H^- centers, 10% of which have been converted to F centers by X-irradiation, the Raman spectrum measured using laser excitation in the F band shows the usual F center Raman spectrum together with weak, sharp Raman lines in the frequency range of the H^- center infrared-active localized modes [4.34]. These lines (Fig. 4.17) are due to F centers associated with substitutional H^- defects in nearest-neighbor sites, i.e., $F_H(H^-)$ centers. Since in this aggregate the F center symmetry is reduced from octahedral to ortho-rhombic, the threefold degeneracy of the local mode of the isolated H^- center is completely raised and the H^- center local modes become Raman active. Now $F_H(H^-)$ centers may be directly converted to F_2 or F_2^+ centers; evidently monitoring of the Raman spectrum of $F_H(H^-)$ centers may be used to follow the progress of the photoconversion process.

Following the pioneering Raman studies of the F center [4.37], it was natural to extend studies to F_A-type centers. The first reported Raman study of F_A centers was by *Fritz* et al. [4.40] for $F_A(Li)$ centers in KCl. In addition to the usual first order defect-induced phonon spectrum, three Raman active local modes at 47 cm^{-1}, 216 cm^{-1}, and 266 cm^{-1} for ^7Li were reported. However, neither the polarized Raman intensities for scattering from localized vibra-tional modes of A_1 or E symmetry in a C_{4v} symmetry defect nor the isotope shifts on ^7Li to ^6Li substitution could be explained. Consequently it was proposed that the Li^+ ion in the $F_A(Li)$ center resides not along the $\langle 100 \rangle$ symmetry axis of the center but is in an off-axis position. This is not surprising since the isolated Li^+ ion in KCl is well known to exhibit tunneling behavior between different $\langle 111 \rangle$ off-center sites. Electro-optic measurements and absorption/luminescence studies confirm that the Li^+ ion lies in a $\langle 110 \rangle$ plane relative to the z axis of the defect and that in response to this there is a tilting of the p-type orbitals of the excited F_A center towards the $\langle 110 \rangle$ plane and away from the $\langle 100 \rangle$ axes. Very recent measurements by *Leblans* et al. [4.41] on $F_A(Li)$ centers in KCl of Raman scattering under resonant excitation in F_{A_1} and

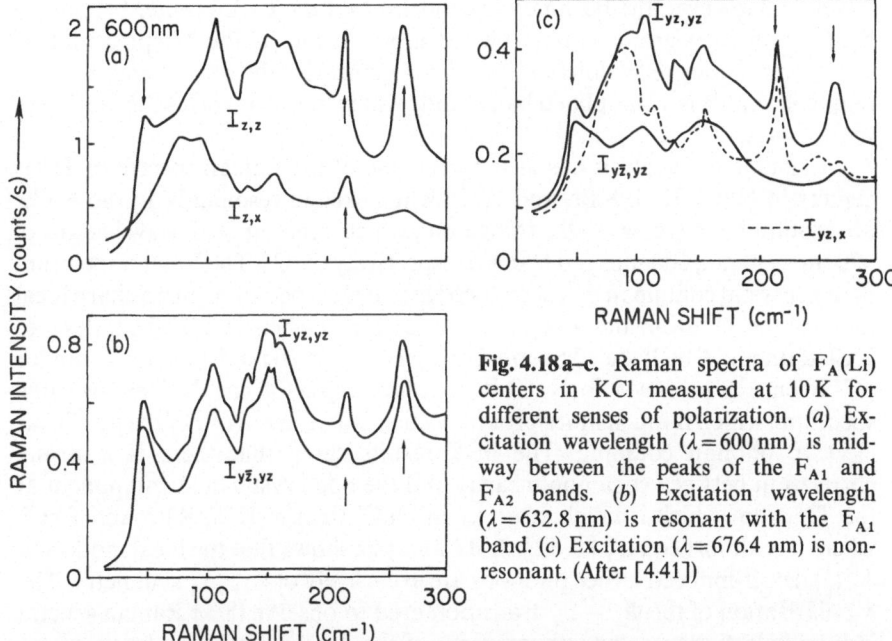

Fig. 4.18a–c. Raman spectra of $F_A(Li)$ centers in KCl measured at 10 K for different senses of polarization. (a) Excitation wavelength ($\lambda = 600$ nm) is midway between the peaks of the F_{A1} and F_{A2} bands. (b) Excitation wavelength ($\lambda = 632.8$ nm) is resonant with the F_{A1} band. (c) Excitation ($\lambda = 676.4$ nm) is nonresonant. (After [4.41])

F_{A2} bands and nonresonant excitation at 676.4 nm were analyzed in terms of a Behavior Type (BT) method. To quote from *Joosen* et al. [4.42]: *"the prominent feature of the BT method is its inductive character: Without any previous knowledge about the defect symmetry and, more generally, about the defect structure, the BT method permits the determination of the point group of the defect and the irreducible representation(s) of the Raman mode(s), which are consistent with the polarized Raman studies."* Figure 4.18 shows the polarized Raman spectra under excitation at 600 nm (midway between the F_{A2} and F_{A1} bands), 632.8 nm (resonant with the F_{A1} band) and 776.4 nm (nonresonantly). Under resonant excitation in the F_{A1} band the $F_A(Li^+)$ center displays each of the localized mode frequencies at 47 cm^{-1}, 216 cm^{-1}, and 266 cm^{-1}. The two high frequency modes at 216 cm^{-1} and 266 cm^{-1} shift to 266 cm^{-1} and 286 cm^{-1} on substitution of ^{6}Li for the ^{7}Li isotope. Under excitation in the F_{A2} band only the two high frequency modes show up. The polarization intensities and the wavelength dependences of Raman components show that only the 266 cm^{-1} mode retains its A_1-mode character of the C_{4v} symmetry group, i.e., it is associated with motions of the Li$^+$ ions in the mirror plane but parallel to the defect axis. As Fig. 4.18 shows, the 216 cm^{-1} mode is much the stronger in nonresonant excitation, which reflects the off-axis position of the Li$^+$ motion vibrating in the mirror plane perpendicular to the defect axis. In fact, the polarization properties of the 216 cm^{-1} mode show that symmetry of the defect contains only a mirror plane. As noted above, this deviation from C_{4v} $\langle 001 \rangle$ symmetry is not observed in the data for the 266 cm^{-1} mode, nor for the

47 cm^{-1}. However, the BT analysis confirms that the $F_A(Li)$ center belongs to the C_{1h} symmetry group within which all modes belong to the A' representation being even under reflections in the mirror plane. On the other hand, the low frequency mode is an amplified band mode which hardly involves the motion of the Li^+ ion.

A similar BT analysis has also been made of the Raman spectra of $Tl^0(1)$ centers in NaCl, KCl, KBr, and RbCl when excited resonantly in the F-like absorption band (i.e. $\Phi_\pm \to \Sigma_\pm$ transition), which occur at peak wavelengths of 470 nm, 540 nm, 600 nm, and 600 nm respectively [4.43]. The Raman spectrum in each crystal contains a defect-induced first-order spectrum, plus a sharp local mode of the Tl^0 atom and several overtones. For the heavier alkali halides (i.e., KCl, KBr, and RbCl) the vibrational energy of the local modes are quite similar (27–30 cm^{-1}). However in NaCl it is rather larger (38 cm^{-1}). The overtone spectrum, which is much more pronounced in NaCl, arises out of the nonlinear electron–phonon coupling. The ESR results show that there is a strong correlation between such nonlinearity and the odd crystal field component at the Tl^0 atom which is much stronger in NaCl than in KCl, KBr, and RbCl, where it is almost equal [4.9, 12]. A BT analysis shows that the local mode and $Tl^0(1)$ defect-induced lattice phonons are exclusively of C_{4v}: A_1 symmetry. The σ-polarization of the $\Phi_\pm - \Sigma_\pm$ transition used to observe these Raman spectra induces selective *resonant* enhancement of the A_1 symmetry modes.

In studies of vacancy-type defects the Raman experiments have been used to elucidate the nature of the modes coupled to the electronic states of the defect. However, this light scattering technique is also capable of revealing details of the structure of defect clusters. The determination of the C_{1h} symmetry of the $F_A(Li)$ center is an elegant example of this particular use. As discussed in Sect. 4.1.6, optical absorption techniques are especially useful in determining the structure of isolated X_2^- and H centers, and even clusters containing (say) two trapped hole centers. Electron microscopy provides evidence of clusters containing more than twenty atoms [4.42, 44]. The intermediate region is a more difficult region to probe. *Lefrant* and colleagues [4.45–48] provided new insight into H center aggregation by using Raman scattering. This work has now been complemented by the work of *Comins* et al. [4.49] and of *Aguilar* et al. [4.50]. The V bands in the blue–near uv region of the spectrum in alkali halides are very broad and complex, so that the structure of "V centers" has been difficult to uncover. Raman scattering from these centers is best under conditions of quasi-resonant scattering. Most studies have involved the use of the resonance lines in Ar^+ and Kr^+ lasers and pure or doped crystals of KBr, KI, BbI, and CsI. Such studies show that the clustering of stabilized interstitials during X-irradiation of KI below 300 K proceeds via the formation of trihalide ions (I_3^-, Br_3^-, …).

The Raman line at 111 cm^{-1} in Fig. 4.19 has been assigned to the symmetrical stretching mode of the linear I_3^- molecular ion. Larger iodide clusters (polyiodide molecular ions) are signaled by the 175 cm^{-1} mode, which has been identified with the vibrational mode of the I_2 molecule in a polyiodide

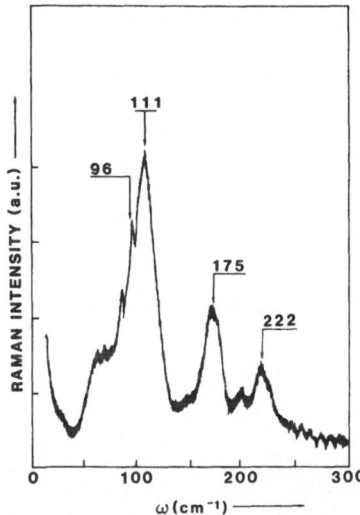

molecular ion (I_n^-, $n = 5, 7, \ldots$). Impurities (M^{2+}) tend to stabilize the interstitial aggregates, leading to enhanced production efficiency of stable F centers during X-irradiation near 300 K [4.49].

4.3.2 Optical Hole Burning Spectroscopy

Very narrow laser lines are ideal light sources for saturation experiments in optical spectroscopy. The earliest examples of such experiments were studies of Doppler-free atomic spectra [4.51]. Inhomogeneous broadening arises when atoms are distinguished by the frequency at which they absorb or emit radiation. The spectral profile is then the sum of separate, independent absorption/emission lines. In solids, atoms oscillate about a mean lattice point in the crystal, and optical transitions are *homogeneously broadened* by the electron–phonon coupling. However, zero-phonon lines in solids are not broadened by the phonon spectrum and are sensitive only to the distribution of internal strains in a crystal and are consequently *inhomogeneously broadened*. It is possible to recover the homogeneous width of an optical transition using the technique of optical hole burning (OHB). Figure 4.20 illustrates an inhomogeneously broadened line of width Γ_{inh}, produced by many narrow components of homogeneous width $\Gamma_{hom} \ll \Gamma_{inh}$, each component being centered at a different frequency within the inhomogeneous line profile. If a narrow laser line of frequency ν_L and bandwidth $\Gamma_L < \Gamma_{hom}$ is incident upon an atomic assembly having an inhomogeneously broadened linewidth Γ_{inh} the resulting absorption of laser radiation depletes only that subassembly of excited centers whose energies are within Γ_{hom} of the laser line frequency ν_L. In other words the absorption signal is determined selectively by some atoms with absorption transitions within the overall line profile but not by others. Consequently there

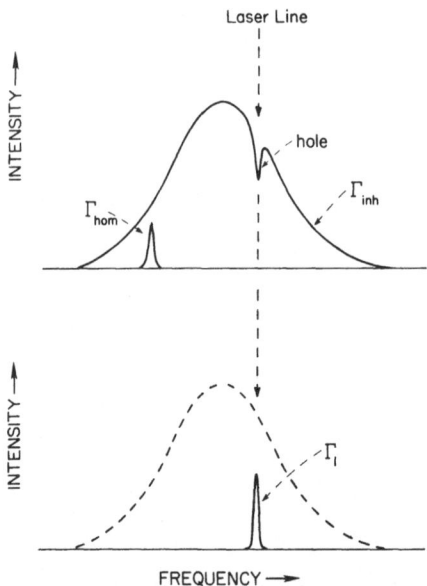

is a distortion of the lineshape in the neighborhood of v_L; a "hole" of width $2\Gamma_L$ is burned in the inhomogeneously broadened line. In order to resolve the homogeneous width, it is required that $\Gamma_L < \Gamma_{hom} \ll \Gamma_{inh}$. In hole burning spectroscopy the narrow laser linewidth and high power make it possible to maintain a significant fraction of those atoms with transition frequency v_L in the excited state, where they no longer contribute to the absorption at this frequency. The experimental arrangements for carrying out hole burning experiments in solids, using cw ring dye lasers or pulsed dye lasers, are fully described by *Selzer* [4.52].

There have been many reports of hole burning spectroscopy on transition metal ions and rare earth ions in inorganic materials [4.52]. Optical zero-phonon lines associated with defects are inhomogeneously broadened by the effects of random lattice strain; they are prime candidates for study by optical hole burning. So far, measurements have been reported for color centers in LiF, NaF, CaF_2, CaO, and diamond. The combination of electron spin resonance, uniaxial stress splitting, Stark and Zeeman effect studies have already resulted in a very detailed understanding of the electronic structure of defects in alkali halide crystals [4.53]. Typical color centers have homogeneous widths of 10–30 MHz and corresponding lifetimes of the order of tens of nanoseconds. For F_3^+ centers in alkali halides, the low-lying transition $^1A_1 \rightarrow ^1E$ is revealed in LiF as a sharp zero-phonon line at 487.4 nm with a halfwidth of 5 cm^{-1} at 4 K: the stress spectra are consistent with the trigonal $^1A_1 \rightarrow ^1E$ transition [4.54]. A zero-phonon line at the same wavelength is observed in emission also. This is also the situation in NaF where the F_3^+ center zero-phonon line of width 1.1 cm^{-1} occurs at 545.6 nm in both absorption and emission. The absorption

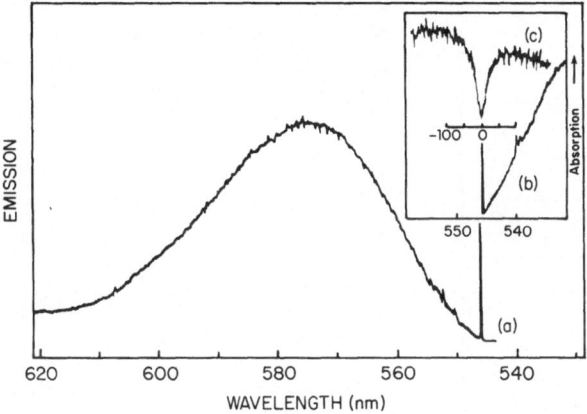

EMISSION

WAVELENGTH (nm)

620   600   580   560   540

Fig. 4.21 a–c. The
$^1A \leftrightarrow {}^1E$ transition of the
F_3^+ centers in NaF. (a)
The emission spectrum;
(b) excitation spectrum
near the zero-phonon
line; and (c) a hole burned
in the zero-phonon line.
(After [4.57])

and emission sidebands of these $^1A \rightarrow {}^1E$ transitions are approximately mirror images of one another in the zero-phonon line. The Stark splittings of this line are linear with the applied electric field, showing that the center lacks inversion symmetry, in accordance with the C_{3v} point symmetry of the F_3^+ center [4.55]. Such narrow linewidths point to dramatic improvements in spectral resolution using OHB techniques in the presence of applied perturbations. The first observation of hole burning in a color center zero-phonon line was for F_3^+ centers in NaF [4.56]. Holes were burned in the zero-phonon line by bleaching for about 2 s with 200 mW/cm^2 of single frequency dye laser light (Fig. 4.21). The homogeneous linewidth of the $^1A \rightarrow {}^1E$ transition, $\Gamma_{hom} = 17$ MHz, corresponds to a dephasing time of $T_2 = 20$ ns. The hole recovery showed two components, one of several seconds associated with a low-lying spin triplet state of the center and the other of order 70 min due to photoionization.

Unambiguous identification of the associated point group would be helpful in determining atomistic models of the various centers. Such identification is possible either by combining high-resolution techniques and perturbation spectroscopy or by combining uniaxial stress and Stark effect measurements. *Macfarlane* et al. [4.57] have taken advantage of such high resolution to demonstrate that assignments of color center symmetry may be incorrect. As an example of such studies we shall discuss the 607 nm line in NaF. This line was assigned by *Baumann* [4.58] to F_4 centers (Fig. 4.3b), on the basis of uniaxial stress effect measurements. *Macfarlane* et al. [4.57, 59] reported the Stark splitting of the hole burned in this zero-phonon line (Fig. 4.22). The measured homogeneous width is only 21 MHz, which is to be compared with the inhomogeneous width of 3 GHz. The Stark splitting patterns shown in Fig. 4.22 are inconsistent with the F_4 center structure. Instead of the C_{2h} symmetry of the F_4 center, they require the symmetry of the center to be C_s with a (110) mirror plane. Obviously, perturbation spectroscopy alone is not always sufficient to determine details of structural models of defects: it may be necessary to have the evidence of very high resolution studies to be able to make confident assignments.

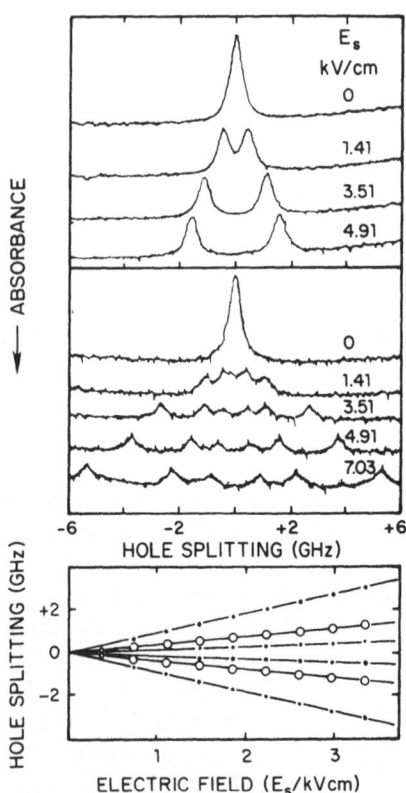

Fig. 4.22 a, b. Stark effect on the 607 nm line in NaF using OHB. The case of E_s parallel to [111] is shown for two polarizations. (a) E_L parallel to [111] and (b) E_L parallel [110]. (After [4.57])

Moerner et al. [4.60, 61] have made a very detailed study of *persistent hole burning* (PHB) of the 833.0 nm zero-phonon line associated with the F_3^- center in LiF. Optical hole burning experiments were carried out at temperatures of 1.4 K by immersing the sample in superfluid He. The shape of the hole varies with the wavelength within the inhomogeneous lineshape of the zero-phonon line. However, close to the peak the most striking feature is the appearance of satellite holes on both sides of a central peak. The satellites are difficult to resolve because of overlap with the central hole. This structure is related to random splittings of the excited state by well-defined local stresses associated with other nearby defects (F, F_2, F_3, ...) introduced by the coloration process. The growth of the depth of holes is a logarithmic function of both laser intensity and exposure time over several decades. Moerner and co-workers explain the results in terms of a phenomenological model in which the F_3^- centers are coupled to a distribution of two-level tunneling states by local strain fields associated with the other nearby defects. These holes are indefinitely stable at low temperatures in the dark. However, there is a purely transient effect associated with the excited state lifetime (~ 10 ms). Over the temperature range 1.4–20 K there is a decrease in hole width of order 15%. The long-lived holes are induced by photoionization, in which the released electron tunnels via the two-level system into a nearby electron trap. Persistent hole burning is of current technological interest because it has the potential for optical storage with terabyte capacity using frequency domain optical storage [4.62]. The F_3^- centers in LiF are an example of single photon PHB. However, destructive "reading" using a laser can be responsible for erasing all the information contained in the PHB pattern. Two-photon PHB can sometimes eliminate this possibility. Essentially photons of wavelength λ_1 are used to burn holes in an inhomogeneously broadened line, then photons of a second wavelength λ_2 are used to photoionize centers raised into the excited state by wavelength λ_1. Only λ_1 is required to read the holes, and this may not of itself produce photochemistry. The first observation of photon-gated OHB was reported for rare earth ions in alkaline earth fluorides [4.63]. Similar results have now been reported for an organic system: carbazole in boric acid glass [4.64]. Given the very many zero-phonon lines in color center spectra and their ready participation in photoionization transitions, such phenomena are to be expected there also.

Other spectral hole burning studies have been reported for color centers in CaF_2 (F_3 centers [4.65], diamond [4.66] and in CaO [4.67]. In both diamond and CaO, triplet excited states are involved either directly or indirectly. The F center in CaO differs from that in the alkali halides in that two electrons are trapped in the oxygen ion vacancy. Since the symmetry is octahedral (O_h) the $1s^2$ configuration leads to a $^1A_{1g}$ ground state and the first allowed optical transition takes the center to the $^1T_{1u}(1s2p)$ state via a broad absorption band peaking at ~ 400 nm. Intersystem crossing then leads to emission via the $^3T_{1u} \rightarrow {}^1A_{1g}$ transition identified via a sharp zero-phonon line at 574.2 nm and vibronic sideband with peak at 601 nm [4.68]. In optically detected magnetic resonance (ODMR) studies *Edel* et al. [4.69] showed that in the excited state there is a strong static Jahn-Teller coupling to E_g modes of vibration leading to a static tetragonal distortion of the center. The shapes of the optical zero-phonon line and the ODMR lines are determined by the same distribution of internal strains of (predominantly) E_g symmetry [4.70–73]. Thus, both ODMR and zero-phonon lines are inhomogeneously broadened and are suitable for OHB measurements. Figure 4.23 shows the hole burned in the 574.2 nm line of the F center [4.67]. The depths and widths of the holes in the zero-phonon line are strongly sensitive to laser power. The data shown in Fig. 4.23 were achieved with a laser intensity of ~ 50 mW for 10 s: the whole linewidth is then about 100 MHz. However, burning with 0.5 mW for 300 s results in linewidths of only 40 MHz. Such widths are much larger than the inhomogeneous width expected from the lifetime of 3 ms [4.68] or the interwell relaxation process [4.71]. *Reddy* et al. [4.67] also report measurements of the Zeeman effect on the hole burned in the 574.2 nm zero-phonon line. These studies were carried out in magnetic fields up to 0.1 T, convincingly confirming the selection rules for linear and circular polarizations determined in the earlier ODMR measurements [4.69]. In fact, ODMR studies using laser excitation also lead to very similar hole burning phenomena, as long as such measurements are carried out detecting at

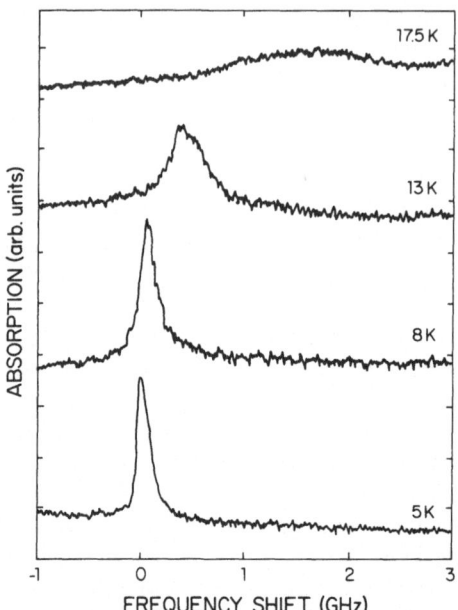

Fig. 4.23. Temperature-induced broadening of holes burned in the 574 nm zero-phonon line of the $^3T_{1u} - {}^1A_{1g}$ transition of the F center in CaO. (After [4.67])

the zero-phonon line only (as opposed to the vibronic sideband). This follows from the fact that the shapes of the zero-phonon line and ODMR lines are determined by the same distribution of strains of E_g symmetry. These hole burning phenomena are long-lived, especially at low temperature, and are associated with photochemical reactions of the type $F \to F^+$ center in which the electron tunnels from the excited $^3T_{1u}$ state to nearby F^+ or F_A^+ centers, thereby creating F and F_A centers in their excited states [4.74]. On decay of "new" F-type centers to the ground state they become extremely stable, so accounting for the very long-lived nature of the optical holes.

4.3.3 Picosecond Spectroscopy of Solids

During the past two decades there have been quite remarkable developments in techniques for generating and measuring ultrashort pulses. Pulses in the 10^{-12} s time domain are routinely available and spectroscopy in the 10^{-14} s domain is becoming comparatively familiar. Such ultrashort pulses are useful in the study of nonradiative deexcitation from optically excited states, involving such processes as ionic relaxations around centers in excited states, which may lead to defect formation or defect reorientations in the case of anisotropic centers. These measurements rely upon methods developed for mode-locking lasers and measuring ultrashort pulses. Mode-locking refers to the process of ensuring that the phases of all the modes are the same, whence the laser output consists of a regular sequence of short pulses. At the heart of the mode-locked laser is the gain medium in an optical resonator of optical path length L, which has a set of characteristic resonant frequencies with intermode spacings $\delta v = c/2L$. In a tunable laser δv is much less than the gain bandwidth so that generally many modes oscillate simultaneously. The repetition rate of the pulse train is then just δv and the pulse duration, $\tau_p = \gamma/\Delta v$ where Δv is the width of the spectrum of excited modes and γ depends on the shape of the mode spectrum and is usually of order unity. A pulse of 10^{-12} s requires a transition with frequency width of about 10^{12} Hz $\equiv 30$ cm^{-1}. Organic dyes, transition metal ions and color centers have sufficient bandwidth for the generation of picosecond or even subpicosecond pulses [4.75]. With the exception of the Auston switch [4.76] photoelectric detectors and electronic recording devices are capable of a resolution of only a few hundred picoseconds. The streak camera is the only electronic detector capable of a resolution of 10^{-12} s. For this reason the most widely used technique for measuring picosecond pulses is the measurement of the intensity autocorrelation function by means of scanning autocorrelation interferometer with second harmonic generation [4.77].

   Many picosecond phenomena, especially nonradiative decay processes, are studied by excite-and-probe techniques in which light pulses at wavelength λ_1 are used to excite a phenomenon of interest and then a delayed optical pulse at wavelength λ_2 interrogates a change of some optical property induced by this phenomenon. Ideally, two picosecond optical sources at two different and independently tunable wavelengths, synchronized on the picosecond timescale, are required. In view of the importance of the excited state relaxation mechanisms of anion vacancy centers in the operation of color center lasers, we will illustrate such pulse and probe techniques by discussing some measurements on the $F_A(Li)$ center KCl. This is an interesting example of configurational relaxation in which the large Stokes shift between the absorption and emission band peaks results from ionic relaxation from a single potential well to a saddle point configuration during the deexcitation, as illustrated in Fig. 4.2. During such deexcitation many phonons are excited in the localized modes coupled to the electronic states, which must be dissipated into the continuum of lattice modes, so that measurement of the relaxation time constitutes a probe of the possible phonon damping. *Mollenauer* et al. [4.78] used the experimental system shown in Fig. 4.24 to carry out measurements of the configurational relaxation time. The mode-locked dye laser, which generated pulses of 0.7 ps duration at 612 nm, was used both to pump into the F_{A2} absorption band of the center and to provide the timing beam. Such pumping leads to optical gain in the luminescence band and prepares the centers in their relaxed state. The probe beam, collinear with the pump beam, is generated by a cw $F_A(Li)$ center laser operating at 2.62 μm. The probe beam and gated pulses from the dye laser are mixed to produce sum and difference frequencies in a nonlinear optical crystal ($LiIO_3$). The filter allows only the sum frequency at 496 nm to be detected. The photomultiplier tube then detects the rise in intensity of the probe beam which signals the appearance of gain when the $F_A(Li)$ centers have reached the relaxed excited state. The pump beam is chopped at low frequency to permit

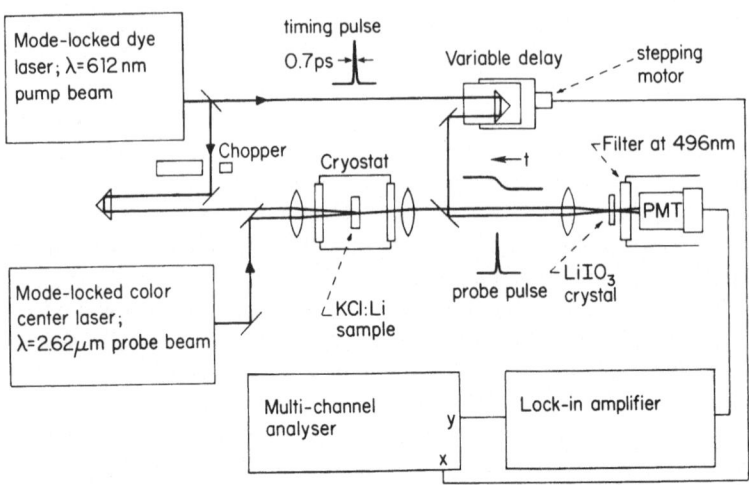

Fig. 4.24. Apparatus for picosecond measurements of configurational relaxation times in color centers. (After [4.78])

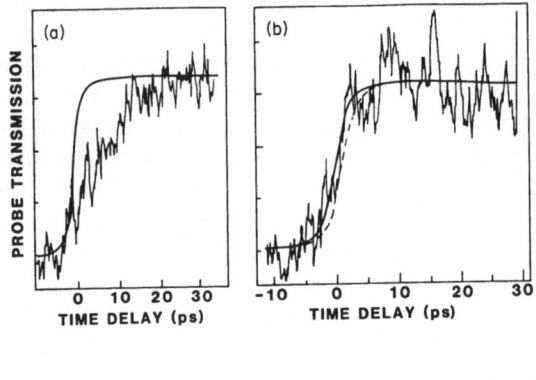

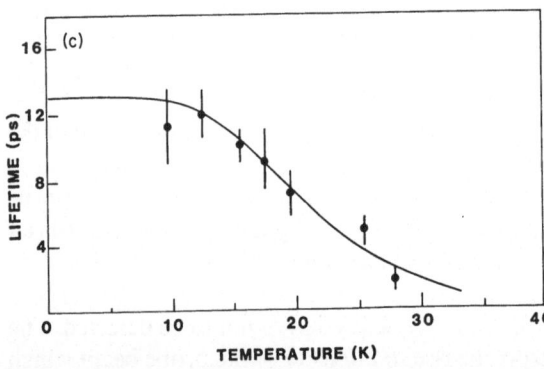

Fig. 4.25a–c. Time-resolved measurements of probe beam transmission for $F_A(Li)$ centers in KCl. The rise of gain is shown at $\lambda = 2.62\,\mu m$ for temperatures of **(a)** 15.6 ± 0.4 K and **(b)** 47.7 ± 0.2 K. The solid line is the instantaneous response of the system, whereas the dashed line in **(b)** is the instantaneous response convolved with a 1.0 ps rise time. Measured rise times are 10 ± 1 ps in **(a)** and <0.5 ps for **(b)** showing the temperature dependence of the relaxation time **(c)**. (After [4.78])

detection with phase-sensitive electronics. The temporal evolution of F_A(Li) center gain is then measured by varying the time delay between pump and gating pulses.

The temporal evolution of gain at 2.62 μm is shown in Fig. 4.25a at $T=15.6$ K and in Fig. 4.25b at $T=47$ K. These results indicate that the relaxation is a very fast process, typically on a timescale of ~ 10 ps at 4 K. Measurements of the temperature dependence of the relaxation times (Fig. 4.25c) show that the configurational relaxation involves a multiphonon mechanism in which of order twenty phonons of energy $E_p/hc \simeq 47$ cm^{-1} are created. This is just the low energy mode which occurred so strongly in the Raman spectrum of F_A(Li) centers in KCl (Fig. 4.18) [4.41]. That only about 20×47 cm$^{-1}/8066$ cm$^{-1} = 0.1$ eV is deposited into the 47 cm^{-1} mode, whereas 1.6 eV of optical energy is lost during the overall ionic relaxation preceding radiative decay indicates that other higher energy vibrational modes must also be involved. *Leblans* et al. [4.41] have shown two high frequency Raman-active modes with energies of 216 cm^{-1} and 266 cm^{-1} are resonantly excited by pumping in the F_{A2} band. The division of relaxation energy between the various modes reflects the coupling resonance between optically prepared and energy accepting phonons [4.79]. This type of measurement could be applied to other color center systems, although the specific experimental details would of necessity be varied to match the particular color center.

4.3.4 Excited State Spectroscopy

This discussion ends with a description of some experiments in which the optical properties of a sample are probed simultaneously by two beams of light. There is considerable flexibility in the experimental arrangement since

    i) both beams may be derived from the same or different lasers,
    ii) one beam may be from a laser and the other from a broad-band lamp.

Furthermore both beams may be cw or pulsed. In general, the presence of color centers in crystals is manifested by the appearance of more-or-less broad absorption bands. This being the case it is often convenient to use tunable lasers to match the excitation to the peak in the absorption band. This will raise a large density of centers to the excited state with considerable efficiency. Such excited centers as there are will decay by phonon emission to the relaxed excited state from which luminescence transitions take place. Since by using a laser we can ensure that the pump rate is always larger than the radiative emission rate, we can have a large steady-state population of the excited state. The probability of exciting absorption transitions out of this relaxed excited state to some higher-lying state is then significant, and their observation may be effected by detecting changes in the absorption coefficient at the pump wavelength or at the emission wavelength (Fig. 4.26). Note that the relaxed excited state studied in excited state spectroscopy may be electronically quite different from the states reached in absorption transitions out of the ground state. For example, the

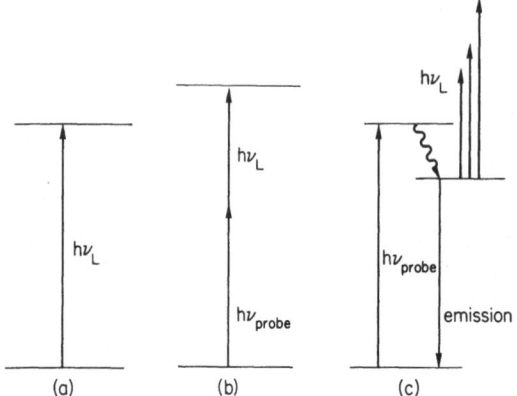

Fig. 4.26 (a) One-phonon absorption between states of opposite parity. **(b)** Two-photon absorption between states of the same parity. **(c)** One-photon absorption followed by nonradiative reloxation and one-photon absorption from the relaxed excited state

ground state of the F_2 center is $^1A_{1g}$ and strongly allowed electric dipole absorption occurs to the $^1B_{1u}$ state. Relaxation from this excited state may lead to an intersystem crossing to a metastable spin triplet state $^3A_{1g}$, which has a lifetime of tens of seconds. In F_2 centers there is a possibility of studying excited state transitions out of $^1B_{1u}$ or $^3A_{1g}$ states. As discussed in Sect. 4.1.3, F_2 centers have several absorption bands, some partially overlapping the F band, with well-defined dichroic properties. These bands are narrower than the F band, indicating a rather smaller electron–phonon interaction. In LiF, NaF, and NaCl there is a single luminescence band originating on the $^1B_{1u}$ state with lifetime of order tens of nanoseconds. However, deexcitation of F_2 centers in KCl and CaF_2 results in the population of low-lying metastable $^3A_{1g}$ state, which have lifetimes of 10–50 s [4.80]. Given the large fraction of centers in the metastable excited state one may study optical absorption transitions to higher-lying triplet states without recourse to laser techniques [4.80].

The long lifetime (10^{-6} s) of the relaxed excited state of the F center in alkali halides made this ideal for studies of excited state absorption. The early work of *Park* and *Faust* [4.81] on F centers in KI indicated laser-induced absorptions for photon energies in the range 0.1–0.5 eV from the relaxed excited $^1T_{1u}^*$ state to other bound levels just below the conduction band. This work correlated well with the measured value of the thermal ionization energy of the excited states determined from photoconductivity measurements [4.82]. Subsequent measurements [4.83, 84] in additively colored KCl, KBr, and KI revealed structure (Fig. 4.27) associated with transient absorption bands from the relaxed excited state to bound 3p-like, 3d-like, and 4p-like states. That such transitions occur with high oscillator strength follows from the mixed $s-p$ character of the relaxed excited state [4.84, 85]. The higher energy structure in Fig. 4.27 is believed to be due to one, two, and three phonon structures.

Although the excited states of F_2 centers have been little studied by laser techniques, both F_2^+ and $F_2^+ : O^{2-}$ centers have been studied in detail [4.86, 87]. In these experiments the F^{2+}-type centers were continuously aligned along a

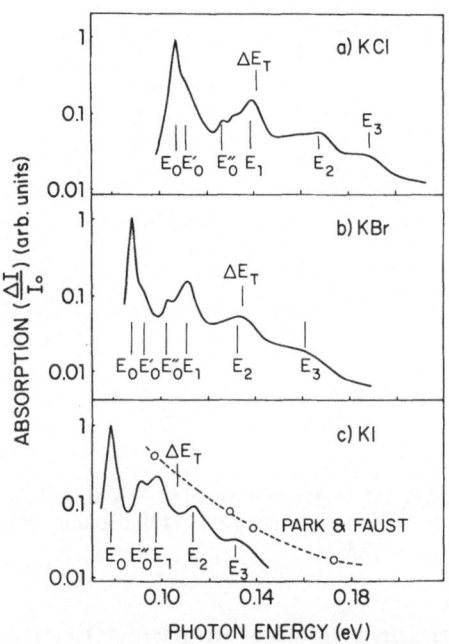

Fig. 4.27. Transient absorption of additively colored crystals induced by F band light at $T = 10$ K. Here ΔE_T represents the thermal ionisation energy of energy of the F center. Results on KI by Park and Faust are shown for comparison. (After [4.84])

particular $\langle 110 \rangle$ direction by irradiation with $\langle 110 \rangle$ polarized F band light, which was the same polarization as the main pump beam. Although this alignment procedure is less than 100% efficient, it is sufficient to significantly improve the signal-to-noise ratio of the recovered signal so that unambiguous determination of the polarization properties of the transition are possible. The main pump beam is $\langle 110 \rangle$ polarized radiation from a Nd:YAG laser operating at 1.06 μm: this pumps the fundamental $^2A_{1g} \rightarrow {}^2B_{1u}$ transition of the aligned centers. To modulate the populations of the $^2A_{1g}$ and $^2B_{1u}$ states of the defect the pump beam is chopped using a high frequency (15 kHz) mechanical chopper, which also provides a reference signal for a lock-in detector. The probe beam may be a broad band source or a tungsten-halogen or xenon arc lamp, depending on the wavelength region of interest. In tandem the chopped pump beam and fast radiative lifetime ($\tau_R \simeq 150$ ns) of the centers create time-dependent populations in phase ($^2B_{1u}$) and out of phase ($^2A_{1g}$) with the pump laser. In consequence, absorption transitions from the ground state are detected in phase with the pump beam whereas absorption signals from the first relaxed excited state are out of phase with the pump beam. In the spectrum shown in Fig. 4.28 positive- and negative-going signals correspond to transitions from the ground state and relaxed excited state respectively. Three ground state and five excited state absorption transitions are observed, their polarization properties are determined and oscillator strengths may be calculated. A comparison of the transition energies and oscillator strengths for these transitions is given in Table 4.3. Theoretical data computed using the dielectric

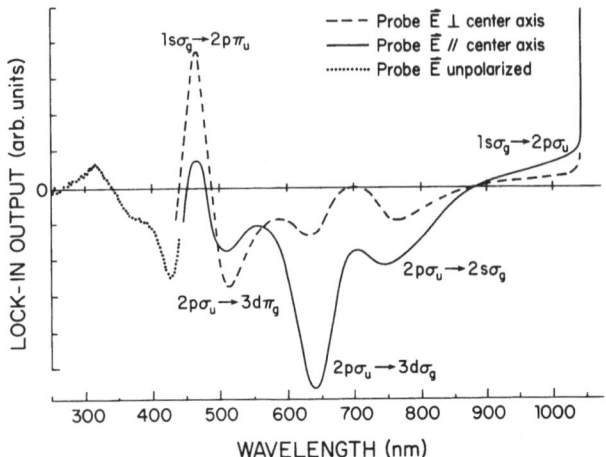

Fig. 4.28. Excited state absorption spectrum of $F_2^+ : O_2^-$ centers measured at 77 K with [110] polarized radiation at 1.06 µm. The alignment of the defects along the [110] direction was maintained by [110] polarized F band light. (After [4.87])

Table 4.3. Calculated and observed transition energies and oscillator strengths of the $F_2^+ : O^{2-}$ center in NaCl. Transition energies for the F_2^+ center are listed for comparison. (After [4.87])

Transition	Calculated		Observed		F_2^+ center ΔE [eV]
	ΔE [eV]	f_0	ΔE [eV]	f_0	
$1s\sigma_g \to 2p\sigma_u$	1.14	0.29	1.14	0.29	1.20
$1s\sigma_g \to 2p\pi_u$	2.70	0.24	2.70	0.13	2.92
$1s\sigma_g \to 3p\pi_u$	a	~0.03	3.91	~0.03	a
$2p\sigma_u \to 2s\sigma_g$	2.0	0.09	1.63	0.14	2.01
$2p\sigma_u \to 3d\sigma_g$	2.19	0.32	1.95	0.39	1.76
$2p\sigma_u \to 3d\pi_g$	2.41	0.15	2.40	0.15	2.48
$2p\sigma_u \to 4d\sigma_g$	a	~0.05	2.88	~0.09	a
$2p\sigma_u \to 4d\pi_g$	a	~0.04	3.29	~0.04	a

a Data unavailable.

continuum model [4.13] (Sect. 4.1.3) are also given. By using (4.1) the vacancy pair separation may be computed using $R_{12} = \kappa r_{12}$, in which r_{12} and κ are chosen to fit the peaks in the experimentally observed transition energies. *Georgiou* et al. [4.87] determine

$\kappa = 2.32, R_{12} = 0.354$ nm,   ground state.

$\kappa = 2.32, R_{12} = 0.418$ nm,   relaxed excited state.

The different values of R_{12} reflect the ionic relaxation of the center following deexcitation from the excited state reached in the absorption transition. The absorption spectra shown in Fig. 4.28 are strongly polarized: the σ and π polarized transitions refer to probe light polarized parallel to or perpendicular to the aligned $F_2^+ : O^{2-}$ defects, respectively. Polarization data was used to help characterize the five lowest energy transitions, whereas the higher energy transitions were assigned according to model predictions of the relative oscillator strengths and transition energies.

Time-dependent perturbation theory in first order shows that one-photon electric dipole transitions involve electronic states of opposite parity; electric dipole transitions between states with the same parity have probabilities for two photon absorption given by second-order perturbation theory:

$$P \sim E_1 E_2 \phi_1 | \sum_i \frac{\langle b | r \cdot \varepsilon_1 | i \rangle \langle i | r \cdot \varepsilon_2 | a \rangle}{E_{ai} - E_2}$$

$$+ \frac{\langle b | r \cdot \varepsilon_2 | i \rangle \langle i | r \cdot \varepsilon_1 | b \rangle |^2}{E_{ai} - E_1} \delta(E_{ba} - E_1 - E_2),$$

where E_1 and E_2 are the energies of photons at wavelengths λ_1 and λ_2 respectively, a and b are the two electronic states separated in energy by $E_{ba} = E_1 + E_2$, and the ε's refer to the polarization of the exciting radiation. Since P involves the photon flux at wavelength λ_1 it is evident that such two-photon transitions are most effectively driven by photons derived from a laser operating at this wavelength. Furthermore, the intermediate states, i, will give the largest contribution to the total matrix element if they are of opposite parity to both states, a and b.

Very little progress with two-photon spectroscopy of defects has been reported since the early work on excitation and band edge absorption reviewed by *Mahr* [4.88]. However, *McClure* [4.89] is making extensive studies of two-photon absorption transitions at transition metal ions, which gives important information about the symmetries of phonons involved in excited state relaxations. Similar work on rare earths is discussed elsewhere in this volume by Downer. As noted above, two-photon absorption processes lead to persistent optical hole burning by photoionization of F_3^- centers in NaF [4.60, 61], and two-step photoionization leads to efficient $F_2 \rightarrow F_2^+$ photoconversion processes in the presence of deep electron traps [4.17, 18].

References

4.1a Much early work is discussed in the classic text by J.H. Schulman, W.D. Compton: *Colour Centres in Solids* (Pergamon, Oxford 1962)

4.1b W.B. Fowler (ed.): *Physics of Colour Centres* (Academic, New York 1968)

4.2 J.M. Spaeth: In *Defects and Their Structure in Nonmetallic Solids*, ed. by B. Henderson, A.E. Hughes (Plenum, New York 1976) and references therein

164   B. Henderson and K. P. O'Donnell

4.3 L.F. Mollenauer: In *Quantum Electronics: Part B*, ed. by C.L. Tang (Academic, New York 1979) Chap 6; and in *Laser Handbook 4*, ed. by M. Bass (North-Holland, Amsterdam 1985)
4.4 E. Sonder, W.A. Sibley: In *Defects in Solids*, ed. by J.H. Crawford, L.F. Slifkin (Plenum, New York 1972) p. 381
4.5 B. Henderson, G.F. Imbusch: In *Optical Spectroscopy of Inorganic Solids* (Oxford University Press 1989)
4.6 F. Lüty: In *Physics of Color Centers*, ed. by W.B. Fowler (Academic, New York 1968) p. 182
4.7 L.F. Mollenauer, D.H. Olson: J. Appl. Phys. **24**, 386 (1974); **25**, 3109 (1975)
4.8 W. Gellermann, F. Lüty, C.R. Pollock: Opt. Commun. **39**, 391 (1981)
4.9 E. Goovaerts, J.A. Andriesson, S.V. Nistor, D. Schoemaker: Phys. Rev. B **24**, 1249 (1981)
4.10 L.F. Mollenauer, N.D. Vieira, L. Szeto: Phys. Rev. B **27**, 5332 (1983)
4.11 F.J. Ahlers, F. Lohse, J.M. Spaeth, L.F. Mollenauer: Phys. Rev. B **28**, 1249 (1983)
4.12 J.M. Spaeth, R.H. Bartram: Private communication (1987)
4.15 M. Aegerter, F. Lüty: Phys. Status Solidi **43**, 227, 245 (1971)
4.14 R. Herman, M.C. Wallis, R.F. Wallis: Phys. Rev. **103**, 87 (1956)
4.15 C.Z. Van Doorn: Philips Res. Rep., Suppl. **4**, 1 (1962)
4.16 R.H. Silsbee: Phys. Rev. **135**, A 180 (1965)
4.17 L.F. Mollenauer, D.M. Bloom: Opt. Lett. **4**, 247 (1979); see also L.F. Mollenauer: ibid **5**, 188 (1980); **6**, 342 (1981)
4.18 D.R. Foster, I. Schneider: In *Tunable Solid State Lasers II*, ed. by A.B. Bugdor, L. Estorowitz, L.G. DeShazer, Springer Ser. Opt. Sci., Vol. 52 (Springer, Berlin, Heidelberg 1986) p. 266
4.19 T.G. Castner, W. Känzig: J. Phys. Chem. Solids **3**, 178 (1957)
4.20 D. Schoemaker: In *Defects and Their Structure in Nonmetallic Solids*, ed. by B. Henderson, A.E. Hughes (Plenum, New York 1976)
4.21 M. Kabler: In *Point Defects in Solids*, Vol. 1, ed. by J.H. Crawford, L.F. Slifkin (Plenum, New York 1972) p. 327
4.22 F. Keller, R.B. Murray, R.A. Weeks, M.M. Abraham: Phys. Rev. **154**, 812 (1967)
4.23 B. Fritz, E. Menke: Solid State Commun. **3**, 61 (1965)
4.24 See e.g. F. Lüty, W. Gellermann: Proc. Int. Conf., Laser '81, ed. by C.B. Collins (STS Press, McClean, Va. (1982)
4.25 J.F. Pinto, E. Georgiou, C.R. Pollock: In *Tunable Solid State Lasers II*, ed. by A.B. Bugdor, L. Esterowitz, L.G. DeShazer, Springer Ser. Opt. Sci., Vol. 52 (Springer, Berlin, Heidelberg 1986) p. 261; J.F. Pinto, L.W. Stratton, C.R. Pollock: Opt. Lett. **10**, 384 (1985)
4.26 D. Wandt, W. Gellermann, F. Lüty, H. Welling: In *Tunable Solid State Lasers II*, ed. by A.B. Bugdor, L. Esterowitz, L.G. DeShazer, Springer Ser. Opt. Sci., Vol. 52 (Springer, Berlin, Heidelberg 1986) p. 252
4.27 I. Schneider, M.J. Marrone: Opt. Lett. **4**, 390 (1979); ibid. **5**, 214 (1980); ibid. **6**, 627 (1981). I. Schneider, C.R. Pollock: J. Appl. Phys. **54**, 6193 (1983)
4.28 I. Schneider, S.C. Moss: Opt. Lett. **8**, 7 (1983)
4.29. L. Isganitis, M.G. Sceats, K.R. German: Opt. Lett. **5**, 7 (1980)
4.30. R. Illingworth, I.S. Ruddock: Opt. Commun. **61**, 120 (1987)
4.31 M.N. Islam, L.F. Mollenauer, K.R. German: in press (1987)
4.32 L.F. Mollenauer, N.D. Vieira, L. Szeto: Opt. Lett. **7**, 414 (1982)
4.33 L.F. Mollenauer, R.H. Stolen: Opt. Lett. **9**, 13 (1984)
4.34 L.F. Mollenauer: unpublished data (1987)
4.35 J. Hegarty, M.D. Sturge: J. Opt. Soc. Am. B **2**, 1143 (1985)
4.36 W. Brya, S. Geschwind: Phys. Rev. Lett. **21**, 1800 (1968)
4.37 J.M. Worlock, S.P.S. Porto: Phys. Rev. Lett. **15**, 697 (1965)
4.38 J.P. Buisson, A. Sadoc, L. Taurel, M. Billardon: In *Light Scattering in Solids*, ed. by M. Balkanski, R. Leite, S.P. Porto (Flammarion, Paris 1975)

4.39 D.S. Pan, F. Lüty: Phys. Rev. B 2, 4252 (1978)
4.40 B. Fritz, J. Gerlach, U. Gross: In *Localized Excitations in Solids*, ed. by R.F. Wallis (Plenum, New York 1968) p. 496
4.41 M. Leblans, W. Joosen, E. Goovaerts, D. Schoemaker: Phys. Rev. B 35, 2405 (1987)
4.42 W. Joosen, M. Leblans, M. Vanhimbeck, H. de Raedt, D. Schoemaker: J. Cryst. Def. Amorph. Solids 16, 341 (1988)
4.43 W. Joosen, E. Goovaerts, D. Schoemaker: Phys. Rev. B 32, 6748 (1985)
4.44 L.W. Hobbs, A.E. Hughes, D. Pooley: Proc. R. Soc. London 332A, 167 (1973); L.W. Hobbs: J. de Phys. 37, C 7–3 (1975)
4.45 S. Lefrant, E. Rzepka: J. de Phys. 41, C 6–476 (1980)
4.46 E. Rzepka, S. Lefrant, L. Taurel, A.E. Hughes: J. Phys. C 14, L 764 (1981)
4.47 E. Rzepka, J.L. Doualan, S. Lefrant, L. Taurel: J. Phys. C 15, L 119 (1982)
4.48 L. Taurel, E. Rzepka, S. Lefrant: In Proc. IVth Europhysical Topical Conf. on Lattice Defects in Ionic Solids, ed. by B. Henderson, J. Corish: Radiat. Eff. 72, 115 (1983)
4.49 J.D. Comins, A.M.T. Allen, P.J. Ford, D.A. Matthews: In Proc. IVth Europhysical Topical Conf. on Lattice Defects in Ionic Solids, ed. by B. Henderson, J. Corish: Radiat. Eff. 72, 107 (1983)
4.50 M. Aguilar, F. Jaque, F. Agullo-Lopez: Radiat. Eff. 71, 215 (1982)
4.51 T.W. Hänsch, I.S. Shahin, A.L. Schawlow: Phys. Rev. Lett. 27, 707 (1972); Nature 235, 63 (1972)
4.52 P.M. Selzer: In *Laser Spectroscopy of Solids*, ed. by W.M. Yen, P.M. Selzer, Topics Appl. Phys., Vol. 49, 2nd ed. (Springer, Berlin, Heidelberg 1988) Chap. 4
4.53 D.B. Fitchen: In *Physics of Colour Centres*, ed. by W.B. Fowler (Academic, New York 1968) p. 293
4.54 A.E. Hughes, W.A. Runciman: Proc. R. Soc. London 86, 615 (1965)
4.55 G. Bauman, F. Lanzl, W. von der Osten, W. Waidelich: Z. Phys. 197, 367 (1966)
4.56 R.M. Macfarlane, R.M. Shelby: Phys. Rev. Lett. 42, 788 (1979)
4.57 R.M. Macfarlane, R.T. Harley, R.M. Shelby: In Proc. IVth Europhysical Topical Conf. on Lattice Defects in Ionic Solids, ed. by B. Henderson, J. Corish: Radiat. Eff. 72, 1 (1983) and references therein
4.58 G. Baumann: Z. Phys. 203, 464 (1967)
4.59 M.D. Levenson, R.M. Macfarlane, R.M. Shelby: Phys. Rev. B 22, 4915 (1980)
4.60 W.E. Moerner, F.M. Schellenberg, G.C. Bjorklund: Appl. Phys. B 28, 263 (1982)
4.61 W.E. Moerner, P. Pokrowsky, F.M. Schellenberg, G.C. Bjorklund: Phys. Rev. B 33, 5702 (1986)
4.62 C. Ortiz, R.M. Macfarlane, R.M. Shelby, W. Lenth, G.C. Bjorklund: Appl. Phys. 25, 87 (1981)
4.63 A. Winnacker, R.M. Shelby, R.M. Macfarlane: Opt. Lett. 10, 350 (1985)
4.64 H.W.H. Lee, M. Gehrtz, E.E. Marino, W.E. Moerner: Chem. Phys. Lett. 118, 611 (1985)
4.65 R.T. Harley, R.M. Macfarlane: Unpublished data (1983)
4.66 R.T. Harley, M.J. Henderson, R.M. Macfarlane: J. Phys. C 17, L 233 (1984)
4.67 N.R.S. Reddy, Z. Hasan, N.B. Manson: J. Phys. C, in press (1987)
4.68 B. Henderson, S.E. Stokowski, T.C. Ensign: Phys. Rev. 183, 826 (1969)
4.69 P. Edel, C. Hennies, Y. Merle d'Aubigné, R. Romestain, Y. Twarowski: Phys. Rev. Lett. 28, 1268 (1972)
4.70 Le Si Dang, Y. Merle d'Aubigné, Y. Rasoalarison: J. de Phys. 39, 760 (1978)
4.71 J. Cibert, P. Edel, Y. Merle d'Aubigné, R. Romestain: J. de Phys. 40, 1149 (1979)
4.72 J. Bontemps-Moreau, A.C. Boccara, P. Thibault: Semicond. Insul. 3, 165 (1978)
4.73 C.J. Krap, M. Glasbeek, J. Van Voorst: Phys. Rev. B 17, 61 (1978); M. Glasbeek, J.W. Van Voorst: Phys. Rev. B 17, 4895 (1978)
4.74 F.J. Ahlers, F. Lohse, J.M. Spaeth: J. Lumin. 24/25, 359 (1981)
4.75 D.J. Bradley: J. Phys. Chem. 82, 2259 (1978)
4.76 D.H. Auston, A.M. Johnson, P.R. Smith, J.C. Bean: Appl. Phys. Lett. 37, 371 (1980)
4.77 See e.g. R.L. Fork, F.A. Beisser: Appl. Opt. 17, 3534 (1978)

166   *B. Henderson* and *K. P. O'Donnell:* Laser Spectroscopy of Color Centers

4.78 L.F. Mollenauer, J.M. Wiesenfeld, E.P. Ippen: Radiat. Eff. **72**, 73 (1983);
J.M. Wiesenfeld, L.F. Mollenauer, E.P. Ippen: Phys. Rev. Lett. **47**, 1668 (1981)
4.79 A.M. Stoneham, R.H. Bartram: Solid State Commun. **21**, 1325 (1978)
4.80 The results of the triplet state absorption of F_2 centers in KCl are discussed by H. Seidel,
H.C. Wolf in [4.16] and in alkaline earth fluorides by W. Hayes, A.M. Stoneham in
Crystals with the Fluorite Structure, ed. by W. Hayes (Oxford University Press, Oxford
1974)
4.81 K. Park, W.L. Faust: Phys. Rev. Lett. **17**, 137 (1966)
4.82 R.W. Swank, F.C. Brown: Phys. Rev. **135**, A 450 (1964)
4.83 Y. Kondo, H. Kanzaki: Phys. Rev. Lett. **34**, 664 (1975)
4.84 G. Spinolo: In *Defects and Their Structure in Nonmetallic Solids*, ed. by B. Henderson,
A.E. Hughes (Plenum, New York 1976) p. 283
4.85 W.B. Fowler: Phys. Rev. **135**, A 1725 (1964)
4.86 L.F. Mollenauer: Phys. Rev. Lett. **43**, 1524 (1979)
4.87 E. Georgiou, J.F. Pinto, C.R. Pollock: Phys. Rev. B **35**, 7626 (1987)
4.88 H. Mahr: In *Physics of Colour Centres*, ed. by W.B. Fowler (Academic, New York 1968)
p. 243
4.89 D.S. McClure: In *Tunable Solid State Lasers II*, ed. by A.B. Bugdon, L. Esterowitz, L.G.
DeShazer, Springer Ser. Opt. Sci., Vol. 52 (Springer, Berlin, Heidelberg 1986) p. 1

5. Dynamics of Molecular Crystal Vibrations

Dana D. Dlott

With 11 Figures

All energy transfer processes in condensed phase molecules, whether chemical or physical in nature, involve vibrational excitations, mechanical fluctuations and dissipation of vibrational energy. These processes are often explored by exciting a molecular vibration with an external optical field and monitoring the subsequent vibrational relaxation (VR) [5.1]. Such measurements are also sufficient to determine the behavior of spontaneous mechanical fluctuations because these are related to the VR by the fluctuation dissipation theorem [5.2]. Many dynamical processes in solids, such as chemical reactions, electron and exciton transport and heat transport, are strongly affected by the VR properties of the particular material.

Molecular crystals have traditionally fulfilled the role of model systems for understanding the dynamics of molecular excitations in condensed phases. It is a great challenge to understand VR in these systems. In the last few years, there has been considerable progress in the theory of interacting anharmonic systems [5.3] and in experimental methods which must overcome the problems of detecting and resolving short-lived vibrations that are difficult to produce in high concentrations.

Vibrational relaxation measurements can be made in the frequency domain with vibrational spectrometers, or in the time domain with ultrafast lasers. Until a few years ago, ultrafast lasers were so cumbersome that it was not possible to systematically study the dependence of VR on excess vibrational energy or the relationship between VR and molecular structure. The situation has changed dramatically with the development of high power, high repetition rate tunable ultrafast lasers. Several groups have used sophisticated nonlinear optical methods such as ps coherent Raman scattering (ps CARS) [5.4, 5] and ps photon echoes [5.6] to investigate VR processes of a variety of pure [5.7–11] or isotopically disordered crystals [5.12, 13], of impurity molecules in host crystal matrices [5.15, 16], of chemically reactive solids [5.17, 18], and of crystalline polymers [5.19].

On the theoretical front, calculations of phonon and vibron relaxation in naphthalene [5.20, 21] and a few other materials have been made. The ultimate goal of these calculations is an understanding of the dissipation of excess vibrational energy in a crystal, and how this process is governed by the chemical nature of the solid.

In this chapter I shall discuss the nature of VR in four model systems, naphthalene, anthracene, benzene, and durene. Naphthalene and anthracene are prototypical molecular crystals of slightly flexible molecules with dense energy level structures, benzene is quite rigid and has a relatively sparse level structure, and durene (1,2,4,5-tetramethylbenzene) is "floppy benzene", a molecule with four nonrigid methyl groups on the periphery.

First I shall describe experimental techniques and the theory of the dynamical lineshape which is required in order to understand these techniques. Then I shall discuss the nature of the vibrational states in crystals and the theory of anharmonic vibrational relaxation. Recent results on VR in these materials are presented with the idea of understanding the relationship between their structure and VR properties. Finally I shall present new calculations on vibrational energy flow in naphthalene which show for the first time how vibrational energy localized in a single high energy mode is dissipated [5.12].

5.1 Experimental Methods

Experiments which measure VR in solids involve either the measurement of the spectral lineshape, or a time-dependent emission from an excited vibration [5.1]. A simplified energy level diagram for organic molecules is shown in Fig. 5.1. Initially, cold molecules are in the ground singlet S_0^0 state. With optical excitation, vibrations are excited in either the ground electronic state S_0^v or excited electronic S_1^v state. In condensed phases, VR from these states is a fast, nonradiative process shown by the wavy arrows in Fig. 5.1. The S_0^v vibrations can be excited indirectly via the Raman process as shown, or directly by absorption of infrared light. Because Raman cross sections are small, Raman excitation is feasible when the concentration is large, for example in a pure crystal. Infrared absorption is much more efficient, but good sources of tunable ultrashort pulses in the vibrational IR are not yet developed. Another approach is to study vibrations in the S_1 state. When the $S_1 \rightarrow S_1$ transition is intense and the Franck-Condon factors are favorable, the absorption strength to the S_1^v mode can be quite large, and these modes can be directly excited. It is possible to achieve high levels of excitation or to study dilute (ppm) impurities in a matrix.

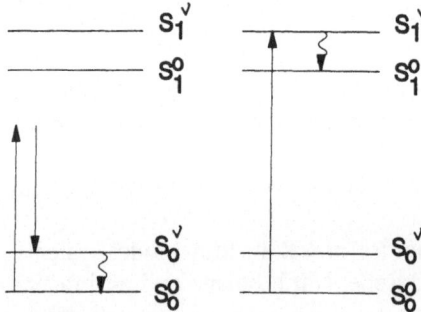

Fig. 5.1. Schematic energy level diagram for VR studies in organic solids. Initially, cold molecules are in the vibrationless electronic ground state S_0^0. Ground state vibrations S_0^v can be indirectly excited by the Raman process (shown), or directly excited with tunable infrared light. The excited vibrations dissipate their excess energy (wavy arrow) in the VR process. Vibrations in the electronically excited state S_1^v can also be directly excited when the $S_0 \rightarrow S_1$ transition is optically allowed

5.1.1 Frequency Domain Experiments

In the 1960s and 1970s a number of workers studied the infrared and Raman spectra of crystalline benzene [5.22] and naphthalene [5.23]. Because the solid state molecular ordering is known from x-ray crystallography, polarized Raman and IR spectroscopy provides a wealth of information about normal mode assignments and perturbations caused by the interacting solid state environment. A number of reviews appeared which described the properties of solid state vibrations, including those by *Robinson* in 1970 [5.24], and *Belusov* [5.25] in 1982. A complete discussion of vibrational spectroscopy in molecular crystals goes beyond the scope of this work. The monograph by *Kitaigorodsky* [5.26] is an excellent source on this subject. I shall restrict the discussion to studies which are primarily concerned with VR processes.

One of the first detailed studies of vibrational dynamics was that of *Prasad* and *Kopelman* [5.27], who used a $1 \, cm^{-1}$ Raman spectrometer to study vibrational perturbations caused by heavy isotopic doping. Subsequently, a $0.1 \, cm^{-1}$ Raman spectrometer was used to measure some phonon and vibron lineshapes in naphthalene [5.28]. It is difficult to achieve better resolution with a spectrometer of conventional design. Resolution of $0.001 \, cm^{-1}$ has been obtained with the CARS technique by *Trout* et al. [5.29], who studied benzene, and by *Decola* et al. [5.30], who studied naphthalene. Equivalent resolution was obtained with a tandem monochromator and scanning interferometer used by *Ranson* et al. [5.31].

Fourier transform IR (FTIR) machines with sub-wavenumber resolution have become quite common [5.32]. Low-temperature FTIR was used by *Ahlgren* and *Kopelman* [5.33] to investigate the IR active phonons and lower frequency vibrations of naphthalene at 77 K and $1 \, cm^{-1}$ resolution. *Kosic* et al. [5.17] studied perylene, a prototype system for solid state chemistry. *Hill* et al. [5.12] studied naphthalene and anthracene vibrational lineshapes and VR with FTIR at 10 K and $0.1 \, cm^{-1}$ resolution.

Vibrational lineshapes can also be measured by persistant hole burning (HB). The resolution is comparable to the megahertz width of the burning and interrogating laser. *Small* [5.34] reviews recent advances in HB. Hole burning experiments are performed on absorbing molecules in a disordered low temperature matrix. Vibrational holes can be created indirectly by irradiation of the electronic origin, or directly by irradiation of an S_1^v state.

De Vries and *Wiersma* [5.35] first used photochemical HB (PHB) to study VR in *s*-tetrazine/benzene mixed crystals. They deduced a VR rate from the (homogeneous) width of the hole. Subsequently *Voelker* and *Macfarlane* [5.36] studied PHB in vibronic bands of free base porphyrin, and *Dicker* and *Voelker* [5.37] deduced a correlation between the VR of porphyrin and the chain length of the *n*-alkane matrix. Vibrational relaxation in chlorophyll was also studied by *Rebane* and *Avarmaa* [5.38]. A recent comparison of the HB process and ps photon echo data [5.39] suggests that the holes are broadened by additional processes related to the mechanism of HB. If this is the case, the lifetimes obtained from HB are probably shorter than the true VR lifetimes.

5.1.2 Time Domain Experiments

The most commonly used time resolved technique is ps stimulated Raman scattering (SRS). In the first experiments, reviewed by *Laubereau* and *Kaiser* [5.1], an immensely intense (>1 GW/cm^2) ultrashort laser pulse at ω_L was used to coherently excite a Raman active mode in a liquid or solid. The mode with the largest Raman cross section is selectively excited. The subsequent decay of coherence is monitored via the intensity of coherent anti-Stokes scattering stimulated by a time delayed ω_L pulse. In complex materials with many vibrations, the selective excitation is a problem. The SRS technique can be improved by using two excitation lasers. In this case, any Raman active vibration can be excited by tuning the frequencies of the two lasers so that $\omega_1 - \omega_2 = \Omega$, where Ω is the frequency of a material vibration. The two-laser technique is usually called ps CARS [5.40]. With ps CARS, it is possible to compare the VR of several modes in the same material [5.4, 5, 7].

Low temperature naphthalene was studied with ps CARS by *Hesp* and *Wiersma* [5.7a], and *Duppen* et al. [5.7b]. Low temperature benzene was studied by *Ho* et al. [5.14]. *Dlott* et al. [5.9] and *Schosser* and *Dlott* [5.10] studied the temperature dependence of VR in naphthalene and anthracene. The effects of isotopic disorder have been investigated in naphthalene by *Chronister* et al. [5.13] and in benzene by *Ho* et al. [5.14] and *Trout* et al. [5.29]. More recently crystals with biological significance such as amino acids, peptides, and a crystalline protein, lysozyme, were studied by *Kosic* et al. [5.11], and co-crystals of nucleic acid base pairs were investigated by *Chronister* and *Dlott* [5.41].

Femtosecond Raman techniques have recently come on to the scene. The theory of femtosecond pulse interactions has been discussed by *Yan* et al. [5.42] and demonstrated in the work of *De Silvestri* et al. [5.18]. Because the frequency spread of a femtosecond pulse is so large (often >100 cm^{-1}), low frequency vibrations can be excited with a single pulse. Unlike ps CARS, the impulsively stimulated Raman technique (ISRS) requires only one laser because both frequencies necessary to excite a vibration are present in the same pulse [5.42].

Under certain circumstances, vibrational lineshapes are dominated by inhomogeneous broadening. This is almost always the case for S_1^{v} states of molecules in host matrices, because the resonance frequency of the $S_0 \rightarrow S_1$ transition is very sensitive to local inhomogeneities. It is possible to remove the inhomogeneous contribution with a photon echo pulse sequence [5.6]. Raman echoes [5.43, 44], which could be used to study S_0^{v} modes with inhomogeneous broadening have been discussed and demonstrated on N_2 vapor and liquid N_2/Ar mixtures [5.44], but not on molecular crystals. This technique could be very useful for disordered systems which show inhomogeneous broadening in the electronic ground state. Such systems include crystalline N_2, H_2, and CO_2 [5.45–47], crystalline polymers [5.19], and glasses.

The photon echo technique was used by *Hesselink* and *Wiersma* [5.15] to study VR of pentacene S_1^{v} modes in a crystalline naphthalene matrix. The effects of deuteration on host and matrix were studied. Subsequently *Hill* et al. [5.16]

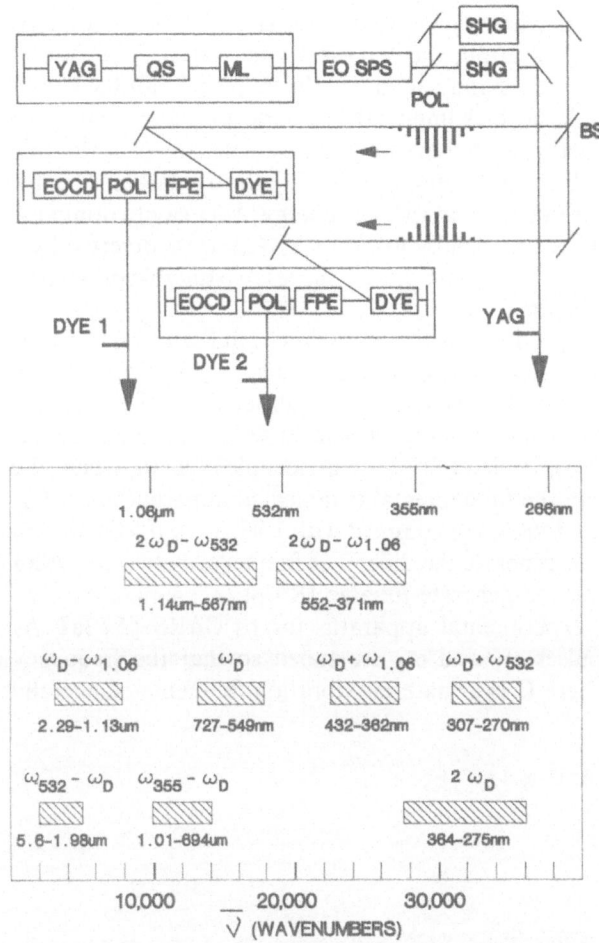

Fig. 5.2. High repetition rate, high power laser system used for VR studies in the author's laboratory. A ps pulsed Nd:YAG laser pumps two synchronous dye lasers, DYE 1 and DYE 2. These two independently tunable pulses and the harmonics of the YAG laser can be mixed in nonlinear crystals to span the frequency ranges indicated in the chart (0.266–5.6 μm), which is reproduced with permission from [5.50]. YAG: (Nd:YAG) laser at 1.064 μm; QS: acoustooptic Q-switch; ML: acoustooptic mode locker; EO SPS: electrooptic single pulse selector; POL: polarizing beamsplitter; SHG: second harmonic generating crystal; BS: thin film beam splitter; DYE: flowing dye cell; FPE: Fabry-Perot etalons; EOCD: electrooptic cavity dumper

used photon echoes to study VR in naphthalene/durene and anthracene/naphthalene mixed crystals.

An advanced laser system used in my laboratory [5.48] to perform ps CARS and photon echo experiments is shown in Fig. 5.2. The system is based on a continuously pumped Nd:YAG laser with acoustooptic mode locker and Q-switch [5.49, 50]. This laser is a highly stable source of Gaussian pulse bursts at 1.06 μm, with a repetition rate variable up to 1000 Hz. Each burst contains about 25 pulses of 100 ps duration, the largest about 100 μJ. The second harmonic at 0.532 μm is generated and one-half of the frequency-doubled burst is used to pump each dye laser. It is important to perform experiments with single pulses, so at a carefully chosen moment, signals are sent to two electrooptic modulators which cavity dump [5.50–52] the dye lasers. The dye lasers then emit single 30 ps pulses of up to 25 μJ energy. After cavity dumping, dye laser action ceases and the next pulse from the YAG laser is diverted by a third electrooptic modulator. The system thus furnishes two tunable pulses and a very intense pulse at 1.06 μm.

The chart in Fig. 5.2 is reproduced with permission from *Patterson* [5.50]. It shows a variety of nonlinear frequency mixing possibilities which are used to frequency shift the pulses from the dye lasers throughout the visible, near-IR and UV with efficiency of at least a few percent. The powerful 1.06 μm pulse can be frequency doubled (70%), tripled (30%), or quadrupled (20%), where the percentage conversion from the fundamental is indicated in parentheses. The dye lasers can be frequency doubled or summed with 1.06 μm or 0.532 μm into the UV. It is also possible to generate the difference frequency between a YAG laser pulse and the dye laser to generate tunable IR out to 5.6 μm.

Figure 5.3 shows the experimental apparatus for ps CARS [5.13a]. As shown in Fig. 5.1, two pulses ω_1 and ω_2 are tuned so that the frequency difference $\omega_1 - \omega_2 = \Omega$, where Ω is a material vibration frequency. Generally

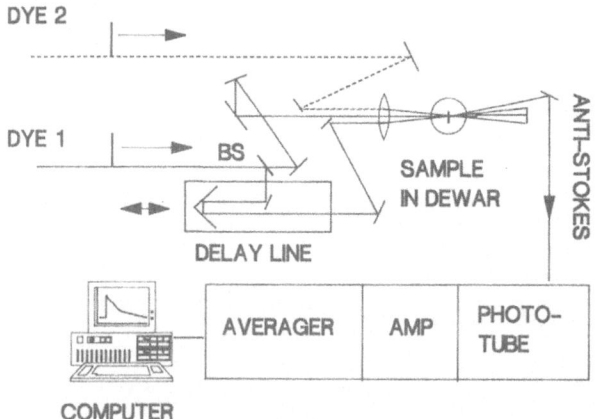

Fig. 5.3. ps CARS apparatus used in the author's laboratory. Two tunable pulses (DYE 1, DYE 2) at frequencies ω_1, ω_2 are tuned so that $\omega_1 - \omega_2 = \Omega$, where Ω is the frequency of a Raman active vibration. The solid sample in the dewar is excited by the pair of simultaneous pulses at zero time. A portion of the ω_1 pulse split and sent down a motorized delay line is used to probe the coherence remaining at time t. The intensity of anti-Stokes emission stimulated by the probe is measured versus delay. The ps CARS decay is given by (5.5)

these pulses originate from the dye lasers, but if resonance enhancement is required, the lasers may be frequency shifted into resonance with a molecular excited state. The decay of the vibrational coherence is probed at time greater than zero by a pulse which travels down a motorized rapid scanning delay line. The experiment involves detection of the intensity of coherent emission stimulated by this delayed pulse versus delay. Because of the high repetition rate and stability of the laser, it is possible to average many hundreds of scans to obtain high-quality data. Photon echo experiments are generally performed with a single laser frequency tuned to excite the S_1^v states in Fig. 5.1. The apparatus is the same as in ps CARS except the ω_2 pulse is eliminated.

A few other time domain techniques have been used to study VR in complex solids. Transient grating vibrational scattering has been used by *Weitekamp* et al. [5.53] to study VR of an S_0^v mode of pentacene in naphthalene. An infrared pump-probe measures VR by saturating an S_0^v mode with a ps IR pulse and probing the recovery of IR absorption. *Heilweil* et al. [5.54] have used this technique to measure VR of OH groups of solids.

For molecules with high fluorescence quantum yields in S_1^v states, it is possible to time resolve the fluorescent emission with a streak camera. *Freiberg* and *Saari* [5.55] observed VR of highly excited luminescing molecules in a low temperature matrix by detecting the time resolved red shift of fluorescence with a synchronously pumped dye laser and synchroscan camera. This technique provides a complete emission spectrum at various delay times and was termed "picosecond spectrochronography". A similar system was used by *Jang* et al. [5.56] and *Brucker* and *Kelley* [5.57] to resolve emission from hot vibrations formed during excited state intramolecular proton transfer.

5.1.3 Steady State Luminescence Experiments

The kinetics of short-lived vibrational states can sometimes be determined in steady state experiments. Typically luminescent molecules in a host matrix are irradiated continuously or quasi-continuously. Relaxation rates are determined from the observed steady state luminescence intensity.

Rabane and *Saari* [5.58] have performed many nice studies of "hot" luminescence in mixed molecular crystals. Luminescing molecules are irradiated in S_1^v states. These states relax to lower energy $S_1^{v'}$ states where the luminescence to S_0 is monitored. Knowing the absorption strength of each state, it is then possible to relate the lifetime to the steady state emission intensity [5.59]. This method was also used by *Hochstrasser* and *Nyi* [5.60] in their study of azulene in naphthalene. The same authors used an analogous technique to measure the rate of excimer formation in perylene [5.61]. Excimer formation is a fast, nonradiative photodimerization in the S_1 state [5.62]. In that work, the relative intensity of steady state fluorescence and resonance Raman were used to determine an ultrafast VR rate of order $10^{13}\,\mathrm{s}^{-1}$.

With intense pumping from long ($>10\,\mathrm{ns}$) laser pulses, it is possible to observe stimulated emission [5.63] from organic guest molecules in crystalline host matrices. The stimulated emission threshold is inversely proportional to the lifetime. Several groups have used this technique to study VR in systems such as perylene or tetracene in anthracene, or anthracene in fluorene [5.64]. The drawback of this technique is that stimulated emission depends a great deal on crystal geometry and quality, and is only observed at extremely large intensities and high concentrations.

5.2 The Dynamic Vibrational Lineshape

The optical transitions shown in Fig. 5.1 are characterized by the center frequency and the lineshape function. Conventional vibrational spectroscopy is mainly concerned with the center frequencies which are characteristic of each normal mode [5.65]. The lineshape function is a source of dynamical information and reflects the coupling between a transition and a "bath" composed of other modes, either on the same molecule or in the surrounding matrix [5.66]. Because vibrations are of primarily internal character, it is usually possible to treat the vibrational dynamics with a model which assumes weak coupling between a mode and the bath [5.1, 67].

A situation which is of great importance in low temperature vibrational spectroscopy occurs when the coupling between the mode and bath is weak and consists of only "fast" and "slow" terms. *Kubo* [5.68] described the problem by assuming that the oscillator has a time-dependent frequency. In the separation of time scales approximation, the time dependence consists of two terms:

$$\Omega(t) = \Omega_0 + \Delta\omega_{\text{fast}}(t) + \Delta\omega_{\text{slow}}(t). \tag{5.1}$$

The time-dependent frequency is caused by stochastically varying interactions between the bath and an ensemble of oscillators. These fluctuations have characteristic relaxation rates Γ_{fast} and Γ_{slow}. The fast and slow limits are realized when $\Gamma_{\text{fast}} \gg \sqrt{\langle\Delta\omega_{\text{fast}}^2\rangle}$, and $\Gamma_{\text{slow}} \ll \sqrt{\langle\Delta\omega_{\text{slow}}^2\rangle}$, where the brackets $\langle\ \rangle$ indicate time average. From the point of view of the oscillator, fast interactions cause a rapid, significant variation of frequency. Because the variation is fast, the time-averaged effect is the same for all oscillators, causing *homogeneous broadening*. The homogeneous lineshape is a Lorentz lineshape function with FHWM (full width at half maximum) $1/\pi T_2$. The slow interactions appear static, and generally different for each oscillator in the ensemble. These environments are distinguishable, and thus slow interactions cause *inhomogeneous broadening*. The inhomogeneous lineshape is typically Gaussian with standard deviation $\sqrt{\langle\Delta\omega_{\text{slow}}\rangle^2} = \Delta$.

The homogeneous broadening parameter T_2 is generally separated into two parts [5.1],

$$(T_2)^{-1} = (\tau_{\text{PH}})^{-1} + (2T_1)^{-1}, \tag{5.2}$$

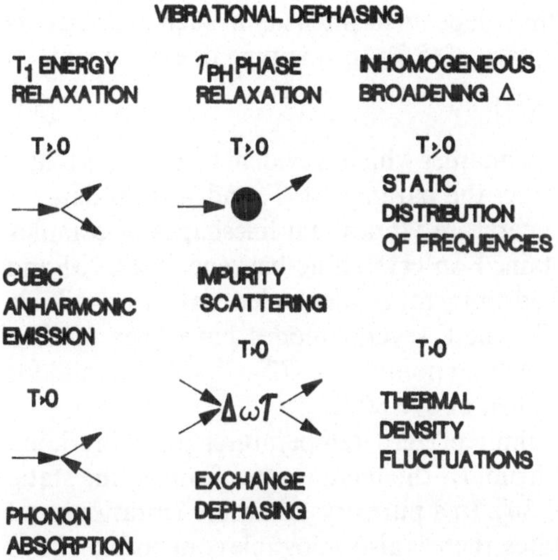

VIBRATIONAL DEPHASING

T_1 ENERGY RELAXATION	τ_{PH} PHASE RELAXATION	INHOMOGENEOUS BROADENING Δ

Fig. 5.4. A summary of the T, τ_{PH}, and inhomogeneous broadening processes

where T_2 is the time constant for all phase relaxation processes, T_1 is the vibrational lifetime, and τ_{PH} is the time constant for pure dephasing due to fast motions of the bath.

Figure 5.4, which is a modified version of Table 1 from *Swiatkiewicz* et al. [5.19], summarizes the dominant T_1, τ_{PH}, and inhomogeneous processes in solids. T_1 processes, described in Sect. 5.5, are caused by anharmonic interactions which permit an excited vibration to relax by simultaneous emission of two or more different vibrations. The τ_{PH} processes are caused by interactions with impurities or thermally excited modes of the bath. Inhomogeneous broadening is caused either by static disorder, or by thermally activated slow fluctuations of local density or vibrational population.

For an oscillator with normal coordinate q and center frequency Ω_0, the time correlation function of the oscillator is given by [5.5, 67]

$$\frac{\langle q(0)q(t)\rangle}{\langle q(0)q(0)\rangle} = \exp\left(-i\Omega_0 t + \frac{t}{2T_1} + \frac{t}{\tau_{PH}} + \frac{1}{2}\Delta^2 t^2\right). \qquad (5.3)$$

The lineshape of the optical transition is proportional to the Fourier transform of (5.3) and is given by

$$I(\omega) \sim \int_{-\infty}^{+\infty} dt \exp\left[-\left(i(\omega-\Omega_0)t + \frac{t}{2T_1} + \frac{t}{\tau_{PH}} + \frac{1}{2}\Delta^2 t^2\right)\right]. \qquad (5.4)$$

The convolution of Lorentzian and Gaussian contributions in (5.4) is called a *Voight* lineshape function [5.67, 69].

In coherent time resolved techniques such as ps CARS, the time domain equivalent of (5.4) is measured [5.4, 5]. Most of these measurements use square-law (intensity) detectors and the time-dependent intensity in such experiments is proportional to the absolute square of (5.3). In this type of experiment,

$$I(t) \sim \exp[-(t/T_1 + 2t/\tau_{PH} + \Delta^2 t^2)]. \tag{5.5}$$

An advantage of time domain techniques which is evident from (5.5) is that it is computationally simpler to extract the parameters T_2 and Δ from data.

The best data on the low temperature vibrational lineshapes of complex molecular crystals have been obtained on crystalline benzene [5.29, 70] and naphthalene [5.7, 9, 10, 12]. Detailed temperature-dependent data are available for naphthalene [5.9, 10, 28, 71], where several modes have been studied between low temperature and the melting point [5.10, 72], $T_m = 353$ K, and for durene [5.73], which has been studied above 80 K.

Figure 5.4 shows that in the limit of zero temperature, the vibrational lineshape contains contributions from T_1, the lifetime, and from Δ, the static spread of oscillator frequencies [5.30]. In a pure crystal, where vibrations and phonons form delocalized band states, there is also a possible contribution from τ_{PH} caused by intraband impurity scattering. Another possible intraband scattering process, the spontaneous emission of phonons, may be neglected in these systems [5.29, 30, 74] at low temperatures.

In the pure crystal there is an interaction between Δ and the amplitude for intermolecular excitation transfer β. When $\beta > \Delta$, motional narrowing of the inhomogeneous linewidth will occur [5.30], yielding a motionally narrowed contribution to the line of magnitude Δ^2/β. The parameter β can be estimated from the factor group splitting; it is typically 0.1–5.0 cm^{-1}. The parameter Δ is estimated by noticing that for electronic transitions, the gas-to-crystal shift is typically $> 10^2$ cm^{-1}, and the value of Δ, which is the variance of the gas-to-crystal shift, is typically 10^0 cm^{-1}. For vibrations, the gas-to-crystal shift is $\sim 10^1$ cm^{-1}, so the value of Δ for vibrations is 10^{-1}–10^{-2} cm^{-1}, and the ratio Δ^2/β is $< 10^{-3}$ cm^{-1} [5.7a, 30].

The motionally narrowed linewidth of about 10^{-3} cm^{-1} corresponds to a time constant of about 5 ns. In the typical case where $T_1 \ll 5$ ns, the low temperature lineshape should be *homogeneously broadened*, with width $2/T_1$. In this case, measurements of the linewidth should yield directly the vibrational lifetime T_1 [5.4, 5, 29, 30].

Ps CARS studies of low temperature naphthalene phonons and vibrations do indeed show nearly perfect exponential decays [5.4–14]. It is possible to observe nonexponential decays in a few cases [5.13, 45–47], for example the long-lived vibrons in diatomic molecules, or when the sample is heavily doped with impurities. Figure 5.5 from *Chronister* and *Dlott* [5.13a] shows ps CARS decays for v_5 ($\Omega = 1385$ cm^{-1}) of naphthalene crystals which are heavily doped with deuteronaphthalene at low temperatures. While exponential decays are observed at "low" dopant concentrations $C_d < 0.5$, nonexponential decays are observed at $C_d = 0.75$, indicating a substantial contribution from inhomogeneous broadening. In this case the vibron is partially localized by impurities. This localization occurs when $\Delta > \beta$ [5.68], where for v_5, $\beta = 0.4$ cm^{-1} [5.30].

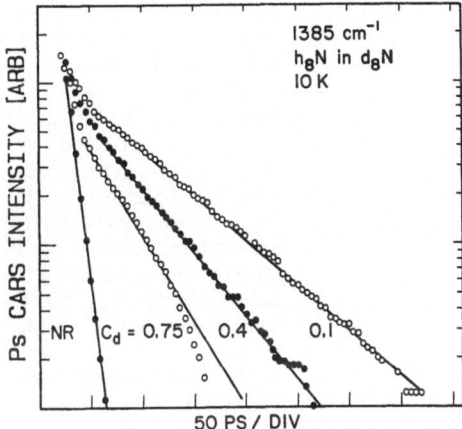

Fig. 5.5. ps CARS decays from [5.13 a] taken on v_5 of h_8N, i. e. naphthalene, ($\Omega = 1385$ cm^{-1}) doped with a fraction, C_d, of d_8N, i. e. perdeuteronaphthalene. At low concentrations of $d_8N(C_d < 0.5)$, the ps CARS decays are exponential, but at high doping levels, a nonexponential contribution due to partially localized states is observed. The data marked NR (nonresonant susceptibility) gives the response of the apparatus, which is about 12 ps

When the optical transition involves S_1^v states of low temperature isolated impurities in a host matrix, the motional narrowing effect does not occur. Because Δ is typically 10^0 cm^{-1} in these systems, conventional spectroscopy will not give VR information, but rather will just probe the inhomogeneous broadening. However, the echo techniques [5.15] remove the effect of Δ, revealing the homogeneous line. As above, at low temperatures the line broadening is dominated by T_1 processes [5.15].

Localization can also occur at high temperatures due to scattering by phonons or low frequency vibrations [5.7a, 10, 28]. As the temperature is raised, slowly varying fluctuations induced by thermal excitations cause Δ to increase. Eventually the limit $\Delta(T) > \beta$ is reached and motional narrowing vanishes. In the limit where $(T_2)^{-1} < \Delta$, the lineshape should change from a Lorentzian to a Gaussian curve of width Δ at the localization temperature T_L where $\Delta(T) = \beta$ [5.68, 75]. In practice, the effect of T_2 is rarely negligible for high temperature vibrations, so the lineshapes are Voightians above T_L. *Schosser* and *Dlott* [5.10] used Raman spectroscopy to measure the lineshapes of naphthalene vibrations at 298 K and near the melting point. They observed Voight lineshapes for several totally symmetric modes with $\Delta \sim 1.0$ cm^{-1} at these temperatures. The inhomogeneous broadening is consistent with the measured factor group splittings which give $\beta < 1$ cm^{-1} for these particular modes. The localization temperature T_L was 200–250 K.

It is possible to study the temperature dependence of the homogeneous linewidth in pure naphthalene up to the melting point. At low temperatures, inhomogeneous broadening is absent, and at high temperature, it can be separated by numerical deconvolution. Figure 5.6 is an Arrhenius plot of the *increase* in linewidth $\Delta v(T) - \Delta v(T=0)$ versus temperature using data for $v_9(\Omega = 511$ cm^{-1}) obtained by *Schosser* and *Dlott* [5.10] (filled squares). When an Arrhenius plot is linear, it indicates that only a single low energy mode is involved in the temperature dependence, and the slope of the plot gives the frequency of that mode [5.73]. In Fig. 5.6, the plot is nonlinear with a slope

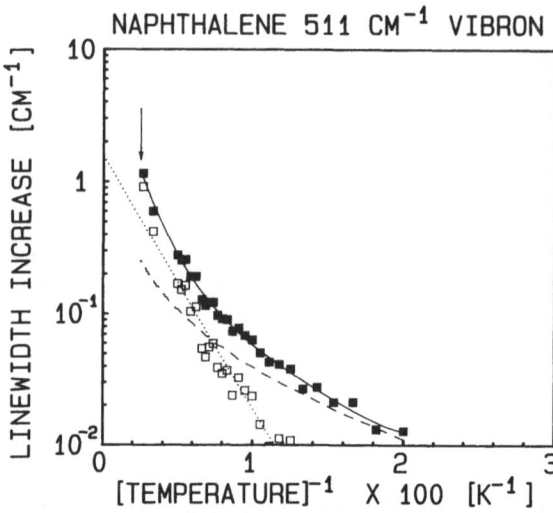

Fig. 5.6. Temperature dependence of homogeneous linewidth of naphthalene v_9 ($\Omega = 511\,\text{cm}^{-1}$). (■) Data from [5.10]; (—) a visual guide; (— —) the calculated contribution of T_1 (energy dissipation) to the linewidth [5.10]; (□) the difference between the linewidth increase and the T_1 process. The difference is caused by the τ_{PH} process which dominates at high temperatures. The arrow indicates the melting point, $T_m = 353\,\text{K}$

which increases with temperature. This type of plot is indicative of a process in which a spectrum of low energy modes participate. The slope increase is caused by the average thermal excitation moving to progressively higher frequency with increasing temperature [5.10].

The *broken line* in Fig. 5.6 is a calculation of the temperature-dependent T_1 using (5.7), the experimental density of states and the observed low temperature value of $T_1 = 140$ ps [5.10]. The data are in agreement with the T_1 model up to about 80 K, but above this temperature the linewidth increases faster than predicted if T_1 were the only homogeneous relaxation mechanism. The remainder of the dephasing, τ_{PH}, found by subtracting the calculated value of T_1 from the data is given by the *open squares*. The τ_{PH} process gives an approximately linear plot *(dotted line)* which yields a thermal activation energy of $300 \pm 50\,\text{cm}^{-1}$. It is evident that the τ_{PH} process, which is negligible at low temperatures, accounts for essentially all of the homogeneous linewidth at ambient temperature. A mechanism for the τ_{PH} process was proposed by *Harris* et al. [5.73].

This model is depicted schematically in Fig. 5.7. The dephasing of high frequency mode Ω is caused by energy exchange with low frequency mode(s) E (for exchanging modes), via quartic anharmonic interactions of the form $Q^2 Q_E^2$. When $\hbar\omega_E/k_B T \sim 1$, thermal population of E modes, which are in equilibrium with the "bath", will occur with excitation rate W_+ and lifetime τ. Whenever an E mode is excited, the anharmonic interaction causes the $G \rightarrow \Omega$ transition frequency to shift by an amount $\delta\omega$ for a time τ. This interaction results in a temperature-dependent frequency shift of Ω and a temperature-dependent line broadening in the intermediate exchange regime where $\delta\omega\tau \sim 1$ given by [5.73a]

$$\Omega^{\text{eff}} = \Omega + \delta\omega/[1 + (\delta\omega)^2\tau^2]\exp(-\hbar\omega_E/k_B T), \tag{5.6a}$$

$$(\tau_{PH})^{-1} = ((\delta\omega)^2\tau/[1 + (\delta\omega)^2\tau^2]\exp(-\hbar\omega_E/k_B T). \tag{5.6b}$$

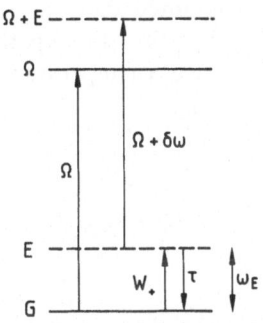

Exchange model:
Durene 2932 cm⁻¹

Fig. 5.7. Vibrational energy exchange model applied to the (crystalline) durene 2932 cm⁻¹ vibration. The $G \to \Omega$ transition is dephased by energy exchange with the E mode(s) which are in thermal equilibrium with the bath. Thermal excitation of an E mode with rate W_+ and lifetime τ frequency shifts the $G \to \Omega$ transition by an amount $\delta\omega$ for a time τ, resulting in a thermally activated frequency shift and linewidth. The data, reproduced with permission from *Harris* et al. [5.73a], show that the same activation energy is observed for the width and shift. The frequency, 227 cm⁻¹, identifies the E mode as a methyl torsion

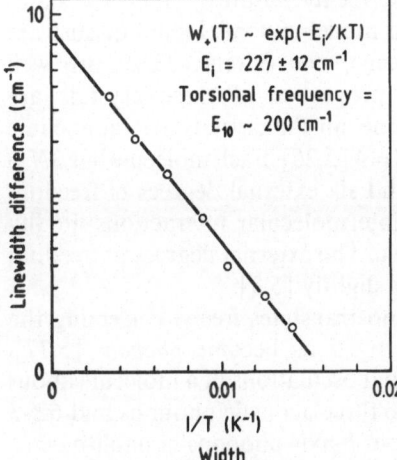

Width

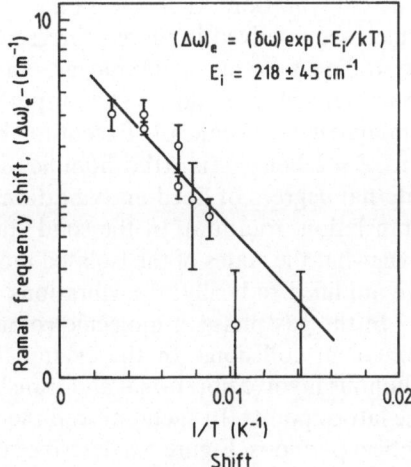

Shift

Equations (5.6) predict that the activation energy $\hbar\omega_E$ observed for the line shift and line broadening should be identical. Experimental observations were made on C–H (methyl) stretching modes of crystalline durene as a function of temperature. The selection of these modes was a very good idea because they are localized on an isolated region of the molecule, and thus would be expected to couple preferentially to the low frequency methyl torsional modes. Figure 5.7 (reproduced with permission from *Harris* et al. [5.73a]) shows Arrhenius plots of the width and shift of the 2932 cm⁻¹ C–H stretching mode of durene. It is clear that the same activation energy, $\omega_E = 227 \pm 12$ cm⁻¹, is observed for both quantities. This energy is almost exactly equal to an observed torsional frequency, which was identified as the dominant E mode.

In extending this energy exchange theory to naphthalene, it is difficult to argue that the skeletal stretching modes of naphthalene should be coupled to only one E mode. Instead one would expect coupling to many or all of the low frequency skeletal deformations. *Hiroke* [5.76] extended the exchange formalism to include the effects of multiple E modes. *Schosser* and *Dlott* [5.10] applied this model to naphthalene. Figure 5.6 shows that the τ_{PH} process has an apparent activation energy of $\omega_E = 300 \pm 50 \text{ cm}^{-1}$. It was noted that naphthalene has no vibrations between 200 and 362 cm^{-1}, but that the equal contribution of all modes between 200 and 500 cm^{-1} yielded a temperature dependence in very good agreement with experiment [5.10].

5.3 Crystal Vibrational States

In the VR process, an excited vibration will relax by emission of two or more different vibrations. In this section I will describe the nature of the crystal states which are involved in these processes. The model systems I shall discuss are crystalline benzene, naphthalene, anthracene, and durene. These are well characterized materials. The important properties of these crystals are summarized in Table 5.1. Except for benzene, all the crystals are monoclinic with $Z = 2$. Benzene is orthorhombic with $Z = 4$ [5.26]. Each molecule has $3N-6$ internal degrees of freedom (vibrations), and six external degrees of freedom (translation, rotation). In the solid state, intermolecular interactions modify somewhat the states of the isolated molecule. The external degrees of freedom are modified radically; the vibrations only slightly [5.3].

In the gas phase, a molecule rotates and translates freely, neglecting the infrequent collisions. In the crystal these motions become *phonons* [5.77]. Phonons involve librational and translational oscillations of a molecule about the lattice points. In each unit cell there are three acoustic phonons and $6Z-3$ optical phonons. Figure 5.8 describes the four *b*-axis phonons in naphthalene. The acoustic phonon involves in-phase translation of the pair in the unit cell

Table 5.1. Properties of aromatic molecular crystals [5.12]

Molecule	Formula	Factor group	Z	$V_0[\text{Å}^3]$	Vibrations/unit cell
Benzene	C_6H_6	D_{2h}^{15}	4	508	120
Naphthalene	$C_{10}H_8$	C_{2h}^5	2	363	108
Anthracene	$C_{14}H_{10}$	C_{2h}^5	2	474	132
Durene	$C_{10}H_{14}$	C_{2h}^5	2	460	132

Molecule	Vibrations [cm^3]	Acoustic phonons [cm^3]	Optical phonons [cm^3]	Ω_{max}
Benzene	2.36×10^{23}	5.91×10^{21}	4.14×10^{22}	135 cm^{-1}
Naphthalene	2.98×10^{23}	8.26×10^{21}	2.48×10^{22}	180 cm^{-1}
Anthracene	2.78×10^{23}	6.33×10^{21}	1.90×10^{22}	170 cm^{-1}
Durene	2.87×10^{23}	6.52×10^{21}	1.96×10^{22}	380 cm^{-1}

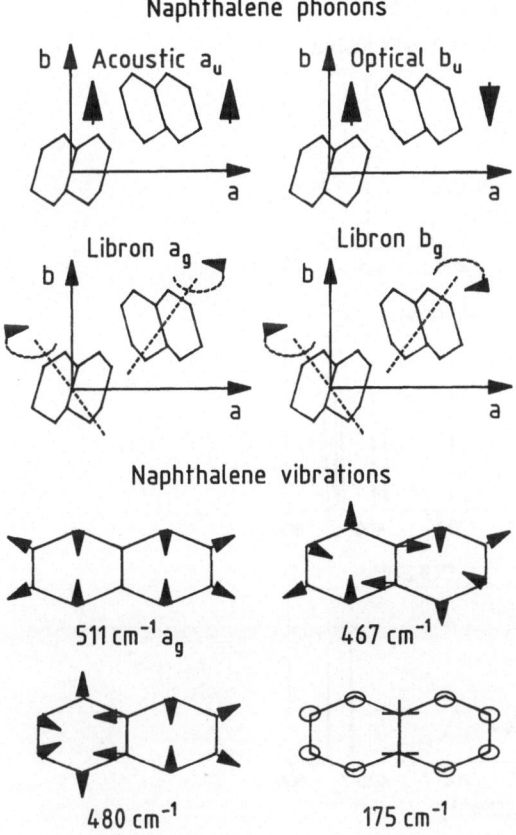

Naphthalene phonons

Naphthalene vibrations

Fig. 5.8. Some phonons and vibrations of crystalline naphthalene. Phonons involve translational and librational oscillations of the entire molecule; vibrations are largely internal deformations of the molecule. The $175\,\mathrm{cm}^{-1}$ "butterfly" motion is a low frequency, large amplitude vibration which is amalgamated into the phonons

while the translational optic phonon involves out-of-phase translation. The two librons involve in- and out-of-phase librations of the pair [5.23d].

Each (nonlinear) molecule of N atoms has $3N$-6 fundamental vibrations. A few of the low frequency modes of naphthalene are shown in Fig. 5.8 [5.23a]. Each vibration of the crystal is split into MZ-fold states, where M is the number of unit cells. However, in molecular crystals where the intermolecular potential is much weaker than the intramolecular potential, these splittings are small, typically 10^{-2}–$10^{-3}\,\mathrm{cm}^{-1}$. The interaction between the Z molecules within the unit cell creates Z *factor group components* which can often be distinguished in polarized light. The interaction between the M translationally equivalent molecules splits each component into a narrow *vibron band*. The factor group splitting is roughly equal to the vibron dispersion.

The $511\,\mathrm{cm}^{-1}$ mode shown in Fig. 5.8 is totally symmetric and has a large Raman cross section. It interacts very weakly ($<0.3\,\mathrm{cm}^{-1}$) with its neighbors and is a good approximation of a localized mode. The $467\,\mathrm{cm}^{-1}$ and $480\,\mathrm{cm}^{-1}$ modes are planar skeletal distortions and are respectively Raman and IR active. Emission of these modes and an accompanying phonon is an important VR

DENSITIES OF STATES

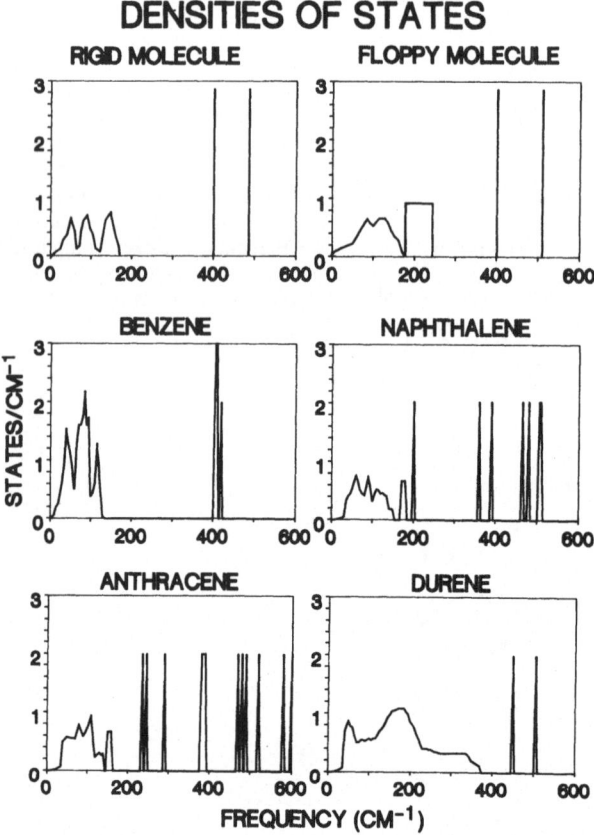

Fig. 5.9. One-phonon density of states calculated from neutron and vibrational spectroscopic data by *Hill* et al. [5.12]. In a very rigid molecule there is a gap separating the phonons (at low frequencies) from the vibrations (sharp spikes). In a "floppy" molecule, the gap narrows. Benzene is highly rigid, naphthalene and anthracene are intermediate, and durene is floppy

mechanism for 511 cm^{-1} [5.10, 21]. The 175 cm^{-1} mode is a "butterfly" motion of the two rings. It is the lowest frequency vibration and its amplitude is large enough that it behaves much like a high frequency phonon. Such a mode is termed an *amalgamated vibration* because it is amalgamated into the continuum of phonons.

The low frequency density of states (DOS) for the model systems is shown in Fig. 5.9 from *Hill* et al. [5.12]. The DOS at zero frequency has a steep rising edge due to acoustic phonons. The Debye frequency is approximately 90 cm^{-1} [5.26]. At about 50 cm^{-1}, the DOS levels out due to optic phonons. When amalgamated vibrations are present, the flat area is extended to higher energy [5.10]. The combination of phonons and amalgamated vibrations comprise the *matrix phonons* [5.12]. At higher energy than the phonon DOS is the vibrational DOS. On the scale of this drawing, the vibrational dispersion and factor group splittings are small, so the DOS consists of spikes. There is a gap between the highest matrix phonon and lowest vibration beginning at Ω_{max}, the matrix phonon cut-off frequency.

The DOS is described with reference to the paradigm of "rigid" and "floppy" molecules. In totally rigid molecules, there are no amalgamated vibrations, so Ω_{max} is small. A wide gap occurs between the phonons and vibrations, and vibrations are localized. The factor group and vibron splittings vanish. In floppy molecules there are many amalgamated vibrations, Ω_{max} is large, the gap is small, and vibrations acquire external character and form delocalized band states [5.3, 12, 26].

Benzene is rigid, so there are no amalgamated vibrations, $\Omega_{max} = 135$ cm^{-1}, and the gap is nearly 300 cm^{-1} wide. Naphthalene, however, is less rigid. The butterfly mode is amalgamated into the phonons, and $\Omega_{max} = 180$ cm^{-1}. The gap shrinks to about 160 cm^{-1}. Anthracene is very similar to naphthalene although somewhat less rigid, and here the gap is only 70 cm^{-1} wide while Ω_{max} is the same as in naphthalene [5.12].

Durene is very different because of the presence of four floppy methyl groups which are much less rigid than the aromatic rings [5.27, 78]. There are more than ten amalgamated vibrations [5.12], arising from methyl torsions, and Ω_{max} is twice as large as in naphthalene, about 380 cm^{-1}. The high frequency vibrations of durene unavoidably involve molecular deformation, and thus they are less harmonic than in the more rigid systems.

5.4 Vibrational Relaxation Processes

It is the anharmonic intermolecular potential which causes VR. When the potential is expanded in a Taylor series, the lowest-order anharmonic term gives rise to $\langle V^{(3)} \rangle$, the cubic anharmonic coupling matrix element, or *V-coefficient* [5.3]. Cubic anharmonicity couples a laser excited mode $(\Omega, k = 0)$ to a pair of different modes (ω_1, k) and $(\omega_2, -k)$. The pair may consist of two phonons, a phonon and a vibron, or two vibrons. When higher order anharmonicity may be ignored, the lifetime T_1 is given by

$$(2\pi c T_1)^{-1} = \frac{18\pi}{\hbar} \sum_{\omega_1} \sum_{\omega_2} \sum_{k} \left| V^{(3)} \begin{matrix} \Omega & \omega_1 & \omega_2 \\ 0 & k & -k \end{matrix} \right|^2$$
$$\times [(n_{\omega_1} + n_{\omega_2} + 1)\delta(\Omega - \omega_1(k) - \omega_2(-k))$$
$$+ (n_{\omega_1} - n_{\omega_2})\delta(\Omega + \omega_1(k) - \omega_2(-k))], \qquad (5.7)$$

where $n_\omega(T)$ is the thermal occupation number of the mode with frequency $\omega: n_\omega(T) = [\exp(\hbar\omega/k_B T - 1)]^{-1}$ [5.3]. Equation (5.7) shows that the excited vibration must relax in a *phase matched* process where energy and wavevector are conserved. There are three distinct relaxation processes. At $T = 0$, VR occurs only through *spontaneous emission* of two lower energy modes as diagrammed in Fig. 5.4. Above $T=0$, *stimulated emission* of the same two modes may contribute to VR. Also above $T=0$, VR can occur by *phonon absorption*, also shown in Fig. 5.4. This process scatters Ω to an ω state located at higher frequency.

For now we will consider (5.7) only $T=0$. In this case T_1 depends on the *number of pairs* of lower energy modes whose summed energy and wavevector are equal to $(\Omega, \mathbf{0})$, and the magnitude of the cubic anharmonic coupling to each pair. The number of pairs is the *two-phonon DOS* [5.3], where we use the term two-phonon to refer to *any state with two excitations*, whether phonons or vibrations.

The VR of a particular mode is controlled by the nature of pairs of states which can participate in the phase matched emission process. The dominant VR mechanism of a particular mode is a function of its frequency. Thus the qualitative nature of the VR process is strongly dependent on the amount of excess vibrational energy. There are five regimes of excess vibrational energy which must be considered to describe crystal VR [5.12]. In order of increasing frequency they are: acoustic regime, optic phonon regime, and vibrational regimes I–III.

The *acoustic regime* extends from zero to about $90 \, \text{cm}^{-1}$. In this regime acoustic phonons relax to lower energy acoustic modes or are absorbed at the boundaries of the crystal [5.79]. The lifetime of acoustic phonons decreases rapidly with increasing frequency, typically as Ω^{-5}. The lower energy acoustic modes (sound) may have very long lifetimes, possibly several milliseconds.

In the *optic phonon regime*, which extends from about $50 \, \text{cm}^{-1}$ to Ω_{max}, optic phonons and amalgamated vibrations relax by emission of a phonon pair consisting of either two optic modes, one optic and one acoustic mode, or two acoustic modes [5.4]. This latter process is the only possible VR mechanism for optic phonons located at the bottom of the optic phonon regime [5.4, 7b, 28a, 74]. At the bottom of this regime, the lifetimes may be as long as a few nanoseconds [5.4]; at the top of the regime they are typically a few picoseconds. The strong dependence of T_1 on Ω in this regime is caused by the steep slope of the acoustic phonon DOS at low frequency [5.11].

In *vibrational regime I*, which extends from the lowest frequency vibration to a cut off at $2\Omega_{\text{max}}$, vibrations relax by emission of two lower energy matrix phonons. The simplest pathway which conserves energy and wavevector is to emit two counterpropagating phonons at frequency $\Omega/2$ [5.12]. The lifetimes are a relatively constant function of the vibrational frequency because the optical phonon DOS is relatively constant. Regime I lifetimes are typically short, perhaps a few picoseconds, because anharmonic coupling to two phonons is large, as is the phonon DOS for this relaxation process [5.15].

In *vibrational regime II*, which extends up from $2\Omega_{\text{max}}$, vibrations relax by emission of a lower energy vibration, which will be termed an *acceptor state*, or *A-state*, and a phonon [5.10, 15]. Emission of two vibrations is usually impossible because of energy conservation. The lifetimes in regime II may range from a few picoseconds to as long as 1 ns [5.7a, 29]. The two-phonon DOS for this process is highly dependent on the number of A-states located between Ω and $\Omega - \Omega_{max}$, and owing to the discrete nature of the vibration DOS, this number is an erratic function of Ω [5.10]. Consequently the two-phonon DOS for this process varies erratically with Ω, as do the lifetimes.

In *vibrational regime III*, which is the high frequency regime extending up to the dissociation limit, vibrations relax into a dense bath of "background" states composed of high-lying combinations and overtones of the fundamentals [5.15]. Regime III VR is basically intramolecular (IVR) and should not be radically different in crystals or isolated molecules. The lifetimes in regime III should be very short, typically no longer than a few picoseconds, and subpicosecond at large values of excess energy.

Cubic anharmonic VR processes depend upon the two-phonon DOS. This DOS is given by [5.3]

$$D(\hbar\Omega) = \frac{1}{\hbar} \sum_{\omega_1\omega_2 k} \delta[\Omega - \omega_1(k) - \omega_2(-k)]. \qquad (5.8)$$

Because it is usually impossible for a mode to relax to two vibrations, calculations of $D(\hbar\Omega)$ neglect this contribution. In this case $D(\hbar\Omega)$ extends from zero frequency to a point $V_{max} + \Omega_{max}$, where V_{max} is the frequency of the highest vibrational fundamental. *Hill* et al. have calculated the normalized two-phonon DOS for several molecular crystals using the experimental phonon DOS and vibrational frequencies obtained from neutron and optical spectroscopy [5.12]. Figure 5.10 taken from that work gives the two-phonon DOS of benzene, naphthalene, anthracene and durene from 0 to 2000 cm^{-1}. The arrow in each plot indicates the position of the regime I – regime II boundary located at $2\Omega_{max}$. In regime I, $D(\hbar\Omega)$ is composed of pairs of phonons, and it is quite smooth, reflecting the dispersed phonon DOS. In regime II, $D(\hbar\Omega)$ is composed of states with one phonon and one vibration, and is a rapidly changing function of Ω, due to the irregular vibrational DOS. In regime II, $D(\hbar\Omega)$ will reach a maximum at a frequency about Ω_{max} above a cluster of vibrations.

In Fig. 5.10, the two-phonon DOS of benzene has deep troughs, going to zero at several frequencies. This is a consequence of the sparse vibrational level structure. Whenever two vibrations are separated by more than $\Omega_{max} = 135$ cm^{-1}, the two-phonon DOS will vanish. In the other materials, gaps in the vibrational DOS never exceed Ω_{max}, and the two-phonon DOS never vanishes below 1700 cm^{-1}. Naphthalene and anthracene are qualitatively similar although the anthracene DOS has more area because there are more vibrations contributing. In durene, the only floppy molecule, the DOS is very different. Durene has many low frequency methyl torsions (150–400 cm^{-1}) which correspond to higher frequency ring twisting modes in naphthalene. In effect, vibrations are changed into phonon-like states, driving the regime I – regime II boundary up to about 800 cm^{-1}, and increasing the area of the two-phonon DOS in regime I at the expense of regime II.

TWO–PHONON DENSITIES OF STATES

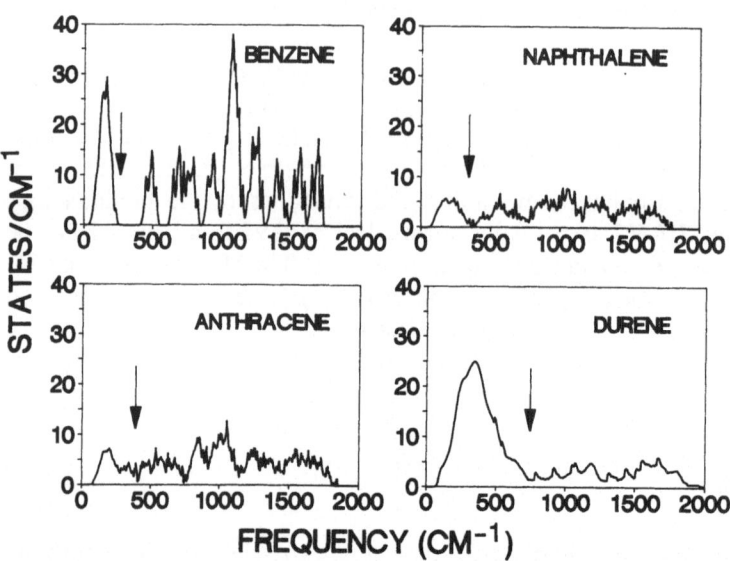

FREQUENCY (CM⁻¹)

Fig. 5.10. Two-phonon density of states calculated from the data in Fig. 5.9 by *Hill* et al. [5.12]. The rate of VR from a state at frequency Ω is proportional to this density, which includes contributions from pairs of lower energy states. The arrow indicates the boundary between regimes I and II where the dominant VR process abruptly shifts from two-phonon emission to emission of one phonon and one vibration

The VR rate is also dependent on the magnitude of the anharmonic coefficients. The cubic coefficients are matrix elements of the form [5.3]

$$\frac{1}{3!}\left.\frac{\partial^3 V(\psi)}{\partial\psi_1\partial\psi_2\partial\psi_3}\right|_{\psi=0}\psi_1\psi_2\psi_3, \tag{5.9}$$

where $V(\psi)$ is the crystal potential surface, and ψ_i is a normal coordinate, $\psi = Q$, a vibrational coordinate, or $\psi = q$, a phonon coordinate. Equation (5.9) shows that the cubic anharmonic matrix element, or V-coefficient, depends on the magnitude of each displacement ψ and the derivative of $V(\psi)$ with respect to each normal coordinate. All things being equal, the largest coefficients are of the form qqq, followed by Qqq, QQq, and QQQ, because the amplitude of vibration is larger for modes with external character than for modes with internal character. However, the value of this coefficient will also depend on whether there is a good geometrical match between the excited mode and the pair of emitted modes. This matching depends on the atom-atom interactions characteristic of each triplet of modes, particularly collisions involving the steeply rising repulsive potential [5.3, 14, 21, 80].

In view of this discussion it would seem that the anharmonic coefficients would vary over a wide range and be highly mode dependent. These coefficients have been calculated using a model anharmonic potential for naphthalene and benzene vibrations by *Righini* [5.21, 80], and for naphthalene phonons by *Delle Valle* et al. [5.20]. It was shown in this latter work that the coefficients for phonon relaxation did not depend much on the type of phonon involved, and if they are treated as equal, the lifetimes at low temperature and the temperature dependence are in good agreement with experiment [5.71]. However *Righini* [5.21, 80] has shown that the anharmonic coefficients in regime II for particular triplets of excited vibration, acceptor mode and phonon are strongly dependent on geometrical factors.

The strong dependence of VR on the V-coefficients makes direct calculation of the state-to-state VR quite difficult. However, in systems composed of large, complex molecules VR can be simply treated in an averaged manner using the average anharmonic coupling approximation [5.10, 12]. In this approximation the low temperature lifetime $T_1(\Omega)$ is given by

$$[2\pi c T_1(\Omega)]^{-1} = \frac{18\pi}{\hbar} \langle V_i^{(3)} \rangle^2 D(\hbar\Omega), \tag{5.10}$$

where $\langle V_i^{(3)} \rangle$ is one of a few averaged anharmonic coefficients and $D(\hbar\Omega)$ is the two phonon DOS. Equation (5.10) has been extended to finite temperature in [5.10].

Equation (5.10) is quite accurate for describing VR in the optic phonon regime. It has also been shown by *Schosser* and *Dlott* [5.10], and by *Hill* et al. [5.12] that this approximation is a good description of VR for regime II modes of benzene, naphthalene and anthracene.

The idea behind this approximation in regime II is that in a complex crystal there is considerable mixing among phonons, and for a particular excited vibration Ω and acceptor A, there will be many different types of phonons present at the frequency $\omega = \Omega - \omega_A$, where ω is the phonon frequency and ω_A is the acceptor frequency [5.80]. In addition, each vibration Ω typically has many A modes [5.10]. Thus as the crystal becomes more complex, the strong dependence of T_1 on the specific character of the anharmonic coefficients lessens, and VR lifetimes can be calculated as a function of the averaged coefficients.

In this approximation, it is possible to derive specific expressions for VR in the optic phonon regime, regimes I and II [5.12]. The behavior of optic phonons has been reviewed recently [5.4]. In regime I, vibrations relax by emission of two phonons. Because the phonons are delocalized states, this process inherently involves terms of the Hamiltonian which annihilate excitation on the originally excited site and create excitations on other sites. Owing to the relatively large dispersions of the emitted phonons, wave vector matching is an important determinant of which modes are emitted. In general the most efficient relaxation involves emission of two counterpropagating phonons at frequency $\Omega/2$ [5.12]. In this case the lifetime in regime I is given by

$$[2\pi c T_1(\Omega)]^{-1} = \frac{18\pi}{\hbar} \langle A^{(3)} \rangle^2 \varrho^2(\hbar\Omega/2), \quad \text{(regime I)} \tag{5.11}$$

where $\langle A^{(3)} \rangle$ is a cubic anharmonic coefficient of the form Qqq, and $\varrho(\hbar\Omega)$ is the phonon DOS. Equation (5.11) predicts a roughly constant dependence of the lifetime on Ω in regime I because the optic phonon density ϱ is also roughly constant [5.4].

In regime II, a vibration relaxes by emission of an acceptor vibration and a phonon. The matrix element is of the form QQ_Aq. However, there are two distinct types of matrix elements. In the first case, which depends on the *B-coefficient*, $\langle B^{(3)} \rangle$, an excitation is annihilated on one site and the acceptor is created on the *same* site [5.12, 14, 81]. This is termed a *one-site* process, although it must be kept in mind that the emitted phonon is inherently delocalized. In the second case, which depends on the *C-coefficient*, $\langle C^{(3)} \rangle$, the acceptor is created on an adjacent site. This is termed a *two-site* process. In the two-site process, intermolecular vibrational energy transfer occurs [5.14, 81]. Wave vector matching is unimportant in regime II because two of the emitted states are narrow dispersion vibrations [5.10]. The lifetime in regime II is given by [5.12]

$$(2\pi c T_1)^{-1} = \frac{18\pi}{\hbar}(\langle B^{(3)} \rangle^2 + \langle C^{(3)} \rangle^2) \sum_{\omega_A} \varrho(\hbar(\Omega - \omega_A))$$

$$= \frac{18\pi}{\hbar}(\langle B^{(3)} \rangle^2 + \langle C^{(3)} \rangle^2) D(\hbar\Omega), \quad \text{(regime II)} \quad (5.12)$$

where $D(\hbar\Omega)$ in this case is a two-phonon DOS containing one vibrational acceptor and one phonon.

Hill et al. [5.16, 82] have treated VR of guest molecules in crystalline host matrices using the averaged anharmonic approximation. In guest-host systems, the emitted phonons are always excitations of the host matrix, and thus in the analog of (5.11, 12) the matrix phonon DOS must be used. Because of a size mismatch between the host and guest, the sum over k in (5.8) does not range over all values in the first Brillouin zone, but rather only over those values which can be mechanically excited by the guest molecule. This effect introduces a factor of $\bar{V}$ in (5.10, 11), where $\bar{V} = 1$ when $V_H/V_G > 1$ and $\bar{V} = V_H/V_G$ when $V_H/V_G < 1$, where V_H and V_G are the molecular volumes of host and guest respectively. In addition to the k-space correction, the anharmonic matrix elements A, B, and C become A', B', and C' in a mixed crystal. The A'-coefficient differs from the A-coefficient in that it describes energy transfer from guest vibrations in regime I to host phonons. The B'-coefficient differs from the B-coefficient in that it describes one site (guest) relaxation with emission of host phonons. Similarly, the C'-coefficient differs from the C-coefficient in that the two-site relaxation now involves intermolecular energy transfer between chemically distinct molecules.

Regime III encompasses all the modes above regime II. Regime III modes relax by emission of high-lying combinations and overtones. In molecules the size of naphthalene, the density of doubly excited states is small, so the most important background states are multiple excitations. The number of these states increases approximately exponentially with Ω [5.12, 83]. However, the anharmonic coupling, which is of the form $QQ_1Q_2 \ldots Q_i$ decreases as i increases [5.83]. As Ω increases, only states with large values of i are resonant. The offsetting nature of these effects gives rise to a gradual boundary between regime II and III often called the IVR "threshold" [5.84]. Above this threshold, regime III modes rapidly relax by emission of regime I and II modes, and possibly one or more phonons.

5.5 Vibrational Relaxation in Low Temperature Crystals

In this section I shall compare the VR behavior of benzene, naphthalene, anthracene and durene pure crystals, and also compare these to the mixed crystal systems naphthalene/durene and anthracene/naphthalene.

Table 5.2 (reproduced from *Hill* et al. [5.12]) summarizes VR information on pure crystals at low temperatures obtained from ps CARS (naphthalene, anthracene, durene), frequency domain CARS (benzene), and low temperature IR (naphthalene, anthracene).

5.5.1 Naphthalene and Anthracene

There is not much data on regime I modes ($\Omega < 360\,\mathrm{cm}^{-1}$) in naphthalene and anthracene, but the existing observations are consistent with an abrupt increase in T_1 occurring at the regime I – regime II boundary as a result of the A-coefficients being larger than B- or C-coefficients. For the naphthalene $362\,\mathrm{cm}^{-1}$ mode, T_1 is about 1 ps [5.12, 33] while one factor group component of the naphthalene $392\,\mathrm{cm}^{-1}$ mode has $T_1 = 21$ ps [5.28c].

The modes listed in Table 5.2 for naphthalene and anthracene are all in regime II. In this regime the lifetimes are not simple functions of excess energy, and some very long-lived ($T_1 \sim 100$ ps) modes and short-lived modes are observed. The VR process is the emission of a phonon and a vibration. The two-phonon DOS, D, at each frequency Ω is given in the table, and using (5.12) and the observed value of T_1, the averaged anharmonic matrix elements $\langle V^{(3)} \rangle$ are determined. For the ps CARS data, which is taken only on totally symmetric modes of a_g symmetry, the average value in regime II is $0.025\,\mathrm{cm}^{-1}$ for both naphthalene and anthracene. For the IR active modes, which are mostly b_{3u} out of plane vibrations, the lifetimes are generally shorter and the anharmonic coefficients about twice as large.

The a_g modes are observed to have small factor group splittings [5.85], and their linewidth increases linearly with impurity doping at small doping levels [5.13]. This behavior indicates that, to a good approximation, they behave as localized states. Thus the only contribution to their VR should be one-site processes governed by the B-coefficient, and the C-coefficient should be small.

190   *D. D. Dlott*

Table 5.2a–g. Vibrational lifetimes in low temperature pure crystals

a) Benzene (S_0^v CARS [5.29, 70])

$\omega[\text{cm}^{-1}]$	$T_1[\text{ps}]$	$D[\text{states/cm}^{-1}]$	$\langle V^{(3)}\rangle[\text{cm}^{-1}]$	Mode
606	2650	0	–	v_6
854	884	0.06	0.042	v_{10}
991	62	2	0.028	v_1
1174	51	2.2	0.029	v_9
1603	50	1.8	0.032	v_8

Average: $\overline{0.033}\pm 0.056\ \text{cm}^{-1}$

$\omega[\text{cm}^{-1}]$	$T_1[\text{ps}]$	$D[\text{states/cm}^{-1}]$ (Four-phonon)	$\langle V^{(4)}\rangle[\text{cm}^{-1}]$	Mode
~ 3042	17.8	37	0.0030	v_7
~ 3063	13.5	21	0.0045	v_2

b) Naphthalene (S_0^v, ps CARS [5.10])

$\omega[\text{cm}^{-1}]$	$T_1[\text{ps}]$	$D[\text{states/cm}^{-1}]$	$\langle V^{(3)}\rangle[\text{cm}^{-1}]$	Mode
511	140	2.32	0.017	v_9
766	62	1.18	0.036	v_8
1023	19	8.00	0.025	v_7
1146	< 10	4.02	> 0.048	v_6
1385	92	2.89	0.019	v_5
1578	14	4.40	0.039	v_4

Average: $\overline{0.025}\pm 0.015\ \text{cm}^{-1}$

c) Naphthalene (S_0^v, IR linewidths [5.12])

$\omega[\text{cm}^{-1}]$	$T_1[\text{ps}]$	$D[\text{states/cm}^{-1}]$	$\langle V^{(3)}\rangle[\text{cm}^{-1}]$	Mode
847	6	4.7	0.058	v_{35}
848	7	4.7	0.055	v_{35}
961	8	5.1	0.049	v_{45}
962	8	5.1	0.049	v_{45}
1007	8	4.2	0.054	v_{43}
1124	4	2.6	0.090	v_{34}
1272	3.5	4.6	0.076	v_{33}
1275	5	4.6	0.065	v_{33}
1389	4	2.9	0.094	v_{32}
1505	5	4.7	0.061	v_{39}
1592	9	2.6	0.064	v_{31}
1593	8	2.6	0.069	v_{31}

Average: $\overline{0.065}\pm 0.015\ \text{cm}^{-1}$

Table 5.2 (continued)

d) Anthracene (S_0^v, ps CARS [5.10])

$\omega[\text{cm}^{-1}]$	$T_1[\text{ps}]$	$D[\text{states/cm}^{-1}]$	$\langle V^{(3)}\rangle\,[\text{cm}^{-1}]$	Mode
395	53	2.81	0.025	v_{12}
753	35	7.2	0.019	v_{10}
1008	<10	28.0	>0.018	v_9
1163	21	8.3	0.023	v_8
1261	15	10.3	0.025	v_7
1404	17	11.2	0.022	v_6

Average: $\overline{0.023}\pm0.003\ \text{cm}^{-1}$

e) Anthracene (S_0^v, IR linewidth [5.12])

$\omega[\text{cm}^{-1}]$	$T_1[\text{ps}]$	$D[\text{states/cm}^{-1}]$	$\langle V^{(3)}\rangle\,[\text{cm}^{-1}]$	Mode
906	9	7.4	0.04	–
907	9	7.4	0.04	–
981	4	7.4	0.05	–
1148	4	3.2	0.09	–
1168	5	3.5	0.07	–
1274	4	6.5	0.06	–
1448	2	2.8	0.14	–
1451	2	2.8	0.13	–
1535	3	6.2	0.07	–
1623	3	4.5	0.08	–

Average: $\overline{0.087}\pm0.021\ \text{cm}^{-1}$

f) Perdeuteronaphthalene (S_0^v, ps CARS [5.10])

$\omega[\text{cm}^{-1}]$	$T_1[\text{ps}]$	$D[\text{states/cm}^{-1}]$	$\langle V^{(3)}\rangle\,[\text{cm}^{-1}]$	Mode
493	74	4.5	0.017	v_9
694	<10	4.0	>0.049	v_8
829	<10	10.4	>0.032	v_7
862	<10	5.9	>0.042	v_6
1390	<10	9.0	>0.034	v_5
1584	<10	3.3	>0.057	v_4

g) Durene (S_0^v, ps CARS [5.12])

$\omega[\text{cm}^{-1}]$	$T_1[\text{ps}]$	$D[\text{states/cm}^{-1}]$	$\langle V^{(3)}\rangle\,[\text{cm}^{-1}]$	Mode
271	<12	21.7	>0.019	–
358	<12	24.7	>0.018	–
506	<12	12.6	>0.025	–
740	<12	1.5	>0.072	–
1265	22	1.5	0.053	–
1390	<12	1.7	>0.068	–
1465	<12	2.8	>0.053	–

The larger V-coefficient for IR active modes indicates that their relaxation includes an additional contribution from two-site processes. This interpretation, put forth by *Hill* et al. [5.12], is consistent with the spectroscopy of the IR active modes, which show much larger factor group splittings. Since the one-site processes are expected to be similar for the Raman and IR active modes, the increased anharmonic coefficient for the IR modes shows that for these modes, $\langle C^{(3)} \rangle$ is about twice $\langle B^{(3)} \rangle$. It was shown that all the measured VR lifetimes in naphthalene and anthracene were determined, within a factor of two, by these two measured parameters.

Beck et al. [5.83] studied naphthalene IVR in isolated molecules and found an IVR threshold at about 2200 cm^{-1}, which decreased to about 1200 cm^{-1} for perdeuteronaphthalene (d_8N). Because this threshold is so high, we expect very little IVR for the skeletal modes of naphthalene, which all lie below 1627 cm^{-1}, but considerable IVR for high energy skeletal modes of d_8N. Table 5.2 shows that while v_9, the lowest a_g mode, has exactly the same anharmonic coefficient in naphthalene and d_8N, the higher d_8N modes decay much faster than in naphthalene. In Table 5.3 it is observed that this effect is independent of the type of crystal matrix [5.16], demonstrating that the increased VR rate in d_8N is caused by IVR. The calculated DOS indicates that anthracene is intermediate between naphthalene and d_8N [5.12]. Since we expect that the anthracene anharmonic coefficients should be very close to those of naphthalene, the somewhat larger value of the coefficient for IR modes suggests that there is some IVR contribution to the relaxation above 1440 cm^{-1}.

Table 5.3a–g. Vibrational lifetimes in pure and mixed crystals at low temperatures

a) Naphthalene (S_0^v, ps CARS)

$\omega[\text{cm}^{-1}]$	$T_1[\text{ps}]$	$D[\text{states/cm}^{-1}]$	$\langle V^{(3)} \rangle [\text{cm}^{-1}]$	Mode
511	140	2.32	0.017	v_9
1023	19	8.00	0.025	v_7
1385	92	2.89	0.019	v_5
1578	14	4.40	0.039	v_4
		Average:	0.025 cm^{-1}	

b) Naphthalene in durene (S_1^v, ps photon echo)

$\omega[\text{cm}^{-1}]$	$T_1[\text{ps}]$	$D[\text{states/cm}^{-1}]$	$\langle V^{(3)} \rangle [\text{cm}^{-1}]$	Mode
503	66.4	21.7	0.0081	v_9
975	<15	23.2	>0.016	v_7
1384	43	10.5	0.013	v_5
1428	38	11.9	0.014	v_4
		Average	$>0.013 \text{ cm}^{-1}$	

Table 5.3 (continued)

c) Naphthalene in perdeuterodurene (S_1^v, ps photon echo)

$\omega[\text{cm}^{-1}]$	$T_1[\text{ps}]$	$D[\text{states/cm}^{-1}]$	$\langle V^{(3)}\rangle[\text{cm}^{-1}]$	Mode
503	63	–	–	–

d) Perdeuteronaphthalene (S_0^v, ps CARS)

$\omega[\text{cm}^{-1}]$	$T_1[\text{ps}]$	$D[\text{states/cm}^{-1}]$	$\langle V^{(3)}\rangle[\text{cm}^{-1}]$	Mode
493	74	4.5	0.017	ν_9
829	<10	10.4	>0.032	ν_7
862	<10	5.9	>0.042	ν_6
1390	<10	9.0	>0.034	ν_5
1584	<10	3.3	>0.057	ν_4

e) Perdeuteronaphthalene in durene (S_1^v, ps photon echo)

$\omega[\text{cm}^{-1}]$	$T_1[\text{ps}]$	$D[\text{states/cm}^{-1}]$	$\langle V^{(3)}\rangle[\text{cm}^{-1}]$	Mode
481	58	25.0	0.0081	ν_9
826	<15	21.5	>0.017	ν_7
835	<15	22.1	>0.017	ν_6
1229	<15	15.6	>0.020	ν_5
1389	<15	7.45	>0.029	ν_4

f) Anthracene

$\omega[\text{cm}^{-1}]$	$T_1[\text{ps}]$	$D[\text{states/cm}^{-1}]$	$\langle V^{(3)}\rangle[\text{cm}^{-1}]$	Mode
395	53	2.81	0.025	ν_{12}

g) Anthracene in naphthalene (S_1^v, ps photon echo)

$\omega[\text{cm}^{-1}]$	$T_1[\text{ps}]$	$D[\text{states/cm}^{-1}]$	$\langle V^{(3)}\rangle[\text{cm}^{-1}]$	Mode
395	48	5.56	0.022	ν_{12}

5.5.2 Benzene and Durene

Benzene data from *Trout* et al. [5.29] were analyzed by *Hill* et al. [5.12], who calculated the DOS and anharmonic coefficients in Table 5.2. Benzene has a sparse level structure compared to naphthalene, and some modes act as extreme bottlenecks in the VR process. In benzene, $\Omega_{\max} = 135 \text{ cm}^{-1}$ [5.81], and there are no regime I modes in the crystal [5.12]. For $\nu_6(\Omega = 606 \text{ cm}^{-1})$, there are no possible two-phonon relaxation mechanisms. This mode probably relaxes via three-phonon emission, which accounts for its long lifetime, $T_1 = 2.65$ ns. Between 850 and 1603 cm^{-1}, four regime II modes were studied, and the average value of $\langle V^{(3)} \rangle = 0.033$ cm^{-1}. It is a bit of a surprise that the averaged anharmonic approximation is so accurate here. Even though each mode has only a small number of acceptors (typically 1 or 2) [5.29], the averaged behavior is observed because there are several types of phonons involved in the relaxation to each acceptor [5.12, 80].

The C–H stretching modes in benzene are very interesting because, unliken naphthalene and anthracene, the IVR threshold of benzene is above the C–H frequencies, at approximately 3400 cm^{-1} [5.84]. It is impossible for these modes to relax to a pair of lower energy modes, and it is predicted that the most likely relaxation pathway is emission of *two* vibrations and one phonon via quartic anharmonic coupling, $\langle V^{(4)} \rangle$. *Hill* et al. have determined the DOS for this process [5.12]. Table 5.2 gives the averaged lifetimes for the factor group components of two modes, v_7 and v_2 from *Hochstrasser* and *Trout* [5.70], the DOS, and the quartic coupling elements, which, as expected, are about an order of magnitude smaller than the cubic coefficients.

One must be careful when interpreting the values $\langle V^{(3)} \rangle = 0.033$ cm^{-1} for Raman active modes in benzene and $\langle V^{(3)} \rangle = 0.025$ cm^{-1} for Raman active modes in naphthalene because the crystal structures are different. Table 5.1 shows that the *density of vibrations* is the same in both while the *density of optic phonons* is twice as large in benzene. With this effect taken into account, the relative anharmonicity of benzene is roughly one-half that of naphthalene.

In durene, the lifetimes are much shorter than in naphthalene or benzene [5.12]. The regime I – regime II boundary is at about 780 cm^{-1}, and the first mode above this value ($\Omega = 1265$ cm^{-1}, $T_1 = 22$ ps) is the longest lived. The short lifetimes result from a combination of a large number of analgamated vibrations, which behave like extra phonons, and the floppy character of the molecule, which mixes anharmonic phonon character into even the high frequency vibrations [5.12].

5.6 Vibrational Relaxation in Mixed Crystals

Naphthalene/durene and anthracene/naphthalene data from *Hill* et al. [5.16] are given in Table 5.3, where comparison between VR in the host matrix and pure crystal is made. In the naphthalene/durene system, the lifetimes of naphthalene S_1^v modes are about half of the same S_0^v modes in the pure crystal. However, since the two-phonon DOS is much larger, the anharmonic coefficients are *smaller* in the durene matrix. *Hill* et al. [5.16] argued that the factor of two difference in the coefficients could not be entirely due to the change in excited state potential surface, especially since it is noticed that anthracene in naphthalene S_1^v has the same coefficient as pure anthracene S_0^v. It was thus concluded that that the durene matrix provided a *more harmonic* environment than the naphthalene matrix [5.12]. Model atom-atom potentials show that H–H interactions are more harmonic than C–C interactions [5.20]. There are considerably more of the former in durene and more of the latter in naphthalene.

Tables 5.2 and 5.3 also show the interesting point that in naphthalene/durene crystals, the VR of the guest is very different from that of the host [5.16, 82]. For example, for naphthalene $v_9(\Omega = 503 \text{ cm}^{-1})$, the lifetime T_1 is roughly an order of magnitude longer than that of the durene 506 cm^{-1} mode. This was taken as evidence that guest and host relax via different channels – naphthalene relaxes by emission of a naphthalene vibration and a durene phonon whereas durene relaxes by emission of durene vibrations and durene phonons. From these data it was possible to conclude that the *intermolecular* potential of durene is more harmonic than naphthalene while the *intramolecular* potential of durene is less harmonic than naphthalene [5.16].

It will be very important to extend these studies to a wider variety of chemical materials in order eventually to develop an understanding of the relation between chemical and crystal structure and the VR rate.

5.7 Vibrational Energy Flow

The experimental techniques described in Sect. 5.1 typically measure the relaxation of the initially prepared state and give little information about the subsequent flow of energy through the crystal. With the reasonably complete data generated by *Hill* et al. [5.12] on naphthalene, it proved possible to perform preliminary calculations of this flow by using measured values of T_1 whenever possible, and when this was not possible, using calculated values of T_1 from the two anharmonic coefficients and (5.12). In the averaged anharmonic approximation, it is also possible to determine the percentage contribution of each possible relaxation pathway to determine where the energy goes in each step [5.10].

Let us consider a cold naphthalene crystal which is excited at the frequency of the highest skeletal mode, which is 1627 cm^{-1} (b_{3g}). This mode will relax by emission of a phonon (probably optical) and an A-mode which is also regime II [5.10]. The A-mode will subsequently emit another phonon and an even lower A-mode. This process is a *vibrational cascade*. The maximum step in frequency space is $\Omega_{max} = 180 \text{ cm}^{-1}$ cm^{-1}, and the average step should be equal to the center of gravity of the phonon DOS, ~ 90 cm^{-1}. Because the boundary between regimes II and I is at 360 cm^{-1}, the average number of steps, or emitted phonons, is fourteen. Once the excitation enters regime I, it relaxes by emission of two phonons; consequently, on the average, at step fifteen there are no more vibrations, only phonons remaining. At an average of 10–20 ps per step, this stage is reached in 150–300 ps in low temperature naphthalene.

During the vibrational cascade, the emitted optical phonons are themselves relaxing via emission of lower energy phonons, either low energy optic phonons or acoustic phonons. It is known that the phonon relaxation rate is a very steep function of phonon frequency. Typically $(T_1)^{-1} \propto \omega^{4-5}$, where ω is the phonon frequency [5.4, 79]. Consequently we would expect the vibrational cooling process to sharply decelerate once the energy has degraded totally into acoustic phonons.

It is possible to treat the vibrational cooling by a master equation for the vibrational excitation density $\varrho_i(t)$, where i denotes the vibrational mode. The master equation is

$$\frac{d\varrho_i(t)}{dt} = -\sum_j W_{ij}\varrho_j(t), \tag{5.13}$$

where W_{ij} is the transition rate from state i to j. Master equations describe the ensemble averaged flow of conserved probability. Because the number of excitations in this problem increases with time as vibrations relax into two

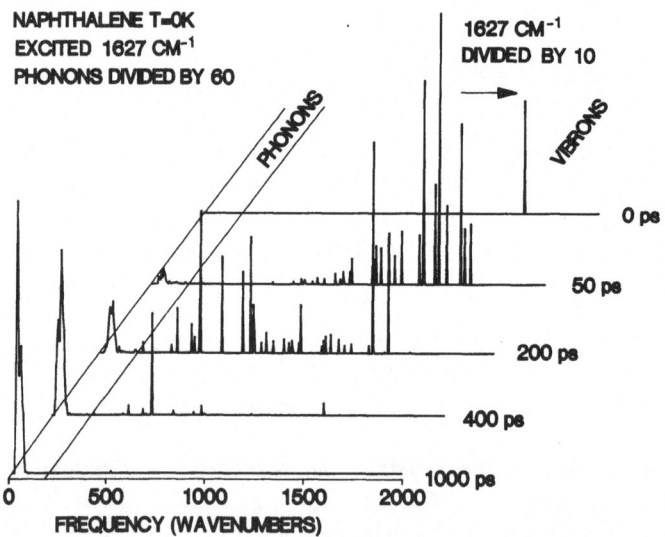

Fig. 5.11. Ensemble averaged vibrational flow through naphthalene crystal after excitation of the highest energy skeletal vibration ($\Omega = 1627$ cm^{-1}). The time dependent populations are calculated using a master equation and a matrix of transition rates containing all the experimentally measured lifetimes from Table 5.2. Where the lifetimes are unknown, calculated lifetimes from (5.11, 12) are used. At $t = 0$, unit excitation is placed at 1627 cm^{-1} (divided by ten for scaling purposes). As time increases, the vibration relaxes by emission of lower energy vibrations and phonons. The long-lived acoustic phonons (0–50 cm^{-1}) accumulate during this process. The phonon population is divided by 60 in the figure

lower energy states, the trick in using (5.13) is to employ composite states which have the same overall energy as the initial state but different occupation numbers [5.12].

Figure 5.11 gives the results of the model [5.12] for unit excitation initially in the $1627 \, \text{cm}^{-1}$ mode (for scaling purposes the initial probability in the figure is divided by 10). Following $t = 0$, the vibrational excitation moves to lower frequency and spreads out. At the same time the phonons accumulate. Note the expanded scale for the phonons, which is necessary because the single vibrational excitation will result in the creation of about 35 high frequency acoustic phonons.

An important result of this calculation is that even though VR lifetimes may be as short as a few picoseconds vibrational cooling occurs on a *much slower time scale*, owing to the finite step size Ω_{max} and the large number of steps. Indeed, even at $t = 1 \, \text{ns}$ a small amount of excitation remains in $v_5 (\Omega = 511 \, \text{cm}^{-1}, \, T_1 = 140 \, \text{ps})$, which is an important bottleneck in the cooling process.

5.A Appendix: List of Abbreviations

ps CARS   picosecond time resolved coherent anti-Stokes Raman scattering
DOS   density of states
FTIR   Fourier transform infrared spectroscopy
HB   hole burning
IR   infrared
ISRS   impulsive stimulated Raman scattering
IVR   intramolecular vibrational relaxation
Nd:YAG   neodymium doped yttrium aluminum garnet
PHB   persistent hole burning
SRS   stimulated Raman scattering
VR   vibrational relaxation

Acknowledgement. This research was supported by the National Science Foundation, Division of Solid State Chemistry under grant NSF DMR 84-15070.

References

5.1   A. Laubereau, W. Kaiser: Rev. Mod. Phys. **50**, 608 (1978)
5.2   B.J. Berne, G.D. Harp: In *Physical Chemistry, an Advanced Treatise*, Vol. 8B, ed. by D. Henderson (Academic, New York 1971)
5.3   S. Califano, V. Schettino, N. Neto: *Lattice Dynamics of Molecular Crystals*, Lect. Notes Chem., Vol. 26 (Springer, Berlin, Heidelberg 1981)

5.4   S.P. Velsko, R.M. Hochstrasser: J. Chim. Phys. **82**, 153 (1985); J. Phys. Chem. **89**, 2240 (1985)
5.5   D.D. Dlott: Annu. Rev. Phys. Chem. **37**, 157 (1986)
5.6   W.H. Hesselink, D.A. Wiersma: "Theory and Experimental Aspects of Photon Echoes in Molecular Solids", in *Spectroscopy and Excitation Dynamics of Condensed Molecular Systems*, ed. by V.M. Agranovich, R.M. Hochstrasser (North-Holland, Amsterdam 1983)
5.7   B.H. Hesp, D.A. Wiersma: Chem. Phys. Lett. **75**, 423 (1980);
       K. Duppen, B.H. Hesp, D.A. Wiersma: Chem. Phys. Lett. **75**, 423 (1981)
5.8   F. Ho, W.-S. Tsay, T.J. Trout, R.M. Hochstrasser: Chem. Phys. Lett. **83**, 5 (1981)
5.9   D.D. Dlott, C.L. Schosser, E.L. Chronister: Chem. Phys. Lett. **90**, 386 (1982)
5.10   C.L. Schosser, D.D. Dlott: J. Chem. Phys. **80**, 1384 (1984)
5.11   T.J. Kosic, R.E. Cline, Jr., D.D. Dlott: J. Chem. Phys. **81**, 4932 (1984)
5.12   J.R. Hill, E.L. Chronister, T.-C. Chang, H. Kim, J.C. Postlewaite, D.D. Dlott: J. Chem. Phys. **88**, 949 (1988)
5.13   E.L. Chronister, D.D. Dlott: J. Chem. Phys. **79**, 5286 (1983);
       E.L. Chronister, J.R. Hill, D.D. Dlott: J. Chim. Phys. **82**, 159 (1985)
5.14   F. Ho, W.-S. Tsay, J. Trout, S. Velsko, R.M. Hochstrasser: Chem. Phys. Lett. **97**, 141 (1983)
5.15   W.H. Hesselink, D.A. Wiersma: J. Chem. Phys. **74**, 886 (1981)
5.16   J.R. Hill, E.L. Chronister, T.-C. Chang, H. Kim, J.C. Postlewaite, D.D. Dlott: J. Chem. Phys. **88**, 2361 (1988)
5.17   T.J. Kosic, C.L. Schosser, D.D. Dlott: Chem. Phys. Lett. **92**, 57 (1983)
5.18   S. De Silvestri, J.G. Fujimoto, E.P. Ippen, E.B. Gamble, Jr., L.R. Williams, K.A. Nelson: Chem. Phys. Lett. **116**, 146 (1985)
5.19   J. Swiatkiewicz, X. Mi, P. Chopra, P.N. Prasad: J. Chem. Phys. **87**, 1882 (1987)
5.20   R.G. Delle Valle, P.F. Fracassi, R. Righini, S. Califano: Chem. Phys. **74**, 179 (1983)
5.21   R. Righini: Chem. Phys. **84**, 97 (1984)
5.22   A. Fruhling: Ann. Phys. (Paris) **6**, 40 (1951);
       M. Ito: J. Chem. Phys. **42**, 2844 (1965);
       M. Ito, T. Shigeoka: Spectrochim. Acta. **22**, 1029 (1966);
       M.P. Marzocchi, H. Bonadeo, G. Taddei: J. Chem. Phys. **53**, 867 (1970);
       H. Bonadeo, M.P. Marzocchi, E. Castellucci, S. Califano: J. Chem. Phys. **57**, 4299 (1972)
5.23   E.R. Lippincott, E.J. O'Reilly: J. Chem. Phys. **23**, 238 (1955);
       W.B. Person, G.C. Pimentel, O. Schnepp: J. Chem. Phys. **23**, 230 (1955);
       A.L. McClellan, G.C. Pimentel: J. Chem. Phys. **23**, 245 (1955);
       M. Suzuki, T. Yokoyama, M. Ito: Spectrochim. Acta. **24A**, 1091 (1968);
       A. Hadni, B. Wyncke, G. Morlot, X. Gerbaux: J. Chem. Phys. **51**, 3514 (1969);
       A. Bree, R.A. Kydd: Spectrochim. Acta. **26A**, 1791 (1970);
       F. Stenman: J. Chem. Phys. **54**, 4217 (1971);
       M. Hineo, H. Yoshinaga: Infrared Phys. **16**, 535 (1975)
5.24   G.W. Robinson: Annu. Rev. Phys. Chem. **21**, 429 (1970)
5.25   M.V. Belousov: "Vibrational Frenkel Excitons", in *Excitons*, ed. by M.D. Sturge (North-Holland, Amsterdam 1982)
5.26   A.E. Kitaigorodsky: *Molecular Crystals and Molecules* (Academic, New York 1973)
5.27   P.N. Prasad, R. Kopelman: J. Chem. Phys. **57**, 856 (1972); ibid. **58**, 126 (1973)
5.28   J.C. Bellows, P.N. Prasad: J. Chem. Phys. **70**, 1864 (1979);
       L.A. Hess, P.N. Prasad: J. Chem. Phys. **72**, 573 (1980); ibid. **78**, 626 (1983)
5.29   T.J. Trout, S. Velsko, R. Bozio, P.L. Decola, R.M. Hochstrasser: J. Chem. Phys. **81**, 4746 (1984)
5.30   P.L. Decola, R.M. Hochstrasser, H.P. Trommsdorff: Chem. Phys. Lett. **72**, 1 (1980)
5.31   P. Ranson, R. Ouillon, S. Califano: Chem. Phys. **86**, 115 (1984);
       R. Ouillon, P. Ranson, S. Califano: Chem. Phys. **91**, 119 (1984);
       P. Ranson, R. Ouillon, B. Halac, S. Califano: J. Chim. Phys. **82**, 169 (1985)
5.32   R.J. Bell: *Introductory Fourier Transform Spectroscopy* (Academic, New York 1972)
5.33   D.C. Alghren, R. Kopelman: Chem. Phys. **48**, 47 (1980)

5.34 G.J. Small: "Persistent Nonphotochemical Hole Burning and the Dephasing of Impurity Electronic Transitions in Organic Glasses", in *Spectroscopy and Excitation Dynamics of Condensed Molecular Systems*, ed. by V. Agranovich, R.M. Hochstrasser (North-Holland, Amsterdam 1983)
5.35 H. De Vries, D.A. Wiersma: Chem. Phys. Lett. **51**, 565 (1977)
5.36 S. Voelker, R.M. Macfarlane: J. Lumin. **18/19**, 213 (1979)
5.37 A.I.M. Dicker, S. Volker: Chem. Phys. Lett. **87**, 481 (1982)
5.38 K.K. Rebane, R.A. Avarmaa: Chem. Phys. **68**, 191 (1982)
5.39 C.A. Walsh, M. Berg, L.R. Narasimhan, M.D. Fayer: J. Chem. Phys. **86**, 77 (1987)
5.40 C.H. Lee, D. Ricard: Appl. Phys. Lett. **32**, 168 (1977)
5.41 E.L. Chronister, D.D. Dlott: In *Laser Applications in Chemistry and Biophysics*, ed. by Mustafa A. El-Sayed. Proc. SPIE, Vol. 620, 1986
5.42 Y.-X. Yan, E.B. Gamble, Jr., K.A. Nelson: J. Chem. Phys. **83**, 5391 (1985);
      Y.-X. Yan, L.-T. Cheng, K.A. Nelson: Adv. Infrared and Raman Spectrosc., in press
5.43 T. Mossberg, A. Flusberg, R. Kachru, S.R. Hartmann: Phys. Rev. Lett. **39**, 1523 (1977)
5.44 V. Brueckner, E.A.J.M. Bente, J. Langelaar, D. Bebelaar, J.D.W. van Voorst: Opt. Commun. **51**, 49 (1984);
      D. Brandt, H.J. van Elburg, B.L. van Hensbergen, J.D.W. van Voorst: In *Time-Resolved Vibrational Spectroscopy*, ed. by A. Laubereau, M. Stockburger, Springer Proc. Phys., Vol. 4 (Springer, Berlin, Heidelberg 1985) p. 86
5.45 I.I. Abram, R.M. Hochstrasser, J.E. Kohl, M.G. Semack, D. White: Chem. Phys. Lett. **52**, 1 (1977); J. Chem. Phys. **71**, 153 (1979)
5.46 I.I. Abram, R.M. Hochstrasser, J.E. Kohl, M.G. Semack, D. White: J. Chem. Phys. **71**, 405 (1980)
5.47 P. Ranson, R. Ouillon, S. Califano: J. Raman Spectrosc. **17**, 155 (1986)
5.48 R.E. Cline, Jr., E.L. Chronister, T.J. Kosic, C.L. Schosser, D.D. Dlott: Proc. Int. Conf. on Lasers '83, ed. by R.C. Powell. (STS Press, Maclean, VA 1984) p. 697
5.49 D.J. Kuizenga, D.W. Phillion, T. Lund, A.E. Siegman: Opt. Commun. **9**, 221 (1973)
5.50 F. Patterson: Ph. D. Dissertation, Stanford University (1986)
5.51 D.E. Cooper, R.W. Olson, R.D. Wieting, M.D. Fayer: Chem. Phys. Lett. **67**, 41 (1979)
5.52 J.C. Postlewaite, J. Miers, D.D. Dlott: IEEE J. Quant. Elect. **24**, 411 (1988)
5.53 D.P. Weitekamp, K. Duppen, D.A. Wiersma: Chem. Phys. Lett. **102**, 139 (1983)
5.54 E.J. Heilweil, M.P. Cassassa, R.R. Cavanagh, R.R. Stevenson: J. Chem. Phys. **82**, 5216 (1985); In *Time-Resolved Vibrational Spectroscopy*, ed. by A. Laubereau, M. Stockburger, Springer Proc. Phys., Vol. 4 (Springer, Berlin, Heidelberg 1985) p. 71
5.55 A. Freiberg, P. Saari: IEEE J. QE-**19**, 622 (1983)
5.56 D.-J. Jang, G.A. Brucker, D.F. Kelley: J. Phys. Chem. **90**, 6808 (1986)
5.57 G.A. Brucker, D.F. Kelley: J. Phys. Chem. **90**, 6808 (1986); **91**, 2856 (1987)
5.58 K. Rebane, P. Saari: J. Lumin. **12/13**, 23 (1976); ibid. **16**, 223 (1978)
5.59 T. Tamm, P. Saari: Chem. Phys. **40**, 311 (1979);
      Ya.Yu. Aaviksoo, P.M. Saari, T.B. Tamm: JETP Lett. **29**, 351 (1979)
5.60 R.M. Hochstrasser, C.A. Nyi: J. Chem. Phys. **70**, 1112 (1979)
5.61 R.M. Hochstrasser, C.A. Nyi: J. Chem. Phys. **72**, 2591 (1980)
5.62 J. Tanaka: Bull. Chem. Soc. Jpn. **36**, 1237 (1963);
      J. Tanaka, T. Kishi, M. Tanaka: Bull. Chem. Soc. Jpn. **47**, 2376 (1974);
      E. von Freydorf, J. Kinder, M.E. Michel-Beyerle: Chem. Phys. **27**, 199 (1978)
5.63 H.S. Avanesyan, V.A. Benderskii, V.Kh. Brikenshtein, A.G. Lavrushko, P.G. Fillippov: Phys. Status Solidi **30a**, 781 (1975);
      Yu.V. Naboikin, L.A. Ogurtsova: Zh. Prikl. Spektrosk. **31**, 189 (1970)
5.64 V.S. Gorobchenko, Yu.V. Naboikin, L.A. Ogurtsova, A.P. Podgornyi: Izv. Akad. Nauk SSSR, Ser. Fiz. **42**, 499 (1978);
      Yu.V. Naboikin, L.A. Ogurtsova, A.P. Podgornyi, F.S. Pokrovskaya: Zh. Prikl. Spektrosk. **27**, 675 (1977)
5.65 E.B. Wilson, Jr., J.C. Decius, P.C. Cross: *Molecular Vibrations, The Theory of Infrared and Raman Vibrational Spectra* (Dover, New York 1955)

5.66 R.G. Gordon: J. Chem. Phys. **40**, 1973 (1964); ibid. **42**, 3658 (1965); ibid. **43**, 1307 (1965)
5.67 K.S. Schweizer, D. Chandler: J. Chem. Phys. **76**, 2296 (1982)
5.68 P.W. Anderson: J. Phys. Soc. Jpn. **9**, 316 (1954);
      R. Kubo: Adv. Chem. Phys. **15**, 101 (1969)
5.69 D.W. Posner: Aust. J. Phys. **12**, 184 (1959);
      B. Debartolo: *Optical Interactions in Solids* (Wiley, New York 1968) p. 366
5.70 R.M. Hochstrasser, J. Trout: In *Dynamics of Molecular Crystals*, ed. by J. Lascombe
      (Elsevier, Amsterdam 1987) p. 61
5.71 C.L. Schosser, D.D. Dlott: J. Chem. Phys. **84**, 1369 (1984)
5.72 G.N. Zhizhin, N.V. Sviridov: Opt. Spectrosc. (USSR) **47**, 227 (1979)
5.73 C.B. Harris, R.M. Shelby, P.A. Cornelius: Phys. Rev. Lett. **38**, 1415 (1977);
      C.B. Harris, R.M. Shelby, P.A. Cornelius: Chem. Phys. Lett. **57**, 8 (1978);
      R.M. Shelby, C.B. Harris, P.A. Cornelius: J. Chem. Phys. **70**, 34 (1979);
      S. Marks, P.A. Cornelius, C.B. Harris: J. Chem. Phys. **73**, 3069 (1980)
5.74 P.N. Prasad: J. de Phys. C **6**, 563 (1981)
5.75 H. Sumi, Y. Toyazawa: J. Phys. Soc. Jpn. **31**, 342 (1971)
5.76 E. Hiroke: Chem. Phys. Lett. **78**, 323 (1981); J. Phys. Soc. Jpn. **51**, 958, 1953 (1982); Chem.
      Phys. Lett. **103**, 54 (1983)
5.77 P. Brüesch: *Phonons. Theory and Experiments I*, Springer Ser. Solid-State Sci., Vol. 34
      (Springer, Berlin, Heidelberg 1982)
5.78 J.J. Rush: J. Chem. Phys. **47**, 3936 (1967)
5.79 P. Klemens: Solid State Phys. **7**, 1 (1958)
5.80 R. Righini: Chem. Phys. Lett. **97**, 308 (1983)
5.81 S.P. Velsko, R.M. Hochstrasser: J. Chem. Phys. **82**, 2180 (1985)
5.82 J.R. Hill, E.L. Chronister, J.C. Postlewaite, D.D. Dlott: Springer Proc. Phys. **46**, 482
      (1986)
5.83 S.M. Beck, D.E. Powers, J.B. Hopkins, R.E. Smalley: J. Chem. Phys. **73**, 2019 (1980);
      S.M. Beck, J.B. Hopkins, D.E. Powers, R.E. Smalley: J. Chem. Phys. **74**, 43 (1981)
5.84 J.D. McDonald: Annu. Rev. Phys. Chem. **30**, 29 (1979)
5.85 D.M. Hanson, A.R. Gee: J. Chem. Phys. **51**, 5052 (1969);
      N. Rich, D.A. Dows: Mol. Cryst. Liq. Cryst. **5**, 111 (1980)
5.86 E.L. Bokhenkov, V.G. Fedotov, E.F. Sheka, I. Natkaniec, M. Sudnik-Hrynkiewicz, R.
      Righini, S. Califano: Nuovo Cimento **44**B, 324 (1978)

6. Laser Spectroscopy of Crystalline Semiconductors

Claus Klingshirn

With 26 Figures

Since their invention in 1960, lasers have been developed in various directions. The range of available laser emission wavelengths extends now from the vacuum ultraviolet through the whole visible part of the spectrum into the infrared. Tunable dye and solid state laser sources are available for large fractions of this spectrum. High power lasers have been designed. For continuously working (cw) lasers the output intensities reach several kilowatts while pulsed lasers go beyond the gigawatt range. Some lasers are designed to emit extremely coherent and monochromatic light, while others produce pulses as short as a few femtoseconds, i.e., only a few cycles of the electromagnetic wave.

One or a combination of several of these properties makes lasers an ideal tool for the spectroscopy of semiconductors. Especially the large and rapidly developing field of nonlinear semiconductor optics cannot be imagined without these unique light sources. Cases where lasers are not essential for the spectroscopic technique but are used just for convenience, as in many cases of linear transmission or reflection spectroscopy, will not be treated in this contribution.

This chapter will be organized as follows: In the first section, we shall outline the linear and nonlinear optical properties of semiconductors. Since the spectral range of available laser photons is centered around the visible part of the spectrum, the primary interaction of laser photons with semiconductors will be by real or virtual excitations in the electronic system. Consequently, this aspect will dominate Sect. 6.1. Phonons will appear in the context of light scattering, e.g., Raman– or Brillouin–scattering, in phonon-assisted absorption and emission or in relaxation processes. We consider mainly an idealized crystalline semiconductor, but we treat also defects and impurities and, briefly, mixed crystals. These disordered materials lead to amorphous semiconductors which are covered in a separate chapter by P. C. Taylor. In Sect. 6.2, we describe the basic methods of laser spectroscopy of semiconductors, without going into the technology of lasers themselves. This aspect is beyond the scope of this contribution and the reader is referred to e.g., [6.1–4]. In Sect. 6.3 we describe examples of laser spectroscopy of semiconductors. Contributions to this field have been made in the last 25 years by many scientists around the world and merely listing all the contributions would fill more than this book, so we decided to present in Sect. 6.3 a limited number of examples of laser spectroscopy of semiconductors which show the main concepts and ideas and which will be discussed in some detail. Concerning the materials, we concentrate on the more commonly known elementary semiconductors, III–V, II–VI, and I–VII compounds with a bandgap in the visible, near IR or near UV which are easily accessible by commercial lasers. We apologize for not being able to refer to the contributions of all groups which are working in this field. Finally, Sect. 6.4 will give a short outlook on some possible applications of the nonlinear optical properties of semiconductors.

6.1 Linear and Nonlinear Optical Properties of Semiconductors

In this section we describe the linear and nonlinear optical properties of semiconductors. We start with the one-particle states, i.e., with the band structure, and proceed then to intra- and interband transitions. The latter will lead to the concept of excitons and polaritons, which govern the optical properties of semiconductors at low excitation. After some excursions to impurity states and mixed crystals, we proceed to the regime of nonlinear optics, where we discuss scattering processes, the spectroscopy of biexcitons and the transitions to an electron hole plasma. Nonlinear optical properties are almost exclusively accessible only if the semiconductors are illuminated with intense laser beams. Therefore, we discuss these properties here in some detail. Finally we discuss some of the changes introduced by reduced dimensionality in quantum well structures.

6.1.1 Linear Optical Properties

In this section we describe the electronic eigenstates and excitations of semiconductors and the resulting linear optical properties, which can be observed under weak excitation.

a) Bandstructure and Intraband Excitations

The eigenstates of crystal electrons can be understood starting from two different limits, the weak and the tight binding approximations. In the weak binding approximation, we start with plane waves

$$\phi(\mathbf{k}, \mathbf{r}) = \Omega^{-1/2} \exp[i(\mathbf{k}\mathbf{r} - Et\hbar^{-1})], \tag{6.1}$$

where Ω is the normalization volume and the relation between the wave vector k and the eigenenergy E is given for nonrelativistic velocities by

$$E(k) = \hbar^2 k^2 (2m_0)^{-1} \quad \text{with} \quad m_0^{-1} = \hbar^{-2} \frac{d^2 E}{dk^2}, \tag{6.2}$$

m_0 being the free electron mass (Fig. 6.1a). If we bring the electron into a weak periodic lattice potential there will be some changes in the dispersion relation

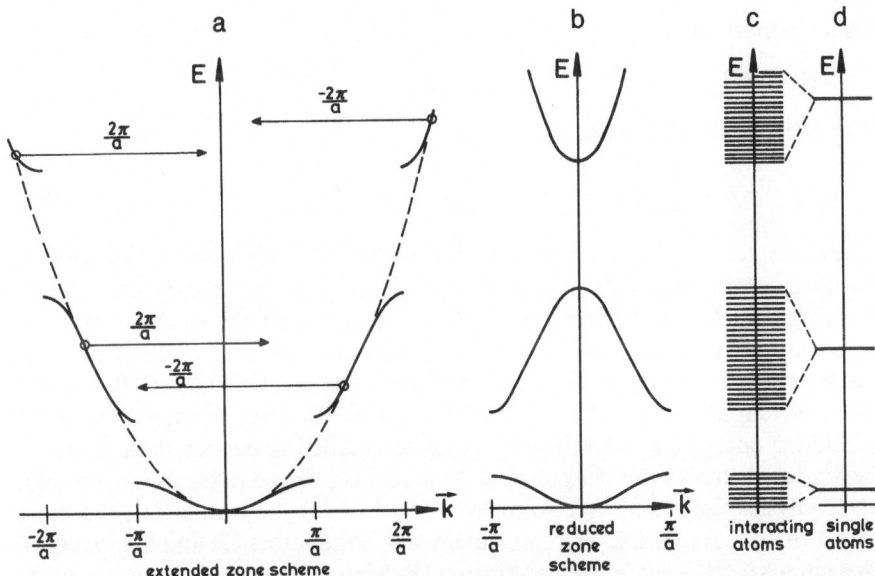

Fig. 6.1. The reduced band structure of an idealized semiconductor (**b**) starting from free and nearly free electrons (**a**) or from atomic orbitals (**c**, **d**)

$E(k)$. The most significant one is the appearance of discontinuities or gaps in the $E(k)$ relation (Fig. 6.1a). They appear for a simple cubic lattice when k is an integer multiple of π/a where a is the lattice constant

$$k_i = n_i \frac{\pi}{a}, \qquad i = x, y, z; \qquad n_i \in Z_0. \tag{6.3}$$

The invariance of the Hamilton operator against infinitesimal translations in space for the free particles leads to conservation of momentum $\hbar k$. This translational symmetry is now reduced to an invariance against translation by integer multiplets of the basic lattice vectors. Consequently, k is conserved only in modulo integer multiples of the vectors of the reciprocal lattice g, in our simple cubic model $g = (2\pi/a)(n_1, n_2, n_3)$ with $n_1, n_2, n_3 \in Z_0$. This allows the outer parts of the $E(k)$ relation to be shifted into the region $-\pi/a \leq k_i \leq \pi/a$; $i = x, y, z$, the first Brillouin zone. In Fig. 6.1 we show schematically on the left-hand side the dispersion relation $E(k)$ for free electrons, for electrons in a weak periodic potential, and the resulting reduced zone scheme (Fig. 6.1b). Arrows indicate which parts of the extended scheme have been shifted by which vectors of the reciprocal lattice.

In the other case, the tight binding approximation, we start with separated atoms and atomic eigenstates (Fig. 6.1d). Moving the atoms together to form the crystal leads to a splitting of the atomic orbitals into bands due to the mutual interaction (Fig. 6.1c). The translational invariance of the lattice leads again to a band structure like the one shown in Figs. 6.1b. The stationary eigenstates are in both cases Bloch functions of the type (6.4), where $u_k(r)$ is the lattice periodic part,

$$\phi(k, r) = \Omega^{-1/2} u_k(r) \exp[i(k, r - Et/\hbar)] . \qquad (6.4)$$

The effective masses m are defined by

$$(m_{ij})^{-1} = \hbar^{-2} \frac{\partial^2 E}{\partial k_i \partial k_j} . \qquad (6.5)$$

They have tensor character and can deviate considerably from m_0. In realistic band-structure calculations, the eigenstates and eigenenergies have to be calculated including the periodic lattice potential and the electron–electron interaction, for example by Hartree-Fock-type methods. We call the uppermost bands which are completely occupied at $T=0$ the valence bands (VB) and the first unoccupied bands, the condution bands (CB). They are separated by a forbidden energy gap of width E_g. A more detailed review of these topics is beyond the scope of this chapter and the reader is referred to textbooks on solid state physics, e.g., [6.5]. We want to take a look at the results, i.e., the band structure of real crystals. The elementary semiconductors Ge and Si crystallize in a cubic structure with the pointgroup O_h. Many compound materials out of the group of III–V, II–VI, or I–VII materials crystallize either in the zinc blende lattice (pointgroup T_d) or in the wurtzite structure (pointgroup C_{6v}). In all three cases the atoms are tetrahedrally coordinated and the band structures are similar. We give some schematic band structures in Fig. 6.2. In Fig. 6.2a, we show schematically the lowest CBs and the upper VBs for Si. This semi-conductor is said to have an indirect gap because the VB maximum and the CB minimum occur at different points in the first Brillouin zone, in this case at the Γ point ($k=0$) and along the direction Δ, respectively. Indirect means that the transition from the VB maximum to the CB minimum is not directly possible with photons, which have a very small k-vector, but involves a third particle, usually a phonon, for momentum conservation. At higher energies there is also a direct gap, i.e., an M_0 critical point and other critical points. The bandstructure of Ge is similar, except that the CB minima occur at the L points, resulting for Si and Ge in 6 and 8 equivalent CB minima, respectively, due to the symmetry of the pointgroup O_h. Many compound semiconductors crystallize in T_d symmetry, e.g., GaAs, GaP, ZnSe, CdTe, CuCl, CuBr. Some of them have a direct gap, others an indirect gap as shown in Fig. 6.2b by solid and dashed lines, respectively. The indirect CB minima occur close to the X or L points. The uppermost p-levels of the cations which contribute mainly to the upper VB (with some admixture of lower-lying d-levels) are split into two subbands at $k=0$ due to spin orbit coupling for O_h and T_d symmetries. The Γ_8 level splits for $k \neq 0$ further into a light and heavy hole band. Many of the II–VI materials and some of the III–V compounds like ZnO, CdS or GaN crystallize preferentially in the hexagonal pointgroup C_{6v} which has a polar c-axis. In this structure the VB is split into three sublevels at $k=0$ due to spin-orbit coupling and the hexagonal crystal field. For more details of the band structure of semi-conductors the reader is referred to [6.5–7].

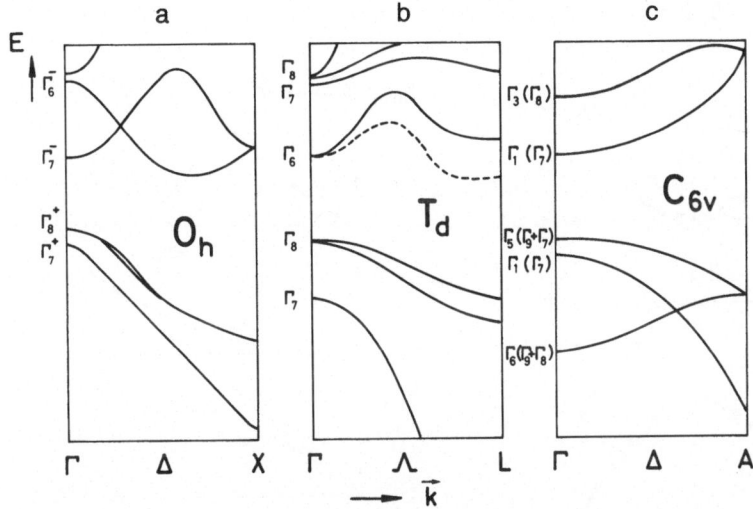

Fig. 6.2. Schematic band structures of (a) a group IV semiconductor, (b) a direct and an indirect gap semiconductor with T_d symmetry, and (c) a direct gap semiconductor with C_{6v} symmetry. Spin is included in (a) and (b); in (c) the respective symmetrics are given in brackets. (Simplified drawings deduced from data compiled, e. g., in [5.5–7])

As already mentioned, the CB and VB describe the one-particle states of the ideal semiconductor (SC). If we bring into a SC with completely filled VB an additional electron, e.g., by ionizing a neutral donor ($D^0 \rightarrow D^+ + e$), this electron will occupy the CB states. If we thermally excite an electron from the VB to an acceptor we create an unoccupied state, i.e., a hole, in the VB ($A^0 \rightarrow A^- + h$). Electrons in the CB or holes in the VB give rise to optical properties due to intraband and collective excitations. They are situated in the IR part and are therefore only of limited interest in laser spectroscopy and will be mentioned only briefly. Let us assume that the semiconductor is doped with donors which are ionized resulting in a density n of electrons in the CB. We can produce a continuous spectrum of one-particle intraband excitations. If we assume low temperatures and a degenerate gas of electrons then the excitation energies start at zero energy over a wave-vector range from zero to $2k_F$ as visualized in Fig. 6.3a. For finite excitation energies only regions with $|k| > 0$ are covered. Since photons have a wave vector close to zero this type of intraband excitation cannot usually be excited optically. Corresponding intraband excitation exists in the VB for p-doped materials. Due to the more complex bandstructure in the VB there are also transitions from the light to the heavy hole subband or from the spin-orbit split subband (Fig. 6.3b). These transitions are optically allowed and can be observed in the IR spectra of strongly p-doped or photoexcited semiconductors [6.9].

Apart from these single-particle excitations there is a collective excitation of the motion of the free carrier gas relative to the lattice, the plasma oscillations.

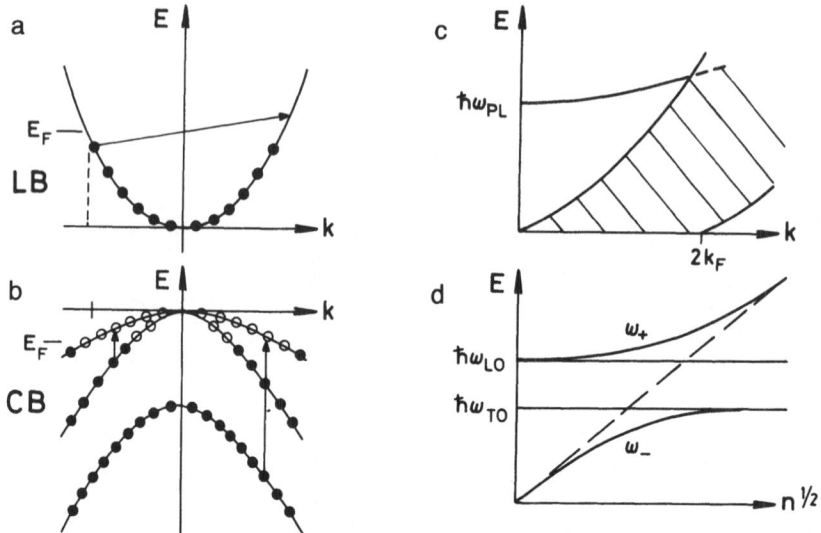

Fig. 6.3. Schematic drawing of (**a, b**) single-particle and (**c**) collective intraband excitations of a semiconductor and (**d**) of the plasmon-phonon mixed modes

The quanta of this excitation are called plasmons. The longitudinal eigenenergy is given for $k=0$ by (6.6) with a quadratic dispersion for $k \neq 0$ (Fig. 6.3c):

$$\omega_{\text{pl}} = \left(\frac{ne^2}{\varepsilon\varepsilon_0\mu}\right)^{1/2}, \tag{6.6}$$

where μ is the effective mass in a one-component plasma, i.e., of either electrons or holes and the reduced mass $\mu^{-1} = m_{\text{e}}^{-1} + m_{\text{h}}^{-1}$ in an electron–hole plasma, which may be formed under intense photoexcitation as will be discussed later.

The frequency of transverse collective excitations of the gas of free carriers is zero. The spectral range below $\hbar\omega_{\text{pl}}$ can thus be considered as the Reststrahlen-bande of the plasma excitation and this explains the high reflectivity of doped semiconductors in the IR [6.10] together with the phonon Reststrahlenbande between the transverse and longitudinal phonon frequencies. Plasmons can be observed in semiconductor laser spectroscopy in Raman scattering. If n is increased, for example by increasing doping or increasing photoexcitation, $\hbar\omega_{\text{pl}}$ may become comparable with the optical phonon energies. Due to the interaction between plasmons and optical phonons there appear plasmon–phonon mixed states. The eigenfrequencies plotted as a function of $n^{1/2}$ show the typical noncrossing rule behavior sketched in Figs. 6.3d. The n-dependent shift of the eigenenergies can be observed experimentally in Raman scattering in doped or photoexcited samples [6.11, 12], and modifies strongly the low frequency dielectric function of semiconductors.

The main topics of laser spectroscopy in semiconductors are connected with interband transitions, i.e., excitation from the VB to the CB. In the simplest

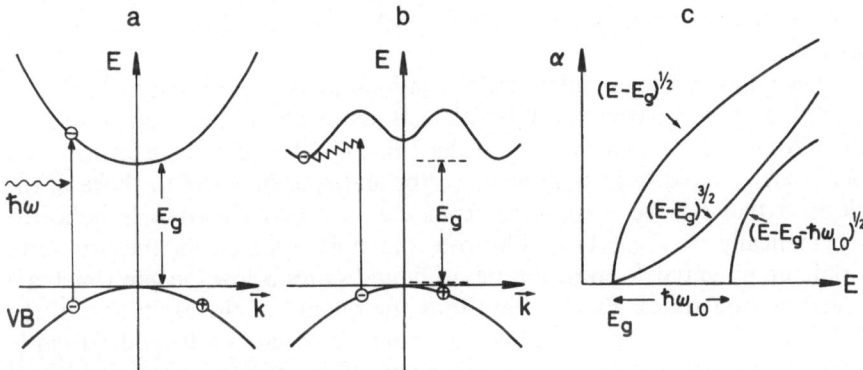

Fig. 6.4a–c. Schematic sketch of interband transitions. (a) direct transition, (b) indirect transition, and (c) the resulting absorption spectra

case, one would expect for a direct gap SC with parabolic bands (i.e., in the effective mass approximation) the following dependences of the absorption coefficient above the gap on the photon energy $\hbar\omega$:

$$\alpha(\hbar\omega) = \begin{cases} \text{const}(\hbar\omega - E_g)^{1/2} & \quad (6.7\text{a}) \\ \text{const}'(\hbar\omega - E_g)^{3/2}. & \hbar\omega > E_g \quad (6.7\text{b}) \end{cases}$$

Equation (6.7a) refers to a semiconductor with a dipole-allowed, and thus in the first approximation energy-independent, band-to-band transition matrix element. The square-root dependence then just reflects the combined or reduced density of states. Most of the direct gap III–V, II–VI, and I–VII semiconductors belong to this group. In contrast (6.7b) describes the dipole-forbidden case where the optical matrix element $\langle c|H_{dip}|v \rangle$ is zero at $k=0$ and becomes allowed proportional to k (Fig. 6.4a). To this group belong materials like Cu_2O, SnO_2 or TiO_2. For indirect materials the participation of a momentum-conserving phonon is necessary (Fig. 6.4b). At low temperatures, only phonon emission is possible, and an allowed transition results in the onset of absorption shifted by integer multiples of the corresponding phonon energy above E_g. These three cases are shown schematically in Fig. 6.4c. The absolute values of the absorption coefficients are much smaller in indirect materials than in direct ones due to the coupling to the phonons. For higher temperatures transitions become possible also under absorption of phonons, i.e., absorption sets in already for $\hbar\omega = E_g - \hbar\omega_{phonon}$.

b) Excitons

Though the considerations and calculations leading to (6.7) and Fig. 6.4c are quite simple and straightforward and are given in every textbook on semiconductor optics, the corresponding behavior is usually never observed in nature.

The reason is that all interband excitations are two-particle transitions: on exciting an electron from the VB to the CB one creates not only an electron in the CB but also an empty state in the VB. This state can be described as a positively charged hole, with wave vector and spin opposite to those of the electron that has been removed from the VB. This situation is indicated schematically in Fig. 6.4a, b. Electron and hole interact via the attractive Coulomb potential, forming a series of bound states below the gap which are called exciton states. Excitons are thus the quanta of the excitation of the VB–CB electronic system of semiconductors. They can also be understood in analogy to positronium atoms, with the main difference being that the widths of the forbidden gaps which separate electrons and holes or electrons and positrons (i.e., holes in the Fermi-Dirac sea) are of the order of 1 eV and 1 MeV, respectively. Excitons are compound particles consisting of two fermions. At very low density and temperature they are expected to behave like bosons [6.13], with increasing deviations at higher densities. Several attempts have therefore been made to observe a Bose-Einstein condensation of excitons or biexcitons (see below). Indications of Bose-type behavior in Ge, Cu_2O, and CuCl [6.14] have been found but a spontaneous condensation into a macroscopic population of one state in phase space has not been observed. Therefore, we shall not follow this aspect further.

In the simplest approximation, i.e., isotropic, parabolic and nondegenerate bands, the exciton problem can be reduced to that of the hydrogen atom. However, the exciton binding energy, E_x^b, and Bohr radius, a_x, are scaled with respect to the Rydberg energy, Ry, and the hydrogen Bohr radius, a_B, by the reduced mass μ and a dielectric constant ε. The wave function consists of a product of a plane-wave factor containing the total momentum $K = k_e + k_n$ and the center of mass coordinate R; an envelope function ϕ, which depends on the relative motion of electron and hole and on the hydrogen quantum numbers n_B, l, m; and electron and hole Wannier functions $\varrho_{e,h}$. These functions are in turn linear combinations of Bloch functions (6.4). In (6.8) we give a set of equations describing the eigenenergies of the excitons and the wave functions: m_0, m_e, m_h, μ, and M are the free electron mass, the effective masses of electrons and holes, the reduced and the translational masses, respectively:

$$E_x(K, n_B) = E_g - E_x^b \frac{1}{n_B^2} + \frac{\hbar^2 K^2}{2M},$$

$$E_x^b = R_y \frac{\mu}{\varepsilon^2}, \qquad a_x = a_B \frac{\varepsilon}{\mu},$$

$$M = m_e + m_h, \qquad \mu = \frac{m_e m_h}{m_e + m_h},$$

$$\phi_x = \Omega^{-1/2} \varrho_e(r_e)\varrho_h(r_h)\phi_{n_B, l, m}(r_e - r_h)e^{iK \cdot R}.$$

(6.8)

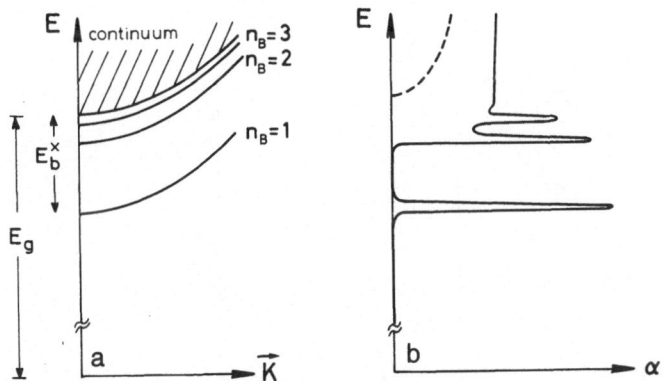

Fig. 6.5. (a) schematic dispersion relation of excitons in a direct gap semiconductor and **(b)** the resulting absorption spectrum. The dashed line gives the square-root behavior expected without the electron-hole interaction

Figure 6.5 give the dispersion relation $E(k)$ of excitons. The interaction between photons and excitons can be described by perturbation theory, i.e., in the weak coupling limit, only for exciton states with small oscillator strength. This is the case, for example, in the indirect transitions. The absorption spectrum of Fig. 6.4c for indirect materials has then to be shifted by E_x^b to lower photon energies. This description is also applicable to exciton states in direct gap materials with weak oscillator strength caused by dipole-forbidden band-to-band transitions, or by spin-flips.

c) Polaritons

If there is a direct, dipole-allowed band-to-band transition, then the excitons with an s-envelope function $(l=0)$ are strongly coupled to the radiation field. Light then propagates as a mixed state of electromagnetic wave (photons) and electronic excitation. The quanta of this mixed state are called excitonic polaritons. The dispersion of the polaritons result from those of photons and of excitons and the noncrossing rule. It is obtained mathematically by solving the polariton equation (6.9) and the equation $\varepsilon(\omega, K)$ of the dielectric function, which is given in (6.10) for one single-exciton resonance. Resonances of higher energy are taken into account by a background dielectric constant ε_b. The functions $f(K)$ and $\Gamma(K)$ are the oscillator strength and the damping, respectively. The latter two may both depend on K. Due to the K dependence of the exciton energy $E(K)$ (the so-called spatial dispersion), (6.10) contains ω and K as independent variables:

$$\varepsilon(\omega, K) = \frac{c^2 K^2}{\omega^2}, \tag{6.9}$$

$$\varepsilon(\omega, K) = \varepsilon_b \left(1 + \frac{f(K)}{E(K)^2 - \omega^2 - i\omega\Gamma(K)} \right). \tag{6.10}$$

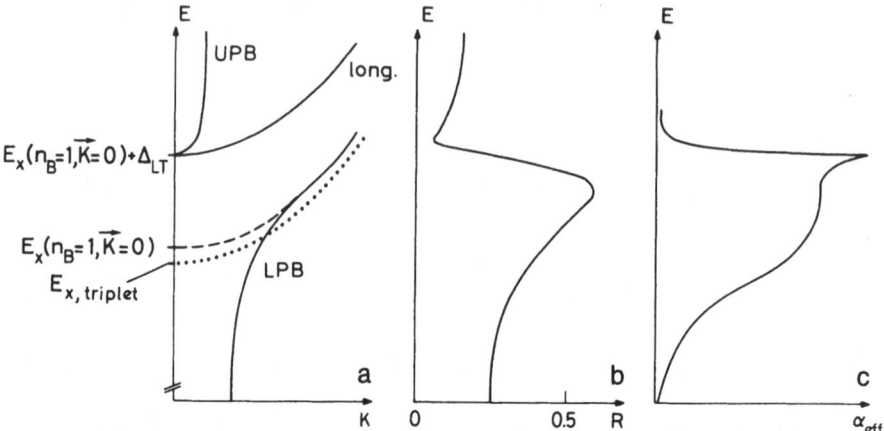

Fig. 6.6. (a) the dispersion relation of an exciton – polariton and the resulting reflection (b) and absorption spectra (c) (schematic drawing). (From [5.15])

The dispersion relation resulting from (6.9) and (6.10) is shown for negligible damping around $E_x(n_B = 1, K)$ in Fig. 6.6a, together with the resulting reflection and absorption spectra (Fig. 6.6b, c). The lower polariton branch (LPB) starts with an almost vertical, "photon-like" dispersion and then bends over to an exciton-like behavior. There is a finite longitudinal–transverse splitting Δ_{LT} which is proportional to the oscillator strength of the resonance and which comes from the nonanalytic or long-range part of the electron–hole exchange interaction [6.16]. There is a longitudinal exciton mode and an upper polariton branch (UPB) which bends over to a photon-like behavior again. We show also the dispersion of an exciton which does not couple to the photon field, e.g. a triplet state with parallel electron and hole spins. This state is shifted with respect to the dipole-allowed singlet by the analytic or short-range part of the exchange interaction [6.16]. It can be seen that for certain energies there are two propagating polariton branches due to the spatial dispersion. This causes a lot of complications, among others the problem of additional boundary conditions. This topic (see, e.g., [6.17]) is however beyond the scope of this contribution. Similar resonances appear for higher s-states, however, with an oscillator strength decreasing like n_B^{-3}. In the continuum, which corresponds to the band-to-band transitions, one has a rather smooth behavior of the real and imaginary parts of the refractive index, $\tilde{n} = n + i\kappa$, or the dielectric function, $\varepsilon = \varepsilon_1 + i\varepsilon_2 = \tilde{n}^2$. The oscillator strength is still enhanced above the ionization continuum by the Coulomb interaction, resulting in a rather constant absorption coefficient, α, instead of the square-root dependence of (6.7a). As a rule of thumb, these modifications extend by about ten times the value of E_x^b into the band-to-band transitions. That high in the bands, there are usually already significant deviations from parabolicity so that (6.7a) is difficult to verify experimentally. There are a lot of improvements to be made in order to describe

Table 6.1. The relation between E_g, E_x^b, and Δ_{LT} for three typical direct gap semiconductors

	E_g [eV]	E_x^b [meV]	Δ_{LT} [meV]
CuCl	3.5	200	5
CdS	2.5	30	2
GaAs	1.5	4	0.1

excitons and polaritons in real crystals. There are degeneracies, anisotropies, k-linear terms or splittings in the valence band, which make a separation into relative and center-of-mass motion (6.8) impossible. The effective polariton masses and the dielectric constant depend on n_B, resulting in deviations from a hydrogen-like series of states. Coupling between the envelope and the electron and hole and other effects lead to a splitting of levels with the same n_B. More detailed information on these topics can be found in [6.17–20] and the references therein.

With increasing temperature T, the damping of the free exciton resonances increases too, depending on $k_B T$ and on the strength of the exciton–phonon coupling. Consequently, the distinct reflection and absorption structures are smeared out and one often observes only a hump on the low energy side of the absorption spectrum. Nevertheless, there is still the Coulomb interaction between electrons and holes and the resulting transfer of oscillator strength towards the band edge.

A low energy wing develops in the absorption edge with temperature according to the Urbach-Martienssen rule

$$\alpha(\hbar\omega) = \alpha_0 \exp[\sigma(\hbar\omega - E_0)/k_B T]. \tag{6.11}$$

The energy E_0 equals roughly the $n_B = 1$ exciton energy; α_0 and σ are material parameters. For the microscopic explanation of (6.11) see [6.21]. A general trend is that the binding energy E_x^b of excitons and their oscillator strength $f \sim \Delta_{LT}$ decrease with decreasing E_g. Examples are shown in Table 6.1.

With decreasing E_x^b and Δ_{LT}, it becomes increasingly difficult to observe distinct exciton resonances even at low temperature and in pure materials. In narrow gap semiconductors ($E_g < 0.5$ eV) like InSb, the lead salts, or $Cd_{1-x}Hg_xTe$, the Coulomb interaction generally only leads to a modification of the shape of the absorption edge, not to resolvable peaks.

It should be stressed that the concept of polaritons is a rather universal one and not a peculiarity of excitons. Whenever light propagates in matter with a refractive index $\tilde{n} \neq 1$ one has actually a mixture of an electromagnetic wave and some excitation of matter, that is, polaritons. Apart from exciton polaritons there are, e.g., phonon polaritons or plasmon polaritons.

d) Impurity States

Apart from the free excitons described above, which are characterized by the plane wave factor $\exp(i\mathbf{K} \cdot \mathbf{R})$, there are excitons bound to defects like neutral acceptors and donors. At low temperatures they show up as extremely narrow absorption and emission lines situated energetically below the free exciton resonances. With increasing temperature they tend to disappear between 20 K and 70 K. Details about bound excitons are found in [6.22]. Increasing doping causes at low temperatures a broadening of both the free and the bound exciton resonances. The free resonances are broadened partly inhomogeneously, partly due to collisions with the impurities. The bound excitons broaden to bands due to mutual interaction.

In mixed crystals one observes at low temperatures a tail of localized exciton states separated by a mobility edge from the delocalized ones. The localization is induced by compositional disorder. In the II–VI mixed crystals one observes localized excitons preferentially in anion-substituted materials like $CdS_{1-x}Se_x$ [6.23]. In this case the fluctuations in x influence mainly the VB, resulting primarily in localized hole states. Binding of an electron to a localized hole leads then to the formation of localized excitons. In cation-substituted materials it is much more difficult to observe localized excitons, since the electrons are less easily localized because of their smaller effective mass [6.23]. Another possibility is to localize excitons as a whole. This process is dominant if the length scale of the potential fluctuations is larger than the exciton Bohr radius (6.8) and in the semimagnetic Mn mixed crystals [6.24].

6.1.2 Nonlinear Optical Properties

In linear optics, one expects the optical properties of matter like the transmission $T(\omega)$ and reflection $R(\omega)$ spectra or the complex dielectric function $\varepsilon(\omega) = \tilde{n}^2(\omega)$ to depend on the frequency of the incident radiation field and, in the case of birefringent materials, also on the direction of polarization and propagation relative to the crystallographic axes, but definitely not on the intensity I or field amplitude E of the radiation field. Consequently, the polarization P of matter oscillates with the same frequency ω as the incident radiation, and two light beams which cross each other in matter will not interact.

All these assumptions are no longer valid in nonlinear optics. This means T, R and ε depend in a reversible way on the light intensity, and light beams start to interact with each other in matter. There are two approaches to describing these types of phenomena. In one case one assumes that the response of the medium to the incident light field, i.e., the polarization, depends only on the instantaneously present field amplitudes. This condition is fulfilled if (electronic) excitations are created only virtually. In this case the polarization $P = \varepsilon_0(\varepsilon - 1)E = \varepsilon_0 \chi E$ can be expanded into a power series of the incident field amplitudes:

$$\frac{1}{\varepsilon_0} P_i = \sum_j \chi_{ij}^{(1)} E_j + \sum_{jk} \chi_{ijk}^{(2)} E_j E_k + \sum_{j,k,l} \chi_{ijkl}^{(3)} E_j E_k E_l + \dots . \tag{6.12}$$

The first term on the right-hand side describes the linear optical properties; $\chi^{(2)}$ gives phenomena like second harmonic-, sum-, and difference-frequency generation or the dc effect; and $\chi^{(3)}$ describes four-wave mixing, hyper Raman scattering (HRS), coherent anti-Stokes-Raman scattering (CARS), etc. [6.25–28]. Some of these terms will be discussed below.

In the second case, the optical properties are modified by real excitations of some quasiparticles with finite lifetime [6.29]. In this case, one has

$$P = \varepsilon_0 \chi(N)E, \tag{6.13}$$

where N stands for an increase of the electron–hole pair density, or of the phonon population number or of the lattice temperature. Due to the finite lifetime, N depends not only on the instantaneous intensity but also on the generation rate, G, in the past, weighted with some decay function:

$$N(t) = \int_{-\infty}^{t} G(t') \exp[-(t-t')/\tau] dt. \tag{6.14}$$

The rate G itself is connected with the intensity I at time t', e.g., in the presence of one- and two-photon excitation by

$$G(t') = \alpha_1 I(t') + \beta I^2(t'). \tag{6.15}$$

The coefficients τ, α, β, may in turn depend on N, leading thus to a rather complex set of equations. In the following, we shall find examples for both types of nonlinearities, i.e., coherent (6.12) and incoherent (6.13) ones.

The variations of the optical properties, such as ε at a frequency ω by excitation at another frequency ω_{exc} and with intensity I_{exc} are often called renormalization effects. By intensity we mean the energy flux density, i.e., the Poynting vector $S = E \times H$ averaged over a period of the light. If one considers the variation of ε or $\tilde{n}$ produced by I_{exc} at $\hbar\omega_{\text{exc}}$ one usually speaks about self-renormalization.

In the following subsections we describe some selected renormalization processes. We start with inelastic scattering processes, then we treat transitions involving biexcitons and finally the electron–hole plasma.

a) Scattering Processes

If we increase the laser intensity falling on a SC and thus the generation rate [e.g., via (6.14) and (6.15)] we may reach densities of excitons or more generally of electron–hole pairs n_p at which these quasiparticles start to interact with each other. The simplest interaction mechanisms, which have been discussed already for many years are elastic and inelastic scattering processes [6.30–32].

Two exciton-like polaritons in the $n_B = 1$ state may collide. One of the inelastic processes involves a scattering of one of the excitons into the photon-like part of the LPB, the other one under energy and momentum conservation into states with $n_B = 2, 3, ..., \infty$. This leads to the well-known luminescence bands P_2, P_3, P_∞. Depending on the losses and on the distribution function of the polaritons in the various exciton-like branches this scattering may result in optical amplification (gain) and/or in excitation-induced absorption [6.32]. The various P bands have been preferentially found in several II–VI semiconductors like CdSe, CdS, ZnO or in GaAs [6.32].

Recently some authors have proposed a different interpretation of the P bands for CdS, mainly in terms of biexciton decay [6.33, 34]. (For biexcitons see below.) The processes discussed especially in [6.34], namely, a decay of a biexciton containing two holes from the uppermost $A\Gamma_9$ valence band into an $A\Gamma_5$ polariton on the LPB and a B exciton, i.e., an exciton containing a hole from the B valence band (Fig. 6.2c), are likely to contribute to the emission in the spectral region of the P_2, P_3, and P_∞ bands in CdS under appropriate excitation conditions. They seem, however, not to be of sufficiently general nature to explain the corresponding results in the other compounds. Furthermore, the process proposed in [6.34] involves a spin-flip of the remaining hole from the $A\Gamma_9$ to the $B\Gamma_7$ state, which reduces the recombination rate significantly. The data in [6.33] give on the other hand a biexciton binding energy which is not compatible with the data of other authors [6.20, 32].

While elastic and inelastic exciton–exciton scattering are predominant at low temperatures, one may imagine similar processes between one exciton-like polariton and a free carrier at higher temperatures, where a certain fraction of the excitons is ionized thermally. This scattering gives rise to a luminescence band and to excitation-induced absorption and gain structures characterized by the fact that the maxima of gain and luminescence shift with increasing temperature faster to smaller photon energies than the band gap does. Models developed to describe this phenomenon are reviewed in [6.32]; a more recent calculation is found in [6.35] which nicely coincides with experiment. In [6.36] it has been claimed that these emission bands may be due to recombination in an electron–hole plasma (see below) but this interpretation has the following shortcoming. At room temperature, the electron and hole populations in the plasma are no longer degenerate. This means, there is no gain due to direct band-to-band recombination in the plasma. On the other hand, stimulated emission has clearly been observed up to room temperature in the emission band presently under consideration [6.37]. As indicated in [6.38] a possible bridge between these two models is to assume inelastic scattering also in a plasma or in a state which is between an exciton gas and a plasma. This idea seems to be favored by the fact that there is a rather continuous transition from the exciton to the plasma state in direct gap materials even at low temperature and in indirect ones above the critical temperature T_c for electron–hole liquid formation.

In Fig. 6.7, we summarize various scattering processes. The upper row shows the exciton–electron scattering process and the appearance of a spectrally broad emission if the thermal distribution of the particles is taken into account. Figure 6.7b inelastic exciton–exciton scattering (the process leading to the P_2 line). The scattering of a polariton from the exciton-like to the

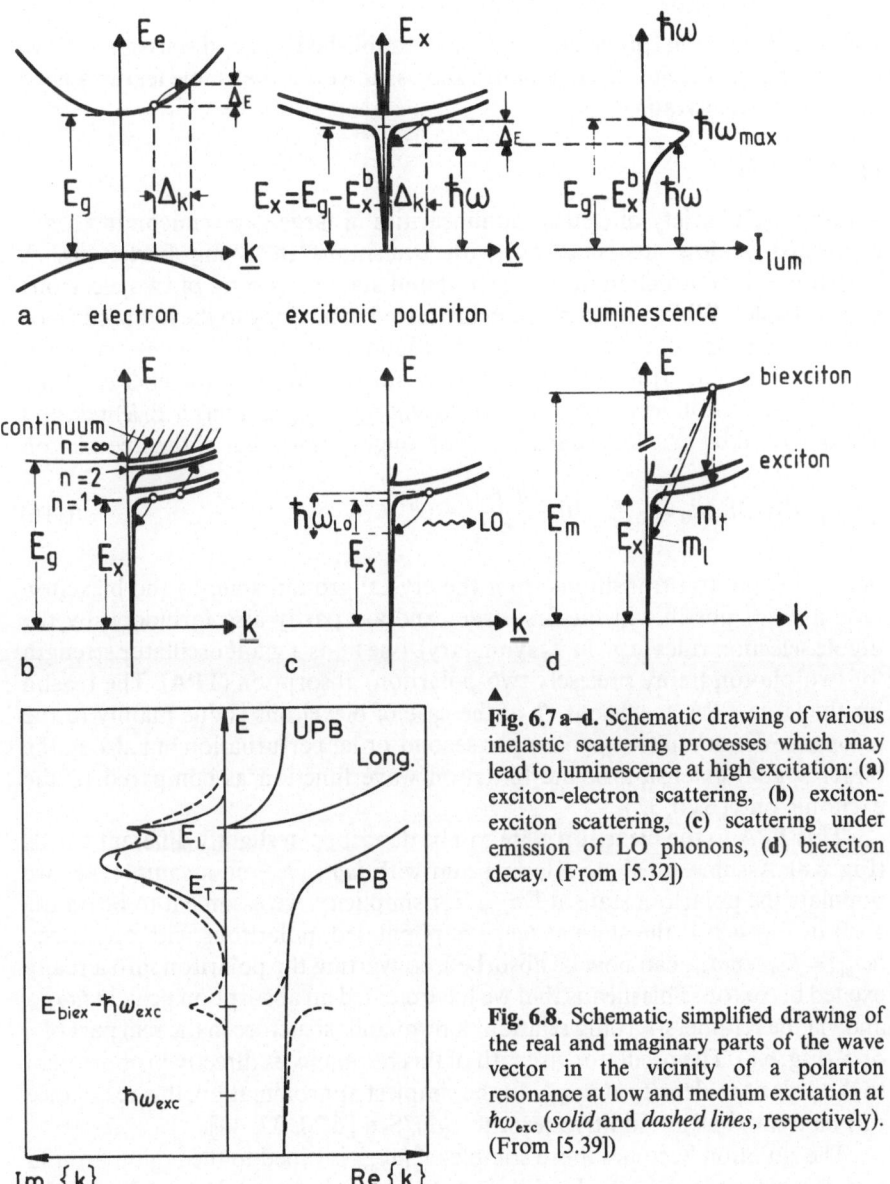

Fig. 6.7 a–d. Schematic drawing of various inelastic scattering processes which may lead to luminescence at high excitation: (**a**) exciton-electron scattering, (**b**) exciton-exciton scattering, (**c**) scattering under emission of LO phonons, (**d**) biexciton decay. (From [5.32])

Fig. 6.8. Schematic, simplified drawing of the real and imaginary parts of the wave vector in the vicinity of a polariton resonance at low and medium excitation at $\hbar\omega_{exc}$ (*solid* and *dashed lines*, respectively). (From [5.39])

photon-like part of the dispersion curve by emission of a LO phonon is shown in Fig. 6.7c. This process occurs already at low excitation but it leads to stimulated emission at high excitation [6.32]. Possible decay processes of a biexciton are shown in Fig. 6.7d and will be described in the next section. All the scattering processes outlined above will contribute to a collision broadening of the free exciton resonance. This excitation-induced collision broadening of the free exciton resonance is shown schematically in Fig. 6.8. The basic ideas of inelastic scattering processes are well established. The discussion above concerning the origins of the P bands shows, however, that some features need more detailed investigation.

b) Biexcitons

A rather rich variety of optical nonlinearities in large-gap semiconductors is connected at low temperatures with transitions involving biexcitons. A biexciton (or excitonic molecule) is a bound state consisting of two electrons and two holes. While the exciton can be treated in analogy to the hydrogen – or the positronium–atom, a biexciton corresponds to the hydrogen – or posi-tronium-molecule. It forms a bound state situated at twice the energy of the lowest free exciton less a certain binding energy $E_{\text{biex}}^{\text{b}}$ and has a $E(K)$ relation characterized by an effective mass M_{xx} of roughly twice that of the free exciton

$$E_{\text{xx}}(K) = 2E_{\text{x}}(n_{\text{B}} = 1, \ K = 0) + \frac{\hbar^2 K^2}{2M_{\text{xx}}} - E_{\text{xx}}^{\text{b}}. \tag{6.16}$$

While one-photon transitions from the crystal ground state to the biexciton level are not possible in the first-order and are partly also forbidden by the dipole selection rules (e.g., in T_{d} symmetry), one finds a giant oscillator strength for two-photon (more precisely two-polariton) absorption (TPA). The reason for the large TPA coefficient, β, in the case of biexcitons is due mainly to the resonance denominator appearing in second-order perturbation but also to the larger spatial extension of the biexciton wave function as compared to the excitonic one [6.20, 32].

This TPA to the biexciton state can be described in slightly different words (Fig. 6.8). Assume we shine a laser beam with $\hbar\omega_{\text{exc}}, I_{\text{exc}}$ on a sample, i.e., we populate the polariton state at $\hbar\omega_{\text{exc}}$. For simplicity, we assume it to be on the LPB of Fig. 6.8. If the state at $\hbar\omega_{\text{exc}}$ is populated, polaritons with an energy $\hbar\omega_{\text{abs}} = E_{\text{xx}} - \hbar\omega_{\text{exc}}$ can now be absorbed, converting the polariton into a really excited biexciton. This means that we have created an absorption peak at $\hbar\omega_{\text{abs}}$, and via the Kramers-Kronig relations a resonance structure in the real part of $\tilde{n}$ or K (Fig. 6.8). The oscillator strength of this resonance is directly proportional to the polariton density at $\hbar\omega_{\text{exc}}$. In the simplest approximation, this resonance can be treated as an additive term to $\varepsilon(\omega)$. See [6.20, 32, 40].

The situation becomes more complex if $\hbar\omega_{\text{exc}}$ is tuned to the region around half the biexciton energy. In this case the laser beam at $\hbar\omega_{\text{exc}}$ falls on the resonance at $\hbar\omega_{\text{abs}}$, i.e., we induce strong self-renormalization phenomena. Various theoretical approaches to this self-renormalization have produced a rather surprising result, specifically, a quadratic pole is predicted in the dispersion relation of the excitonic polariton, together with a strong deform-ation at the exciton resonance itself [6.41]. The solution to this problem was found only recently. It is necessary to include the spatial dispersion, i.e., the K dependence of the eigenenergies (here of excitons and biexcitons) and to treat the electronic excitations and the radiation field on an equal footing. In doing so, one gets a "normal" resonance with some additional polariton branches [6.42].

In cubic materials (T_d symmetry) left and right circularly polarized polaritons are "good" eigenmodes of the system. The crystal ground state and the biexciton ground state both have symmetry Γ_1. These two facts produce induced circular dichroism by TPA. If σ^+ polarized light at $\hbar\omega_{exc}$ is incident on the sample, then a TPA process is possible only if the other quantum is polarized σ^-. This means one gets the additional resonance at $\hbar\omega_{abs}$ (Fig. 6.5) only for this polarization while the dispersion for σ^+ light is not influenced in this region. If one shines linearly polarized light around $\hbar\omega_{abs}$ on the sample, the beam is decomposed into two beams of initially equal intensities polarized σ^+ and σ^-, respectively. The σ^+ polarized polaritons propagate in the sample according to the unrenormalized dispersion curve, while the σ^- ones are absorbed and propagate with a different velocity. As a consequence, the probe beam around $\hbar\omega_{abs}$ becomes elliptically polarized after passing through the sample, with the long axis of the ellipse being turned with respect to the original linear polarization. These phenomena have been investigated in CuCl experimentally [6.43] and theoretically [6.40,44]. A similar effect should be observable also in uniaxial materials (e.g. C_{6v} symmetry) when both beams propagate parallel to the crystallographic axis.

Apart from the TPA processes described above, there is also induced absorption. By this term one means transitions to the biexciton state from exciton-like polaritons which are distributed at the bottleneck or thermally on the exciton-like part of the LPB, the UPB or the longitudinal exciton branch. Obviously one needs some resonant or nonresonant pump source to produce the excitons. The spectral position of the induced absorption is rather independent of $\hbar\omega_{exc}$ and occurs around $\hbar\omega = E_{xx} - E_x$. The shape of the absorption peak is determined by the distribution of the excitons on the various branches [6.45]. The inverse process occurs also, i.e., the decay of biexcitons into a photon-like polariton appearing as luminescence and an exciton-like polariton (Fig. 6.7d). The shape of this luminescence band depends on the distribution of the biexcitons. The gain and/or loss spectra are determined in addition by the population of the various exciton and biexciton branches [6.32,45].

In TPA and induced absorption, biexcitons are created at their proper energy. It is also possible to create virtual biexcitons with two incident quanta at an energy different from their eigenenergy. Such an excitation can exist only for times τ given by the uncertainty principle, following which it must disappear. One such radiative decay process is a decomposition into a photon-like polariton, $\hbar\omega_R$, and a longitudinal exciton at energy $\hbar\omega_f$. Energy and momentum have to be conserved in the hole process. If the virtual biexciton is created from two identical quanta $\hbar\omega_{exc}$, one finds

$$\hbar\omega_R = 2\hbar\omega_{exc} - \hbar\omega_f,$$

$$K_R = 2K_{exc} - K_f. \qquad (6.17)$$

Since the eigenenergy of the longitudinal branch is only weakly dependent on K, a plot of $\hbar\omega_R$ against $\hbar\omega_{exc}$ gives a straight line with slope two. Therefore, this process is called two-photon- or hyper-Raman scattering (HRS). If the final state $\hbar\omega_f$ is a polariton, the energy can depend sensitively on K (Fig. 6.6a). As a consequence, one finds values of the slopes ranging from zero to two in a plot $\hbar\omega_R$ versus $\hbar\omega_{exc}$ depending on the scattering geometry. The relation between $\hbar\omega_{exc}$ and $\hbar\omega_R$ is usually a rather smooth and monotonous one. If, however, $\hbar\omega_{exc}$, $\hbar\omega_f$, or $\hbar\omega_R$ fall on a sharp resonance, then discontinuities appear. Thus, HRS can be used as a means to detect and analyze excitation-induced resonances like the one at $\hbar\omega_{abs}$ shown in Fig. 6.8. The HRS process and its inherent possibilities of K-space spectroscopy are discussed in more detail in [6.20, 32, 46], for example.

If the sample has a resonator-like shape, formed for example by two plane parallel surfaces, the HRS process can result in laser emission. The emission wavelength can be tuned by changing the scattering geometry [6.47]. On the other hand, it is also possible to stimulate the decay of the virtually excited biexcitons by sending an additional, monochromatic beam with energy $\hbar\omega_f$ and wave vector K_f into the sample. The virtually excited biexciton then decays under energy and momentum conservation into a quantum $\hbar\omega_f$, K_f and another one $\hbar\omega_R$, K_R. This process is identical with degenerator or nondegenerate four-wave mixing depending on whether $\hbar\omega_{exc} = \hbar\omega_R = \hbar\omega_f$ or not. Alternatively, this $\chi^{(3)}$ process can also be considered as diffraction from a dynamic, laser-induced grating [6.20, 48–50].

c) Electron–Hole Plasma

Until now we considered optical nonlinearities in a range of excitation densities where excitons and biexcitons are still good quasiparticles [6.51]. With increasing electron–hole pair density this picture breaks down, as shown schematically in Fig. 6.9. In the upper half of the "sample" we show free and bound excitons, a scattering process and a biexciton. In the lower half, the excitation is so strong that the mean distance between excitons is comparable to their Bohr radius. Under this condition, it is no longer possible to say that a given electron is bound to a given hole. Instead one gets a new collective phase of electrons and holes, the so-called electron–hole plasma (EHP). Another point of view is to state that the Coulomb interaction between an electron and a hole is screened by the other carriers to an extend that there are no more bound states. The formation of an EHP has drastic consequences on the optical spectra of semiconductors. The effects are illustrated for low and high temperatures in Fig. 6.10.

Figure 6.10a shows the low temperature case. On the left-hand side, one sees the absorption spectrum at low excitation, revealing the series of exciton states and at higher energies the transitions into the continuum. The right-hand part gives various energies as a function of the electron–hole pair density n_p. The width of the forbidden gap E_g is a monotonically decreasing function of n_p

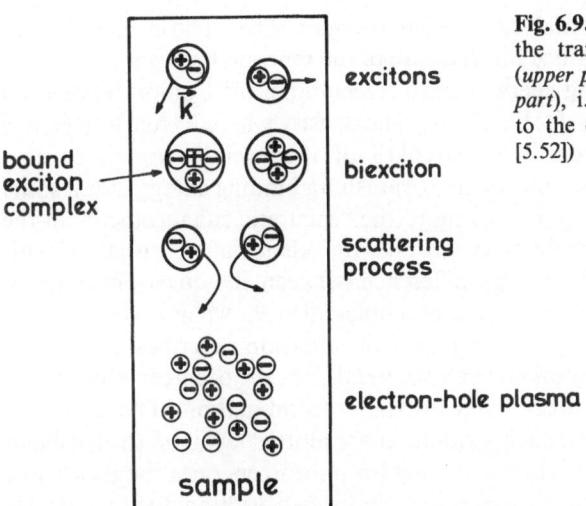

Fig. 6.9. Schematic representation of the transition from low excitation (*upper part*) to high excitation (*lower part*), i.e., from individual excitons to the electron-hole plasma. (From [5.52])

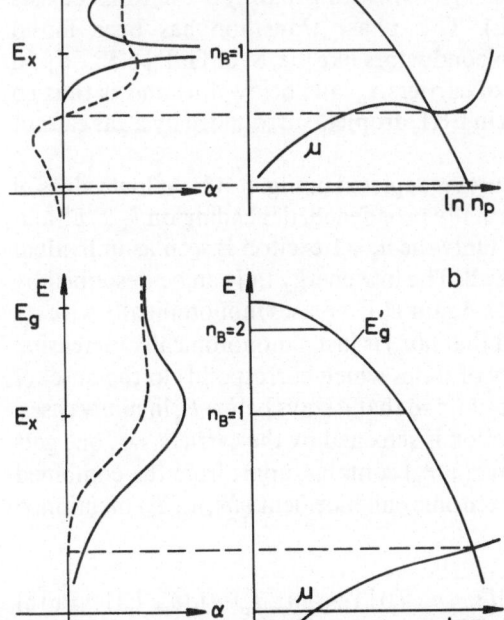

Fig. 6.10. Sketch of the variation of the absorption spectra (*left*) in the transition from excitons to the electron – hole plasma for (**a**) low and (**b**) high temperatures. The right-hand side gives various energies as a function of the electron-hole pair density n_p. (From [5.52])

mainly due to exchange and correlation energies. The exciton energies are nearly independent of n_p since the decrease of the renormalized gap $E_g(n_p)$ and the decrease of the binding energy due to screening of the Coulomb potential compensate for each other [6.32, 53, 54]. The density where E_x^b tends to zero is the Mott density. The oscillator strength of the exciton resonance also decreases with increasing n_p but even for vanishing binding energy is still some electron–hole pair correlation, leading to the "excitonic enhancement" in the plasma recombination probability around the chemcial potential μ. This chemical potential μ is the energy difference between the quasi-Fermi-levels describing the population of electrons and holes, also shown in Fig. 6.10. With increasing n_p, μ can move above $E_g(n_p)$. This situation describes population inversion between conduction and valence bands, i.e., the occurrence of optical amplification (gain) by direct band-to-band recombination. The gain and absorption (loss) spectra for such a situation are plotted again on the left-hand side of Fig. 6.10. The sample is transparent for photon energies roughly below $E_g'(n_p)$. Between E_g' and μ one has gain and above μ absorption due to band-to-band transitions. The fact that the gain spectrum extends somewhat below E_g' is due to a broadening connected with final state damping [6.32]. The structure of $\mu(n_p)$ as a function of increasing n_p shows a maximum and then a minimum and is thus similar to the van der Waals equation for real gases. One can apply Maxwell's construction, resulting in a first-order phase transition below a critical temperature T_c. A low density gas consisting mainly of excitons coexists with a liquid-like plasma (EHL). This phase transition has been found experimentally in indirect gap semiconductors like Ge, Si or GaP [6.55, 56]. In direct gap materials, the lifetime of carriers τ_p falls below 1 ns and is thus so short that a clear phase separation in EHL droplets surrounded by a gas cannot develop [6.57–59].

The situation at high temperatures is depicted in Fig. 6.10b. At low values of n_p, the exciton resonances are thermally broadened, depending on $k_B T/E_x^b$ and on the exciton–phonon coupling. Only the $n_b = 1$ exciton is seen as individual structure, if there are structures at all. The low energy tail can be described by the Urbach-Martienssen rule (6.11). Again E_g' decreases monotonically with n_p. Since we are now above T_c, we find that $\mu(n_p)$ is just a monotonically increasing function of n_p. The effective density of states which corresponds to the onset of population degeneracy increases as $T^{3/2}$ so that μ stays below E_g' in many cases. Nevertheless, the Coulomb interaction is screened by the carriers, i.e., one gets an EHP. The absorption spectrum $\alpha(\omega, n_p)$ contains, apart from the combined density of states $D(\omega, n_p)$ and the excitonic enhancement $\varrho(\omega, n_p, T_p)$ mentioned above, a population term:

$$\alpha(\omega, n_p, T_p) \sim D(\omega, n_p)\varrho(\omega, n_p, T_p)\Gamma(\omega, n_p, T_p)\,[1 - f_e(n_p, T_p) - f_h(n_p, T_p)],\quad (6.18)$$

where f_e and f_h are the quasi-Fermi-functions of electrons and holes, respectively. Since we have $\mu \approx E_g$ in the EHP, we find for $\hbar\omega \approx E_g$, $1 - f_e - f_h \approx 0$. Since at high temperatures the Fermi functions are quite smoothly varying functions of energy, we get weak absorption over a rather large range of energies around E'_g, as is shown on the left-hand side of Fig. 6.10b (dashed line). This means that we get a blue-shift of the absorption edge as compared to the low density case. At low temperatures we have a red shift as long as μ stays below the exciton energy. For $\mu > E_x$ we again find a blue shift. The blue shift of the absorption edge, which occurs at high temperatures and which may appear also at low temperatures depending on the generation rate and the material parameters, has sometimes been interpreted in terms of a simple dynamic Burstein-Moss shift, i.e., as a band-filling effect. Actually for a quantitative description one always has to include band filling and many-body effects like band-gap renormalization and screening of the Coulomb interaction [6.60–62].

Until now, we have discussed the EHP formation mainly with respect to direct gap materials. In indirect gap materials there is no measurable gain even in the case of population inversion. The participation of momentum-conserving phonons reduces the transition probabilities and thus the gain to values below 1 cm^{-1} [6.55, 56]. This is below the detection threshold. Information about the plasma is obtained mainly from luminescence spectroscopy. The indirect absorption edge is modified by the creation of a plasma, and the direct exciton behaves in indirect gap materials in the same way as in direct gap materials [6.40, 63, 64].

Due to the high hole concentration in an EHP, inter-valence band transitions become possible (Fig. 6.3b). The corresponding induced absorption bands are situated in the IR part of the spectrum [6.9, 60, 65].

The electron–hole pair density, n_p, which is the crucial parameter in Fig. 6.10 apart from the temperature can sometimes be assumed to increase linearly with I_{exc}. Often one finds a saturation of the curve $n_p = n_p(I_{exc})$, e.g., connected with the onset of stimulated emission. In some materials the dominant recombination process is of the Auger type. In this case the recombination rate increases with the third power of the carrier density. By simple reaction kinetics it can be shown that this results in a dependence $n_p \sim I_{exc}^{1/3}$ [6.66].

d) Photothermal Optical Nonlinearities

In the last part of this section, we discuss briefly changes of the optical properties connected with sample heating due to light absorption. In most semiconductors one has a red shift of the gap with increasing temperature; this shift starts quadratically with T and then evolves to a linear T dependence [6.7, 67]. Free and bound exciton energies tend to shift parallel with the gap. If the photon energy $\hbar\omega$ is below the free exciton resonance, one finds an increase of α and n for constant $\hbar\omega$ due to this phenomenon and the Urbach-Martienssen rule (6.11). Exceptions may occur at very low temperatures where narrow absorption structures connected with bound exciton complexes may shift over $\hbar\omega$ as the incident intensity increases, thus increasing the sample temperature, resulting in

$$\frac{\partial \alpha}{\partial I} = \frac{\partial \alpha}{\partial T} \cdot \frac{\partial T}{\partial I} \gtreqless 0$$

depending on the relative position of $\hbar \omega$ to the eigenenergy of the BEC.

There are some semiconductors which show an increase of the gap with temperature, e.g., CuCl, CuBr, and some of the narrow gap materials (the lead salts and $Cd_{1-x}Hg_xTe$). In these cases an increase of the incident intensity can result in a decrease of the absorption and of the refractive index [6.7].

Photoelectronic and photothermal optical nonlinearities can be distinguished in direct gap materials by their different decay times. They are of the order of nanoseconds for electronic nonlinearities and in the microseconds to milliseconds range for thermal ones, depending on sample temperature and spot size of the excitation.

In sample with relatively long carrier lifetimes (indirect or narrow gap materials) or which are illuminated with long pulses, optical nonlinearities due to electronic and thermal excitation can be present simultaneously. Depending on the material and the excitation conditions both effects may have the same or opposite signs.

6.1.3 Influence of Dimensionality

The considerations about linear and nonlinear optical properties of SC assumed until now a three-dimensional SC. In the last decade, SCs of reduced dimensionality have become of increasing interest both in basic research and with respect to applications [6.68–70]. In the simplest case, we consider a two-dimensional quantum well (QW) of thickness L_z with barriers of infinite height on both sides, i.e.,

$$V(x, y, z) = \begin{cases} 0 & \text{for} \quad 0 < z < L_z \\ \infty & \text{for} \quad z < 0 < L_z < z. \end{cases} \tag{6.19}$$

The wave function and eigenenergies are then given by

$$\varphi_n(x, y, z) = \left(\frac{2}{L_z}\right)^{1/2} \sin\left(\frac{n\pi z}{L_z}\right) \exp[i(k_x x + k_y y)],$$

$$E_n(K) = \frac{\hbar^2 \pi^2}{2} \cdot \frac{n^2}{L_z^2 m_{eff}} + \frac{\hbar^2(k_x^2 + k_y^2)}{2m_{eff}}. \tag{6.20}$$

This means one has plane waves in the x and y directions and quantization along the z axis. The density of states as a function of energy is a step-like function:

$$D(E) = \sum_n D_n \theta(E - E_n). \tag{6.21}$$

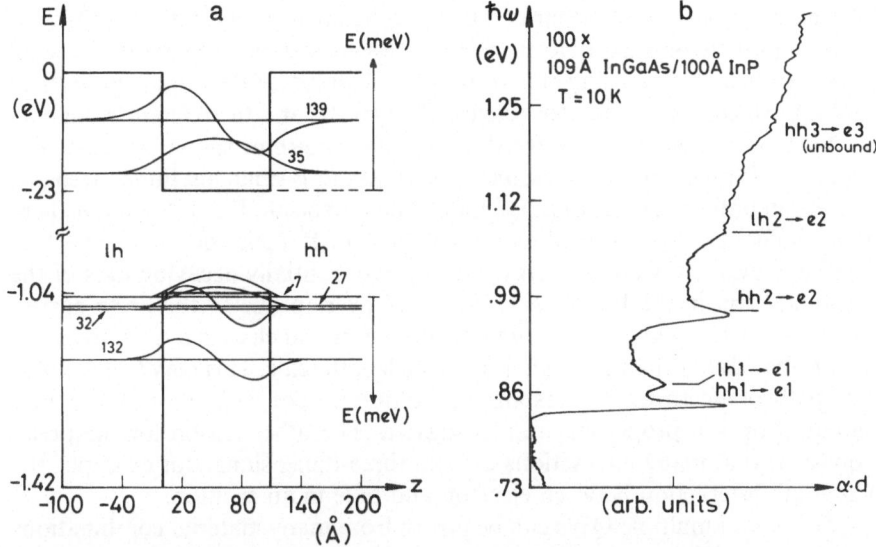

Fig. 6.11. (a) The first two bound states of electrons, and light and heavy holes in a quantum well. Parameters chosen for InGaAs/InP. **(b)** the absorption spectrum of a InGaAs/InP MQW. The band-to-band transition energies from **(a)** are indicated [5.71]. (Copyright: Bell Telephone Laboratories, Inc., 1987; reprinted by permission)

Such a system can be realized to a certain approximation by imbedding a thin slice of a SC with forbidden gap width E_{gw} in a SC with gap E_{gb}, where $E_{gw} < E_{gb}$ in the way shown in Fig. 6.11a. Due to the finite barrier height, the wave functions extend into the barriers. The levels consist of a finite number of bound states and then a continuum. The first two bound states for electrons, and light and heavy holes are shown in Figs. 6.11b. The existence of light and heavy holes results in two series of bound states in the VB. In these cases, the plane wave factors in (6.20) have to be changed to Bloch-type functions.

If one has a situation as in Fig. 6.11a where the states are confined both in the CB and the VB, one speaks about a QW of type I. They are the only ones which are considered in this contribution. Other types of QW, where there is confinement only for one type of carrier or where the forbidden gaps do not overlap, will not be treated here. The reader is referred to [6.70].

The values of L_z are usually chosen between 20 Å and 200 Å. To enhance the effect of a QW in transmission one often grows several of them in one sample. If the thickness of the barriers between the wells is so large that the wave functions of the confined states do not overlap significantly, one speaks about multiple quantum wells (MQW). Their properties are identical with the properties of a single QW. If there is significant overlap, one has a superlattice. The new periodicity with a length larger than the lattice constant leads to a reduction of the Brillouin zone in the k_z direction and to a corresponding folding back of the dispersion curves of phonons or electrons [6.70]. Periodic alternate doping of a SC separated by periodic intrinsic layers leads to the so-called nipi structures; here electrons and holes are separated spatially. We shall concentrate in the following only on single and multiple QWs. For superlattices and nipi structures the reader is referred to [6.70] and the references therein.

Generally QWs are considered to be quasi-two-dimensional systems. The addition of "quasi" has two reasons. One is that L_z is small but finite. Actually, there is an optimum value of L_z for confinement to occur. If L_z is too large, there is no difference to a three-dimensional system. If L_z is too small, the first confined level shifts up in energy and the exponentially decaying tails of the wave function in the barrier get wider and overlap due to the finite barrier height. In order to maintain a quasi-two-dimensional character, it is necessary that L_z be smaller than some characteristic length scale. This may be the mean free path of electrons, if transport properties are considered, or the exciton radius, if optical properties are investigated. The other reason for the prefix "quasi" is that many interactions are still three-dimensional, for example, the Coulomb-interaction between electron and hole in an excition.

Single and multiple QWs can be grown from many material combinations consisting of III–V and/or II–VI compounds [6.70]. The system which was investigated first consisted of GaAs wells with $Ga_{1-x}Al_xAs$ barriers. The reason for this is that these materials are almost perfectly lattice matched and form mixed crystals for all values of x. Other possibilities are indicated in [6.70].

In the following, we shall outline the linear and nonlinear optical properties of SQWs and MQWs.

Due to the step-like density of states (6.21), for band-to-band transitions one would expect an absorption spectrum consisting of a step-like function. This approximation corresponds to the square-root dependence in 3d systems (6.7a). Again there are significant modification due to exciton effects. The spatial confinement of electron and hole in the QW leads to an increase of both the oscillator strength and the exciton binding energy as compared to the 3d case. In a strictly 2d confinement, the hydrogen series $E_x^b(1/n_B^2)$ of (6.8) becomes

$$ E_x^b \, \frac{1}{(n_B - \frac{1}{2})^2}, \tag{6.22} $$

i.e., the binding energy of the $n_B = 1$ state increases by a factor of four. In actual quasi-2d systems values of about 3.5 can be reached. The series of states in (6.22) cannot usually be resolved because there is considerable inhomogeneous broadening of the exciton resonances even at low temperatures. One contribution to this broadening comes from spatial fluctuations of L_z by at least one atomic layer. The other concerns spatial concentration fluctuations if the well and/or the barriers are made from an alloy (e.g., in the systems $GaAs/Ga_{1-x}Al_xAs$ or $Ga_{1-x}In_xAs/Ga_{1-x}Al_yAs$) since fluctuations in the composition result in fluctuations of E_g. Both effects may lead to localization effects as discussed in some detail in [6.69]. Figure 6.11b gives an absorption spectrum of a $Ga_{1-x}In_xAs/InP$ MQW at low temperature. The various exciton resonances situated in front of the band-to-band transitions can be seen nicely.

Concerning optical nonlinearities, not much is known about biexcitons in QW, since biexciton binding energies tend to be small compared to the broadening of the exciton resonances in most of the materials investigated so far.

The EHP is known in MQWs [6.72]. There is a reduction of the gap due to many-particle effects in the EHP, mainly arising from exchange and correlation. The screening of the Coulomb potential between electron and hole is less pronounced in quasi-2d systems compared to 3d ones. The reason is that only the electric field lines in the QW, not the fields in the barrier, can be easily screened by additional carriers. Consequently, some finite exciton binding energy exists in the 2d EHP. Generally one observes a blue shift of the absorption edge in an EHP in a (M)QW. This is due to the fact that the filling of the phase space under excitation, i.e., the Burstein-Moss effect, overcompensates the renormalization of the gap [6.72–74].

There is another nonlinear effect which may occur in both 3d and quasi-2d SCs, namely the ac Stark effect [6.75–77]. This is a Stark effect introduced by the electric ac field of an intense light beam propagating through the sample. This effect can be described qualitatively in the following way using perturbation theory, i.e., in the limit of weak exciton–photon interaction [6.74, 77]. Assume an intense light beam is sent into the sample with an energy slightly below the exciton energy. Then excitons are virtually excited. These virtual excitations lead to a filling of the phase space for electrons and holes. Consequently the absorption in the exciton resonance is partly bleached and the absorption peak shifts to higher photon energies [6.76, 77]. In other words, this may be considered as a "repulsion" between photons and electronic excitations similar to that found in atomic transitions [6.26, 76, 78]. In [6.74, 77] the approximate shift of the $1s$ exciton δE_{1s} in a quantum well is given by

$$\delta E_{1s} = 2\frac{|er_{cv}E_p|^2}{E_{1s}-\hbar\omega_{exc}} \cdot \frac{|U_{1s}(r=0)|^2}{N_S^{PSF}},\tag{6.23}$$

where r_{cv} is the band-to-band transition matrix element, $|U_{1s}(r=0)|^2$ gives the excitonic enhancement, E_p is the field amplitude of the pump, $E_{1s}-\hbar\omega_{exc}$ is the detuning, and N_S^{PSF} the saturation density due to excitonic phase space filling. Since all excitations are only virtual ones, the effect is ultrafast with a time scale depending on $E_{1s}-\hbar\omega_{exc}$.

6.2 Methods of Laser Spectroscopy

In this section we present some typical methods of laser spectroscopy in semiconductors. We present first linear techniques and in the second part nonlinear ones. The last part is devoted to some aspects of time resolved spectroscopy.

6.2.1 Linear Spectroscopy

In linear optics, the use of lasers is necessary if processes are investigated which need spectrally extremely narrow light sources with high power density per spectral interval or, in other words, high occupation numbers per unit volume in phase space. Such processes are Raman scattering with optical phonons, which has a low transition probability, Brillouin scattering with acoustic phonons, which also has low probability and requires the detection of small frequency shifts of the order of 0.1 meV. Luminescence excitation spectroscopy of bound or localized excitons is another field where spectrally narrow, tunable light sources are necessary. In this type of experiment one keeps the incident intensity constant and varies the photon energy. The intensity of a certain emission line such as the recombination from the ground state of the bound exciton complex is recorded as a function of $\hbar\omega_{exc}$. Excited states of these complexes show up as resonances in the excitation spectrum $I_{lum}(\hbar\omega_{exc})$ if the excited state can be created optically and relaxes with reasonable probability into the ground state. For a quantitative interpretation, additional details have to be considered such as the reflectivity as a function of $\hbar\omega_{exc}$, the excitation depth, which is roughly given by $[\alpha(\hbar\omega_{exc})]^{-1}$ or by the diffusion length of the excited species and the escape depth of the luminescence $[\alpha(\hbar\omega_{lum})]^{-1}$.

Raman and Brillouin excitation spectra reveal interesting features of excitonic resonances. As we shall see in Sect. 6.3, the Stokes and anti-Stokes shifts may vary with $\hbar\omega_{exc}$ in the latter cases. The frequency shift depends strongly on the scattering geometry; this is also the case in Raman scattering if photon-like phonon-polaritons are involved [6.9]. A typical setup for these types of measurements is shown schematically in Fig. 6.12a. It consists of a narrow, tunable dye laser (e.g. a ring laser) pumped by a cw gas laser (e.g., an Ar$^+$ laser). The sample, which is mounted in a cryostat, is illuminated and the luminescence and/or the scattered light are dispersed by a spectrometer. For the suppression of stray light, usually double monochromators are used and for extremely high spectral resolution Fabry-Perot devices are employed. The light is detected and the electronic data are handled by data processors which may also operate the dye laser. In Fig. 6.12b we show two scattering geometries which correspond to backscattering in the sample. The direct reflection of the incident beam is blocked by a diaphragm.

Figure 6.12c shows a scattering geometry which is very simple and allows a precise determination of the polariton dispersion in the vicinity of an exciton

a

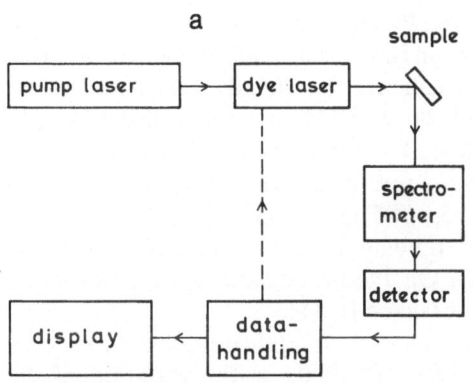

sample

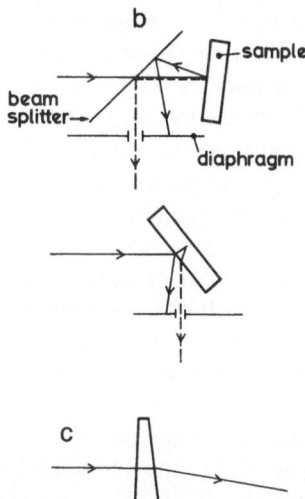

Fig. 6.12. Schematic drawing of a setup for photo-luminescence (excitation) spectroscopy, (**a**) Raman or Brillouin scattering, (**b**) some backward scattering configurations, and (**c**) diffraction in a thin prism

resonance. One needs a thin sample ($d < 10$ μm) which is prism shaped. The prism angle can be smaller than 1°. The narrow, tunable laser beam is deflected by the prism. The deflection is measured as a function of $\hbar\omega_{exc}$ and this gives the real part of the refractive index $n(\hbar\omega_{exc})$ and via $K_{medium} = \hbar\omega_{exc} n(\hbar\omega_{exc}) (c\hbar)^{-1}$, the dispersion relation $E(K)$. This technique has been successfully applied to CdS where platelet-type samples sometimes grow in the desired prism-like shape [6.80, 81].

6.2.2 Nonlinear Spectroscopy

The luminescence from high-density phenomena like biexciton decay or EHP luminescence can be investigated under cw excitation only in materials with long carrier lifetime as in Ge or Si. Some results in this direction are found in [5.55, 82, 83]. In direct gap materials the carrier lifetime is usually below 1 μs. Consequently, incident intensities in excess of 1 kW/cm^2 and up to GW/cm^2 have usually to be applied to get nonlinear optical phenomena and/or high excitation effects. These power densities can be produced only by (generally pulsed) lasers and can be applied to samples without irreversibly damaging them only for periods below 100 ns. Excimer and N$_2$ lasers and tunable dye lasers pumped by them emit pulses of 1–20 ns duration. Mode-locked Nd:YAG and Ar$^+$ lasers and dye lasers synchronously pumped by them reach pulses in the picosecond range. With special techniques it is possible to produce pulses as short as a few femtoseconds either in dye or in color center lasers [6.2, 3].

The simplest technique to investigate high density phenomena is to excite the samples by band-to-band excitation at a fixed frequency and to investigate the luminescence spectra as a function of excitation intensity and sample temperature. High excitation effects usually show up by the appearance of new emission bands with increasing excitation [6.32]. The information which can be obtained in this way is often not sufficient to identify the processes, as mentioned in Sect. 6.1 for the P bands or in [6.32] for the so-called M bands. Additional information can be obtained by varying $\hbar\omega_{exc}$, i.e., by performing photoluminescence excitation spectroscopy. In a similar way hyper-Raman scattering can be observed [6.30, 32, 85, 86]. High excitation effects in SCs with a direct gap are often connected with stimulated emission. This effect is on one hand proof of optical nonlinearities. On the other hand, stimulation often distorts the characteristic emission lineshape [6.84].

It is therefore better to investigate the spectra of optical amplification (gain) or of induced absorption directly. One possibility is the variation of the excitation strip length, L, which is shown schematically in Fig. 6.13a. By measuring the emission spectra as a function of L it is possible to determine the gain due to the amplification of the spontaneous emission along the excitation strip [6.32, 85, 86]. The technique is useful if processes with low gain values (up to 10^2 cm^{-1}) are involved. Otherwise, problems appear with gain saturation. Gain values up to 10^4 cm^{-1} (i.e., 1 μm^{-1}) can be measured in thin platelets by the pump and probe beam technique, which is very versatile and which will be discussed below in more detail.

The basic idea of the pump and probe beam technique is to measure the transmission – or reflection – spectrum of a thin platelet-type sample with a spectrally broad, weak probe beam with and without the presence of an additional, intense, spectrally narrow pump beam. If the duration of the pump beam, τ_p, is longer than the lifetime of the excited species, τ, and if the probe beam (with $\tau_p > \tau_{probe} > \tau$) coincides temporally and spatially with the pump, one can measure optical properties under quasi-stationary conditions. (See e.g. [6.57, 87].)

If the probe beam diameter is larger than the pump beam, it is possible by spatial resolution to get information about the lateral expansion of the excited species [6.87]. If the pump beam is absorbed in a layer which is thinner than the

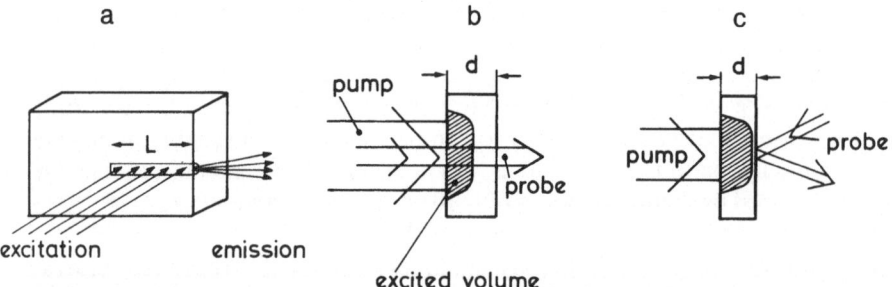

Fig. 6.13. Schematic drawing for a setup suitable for measuring the optical amplification or induced absorption by **(a)** the variation of the excitation strip length or by **(b)** pump and probe techniques. In **(c)** a setup is shown which allows measurement of the longitudinal expansion of excited species

diffusion or drift length of the excited species, l_D, e.g., of an EHP, it is possible to determine l_D into the depth of the sample by varying d and measuring the excitonic reflection spectra on the unexcited surface, which is altered as the EHP reaches it (Fig. 6.13c and [6.87]). Similar information is obtained if the integral EHP gain is measured as a function of d [6.87]. The variation of $\hbar\omega_{exc}$ at constant I_{exc} gives information about gain – or reflection – excitation spectroscopy [6.57]. While the technique in Fig. 6.13b is primarily sensitive to variations of the absorption spectrum, it is also possible to use it for the determination of changes in the real part of $\tilde{n}$ if the platelet has plane-parallel surfaces. In this case, Fabry-Perot modes appear in the transparent spectral range both in the transmission and reflection spectra and their shift with excitation contains information about Δn [6.88].

Usually one determines with the pump and probe beam changes of α and n at a photon energy $\hbar\omega_{probe}$ as a function of $\hbar\omega_{exc}$ and I_{exc}, i.e.,

$$\Delta\alpha(\hbar\omega_{probe}, \hbar\omega_{exc}, I_{exc}),$$
$$\Delta n(\hbar\omega_{probe}, \hbar\omega_{exc}, I_{exc}). \tag{6.24}$$

By extrapolating $\Delta\alpha$ and Δn for $\hbar\omega_{probe}$ towards $\hbar\omega_{exc}$ this technique also gives information about self-renormalization effects [6.88]. The pump and probe beam technique is also useful for two-photon absorption spectroscopy of excitons or biexcitons [6.20, 45]. Sometimes the probe beam can be replaced by the luminescence of the sample itself. This technique is called LATS (luminescence assisted two photon spectroscopy [6.89, 90]).

The basic idea of the spectroscopy with laser-induced gratings (LIG) [6.27] is the following. A spectrally narrow, tunable laser beam is split into two parts which are made to interfere in the sample with small crossing angles. This results in a periodic modulation of the intensity in the sample (Fig. 6.14a). If there are any dependences of the optical properties on the intensity due to real or virtual excitations in the sample (see e.g. Fig. 6.8 or Fig. 6.10) one gets an amplitude or a phase grating depending on whether the variation of the imaginary or of the real part of $\tilde{n}$ is predominant. This gives rise to diffraction. In the simplest case, the beams which are producing the grating are themselves

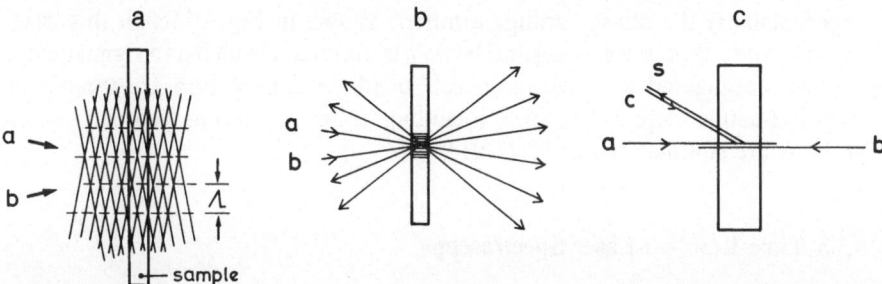

Fig. 6.14. Spectroscopy with laser – induced gratings: **(a)** principle and **(b, c)** two scattering geometries

diffracted from it. This self-diffraction experiment probes the self-renormalization effects. One distinguishes thick and thin gratings, depending on the inequality

$$d \gtrless 2\Lambda^2/\lambda, \tag{6.25}$$

where d is the thickness of grating, Λ the grating period, and λ the wavelength of the light. In the case of thick gratings, the Bragg condition allows only diffraction from one incident beam in the direction of the other and vice versa. In the case of thin gratings, which we discuss in the following, several diffracted orders are possible. Measuring the diffraction efficiency for the first order as a function of $\hbar\omega_{exc}$ usually reveals resonances in the nonlinear optical properties, e.g., of the type discussed in connection with Fig. 6.8.

Another possibility of LIG spectroscopy consists in producing the grating with two beams of frequency $\hbar\omega_{exc}$ and reading it with a probe beam $\hbar\omega_{probe}$. This reveals information on renormalization processes at $\hbar\omega_{probe}$ induced by light at $\hbar\omega_{exc}$, or coherent wave mixing processes of the type $\chi^{(3)}(\omega_{probe} : \omega_{exc}, \omega_{probe}, -\omega_{exc})$. The beams producing the grating can have identical or orthogonal polarizations (e.g. [6.27, 91–93]). If the modulation is due to a real excited species (population grating) it decays by recombination and by diffusion. Both contributions can be separated by variation of Λ [6.27], since the recombination is independent of Λ while the diffusion destroys the grating if $2\pi l_D > \Lambda$, with l_D being the diffusion length. In the case of virtual or coherent excitation, the diffracted orders can also be described in the framework of $\chi^{(3)}$ and higher susceptibilities (wave mixing) [6.26, 27, 91]. This theory describes also the frequency shift of the diffracted orders if the two incident beams have different photon energies $\hbar\omega_{exc\,1}$ and $\hbar\omega_{exc\,2}$ via terms like $\chi^{(3)}$ or $\chi^{(5)}$

$$\chi^{(3)}(\omega_4 : \omega_{exc\,1}, \omega_{exc\,1}, -\omega_{exc\,2}). \tag{6.26}$$

In a more classical approach, one notices that two beams with different frequencies produce a moving grating and the diffracted orders are frequency shifted by the Doppler effect. Exhaustive reviews on LIG are found in [6.25, 27, 91].

Some of the diffracted orders in LIG are phase-conjugate with respect to the incident ones. A geometry which is especially useful for these types of experiments is the phase-conjugate mirror shown in Fig. 6.14c. In this case, beams a and c form the grating and beam b is diffracted from it as a signal beam, s, phase-conjugated to c. More aspects of phase conjugation are treated in [6.94]. Further aspects of LIG including other scattering geometries for gratings are summarized in [6.27, 91].

6.2.3 Time-Resolved Laser Spectroscopy

To conclude this section we give some information on the techniques of time-resolved laser spectroscopy in SCs. The aim of this type of measurement is either to determine various relaxation time constants after excitation of electron–hole pairs or to determine switching times in connection with optical bistability. The second point will be discussed in Sect. 6.4. Concerning the first one, we have to distinguish between various time constants which are outlined below. If we excite some species, they will undergo scattering processes which perturb the phase by changing either φ_0 or the direction or the magnitude of K in the plane wave factor

$$\exp[i(K \cdot R - Et\hbar^{-1} + \varphi_0)] \tag{6.27}$$

with a characteristic phase relaxation time, τ_2. Additionally, the species will relax energetically in their band with an intraband relaxation time τ_i. Eventually, the ensemble will thermalize to a temperature T. Often T is above the lattice temperature T_L, especially at low values of T_L and relaxes towards T_L with a time τ_T. Finally, the density of the species will decrease after pulsed excitation by radiative and radiationless recombination with a constant τ_1. One often finds

$$\tau_2 \lesssim \tau_i \lesssim \tau_1. \tag{6.28}$$

It should be mentioned that it is only possible to speak about time constants in the case of exponential decays. Often one has a nonexponential behavior. The various τ's have then to be considered as "effective" values only. They may depend on the excitation conditions like the density, the excess energy over the bottom of the respective band, etc.

The time-resolved experiments are often carried out in the following way. The sample is excited by a short pulse of duration τ_{exc} and the decays, for example of the luminescence or of the grating efficiency, are monitored. For times down to 1 ns, fast vacuum or semiconductor photodiodes or photomultipliers together with fast oscilloscopes are useful. Streak cameras can be used down to 1 ps. In the picosecond region and below, various types of pump and probe techniques are applicable with optical delay lines (1 ns $\cong$ 30 cm). In the simplest case the sample is excited by a pulse and the temporal variation of the transmission or reflection is monitored by a delayed probe pulse of the same or a different photon energy. See contributions to [6.28] and the literature cited therein. Luminescence decay measurements in the picosecond and sub-picosecond range can be performed by either transmitting the luminescence through a Kerr medium, which is placed between crossed polarizers and which is opened by the electric field of a short laser pulse, or by upconverting the luminescence by frequency mixing with a short laser pulse and detecting $\hbar(\omega_p + \omega_{lumin})$ as a function of the time delay between the excitation and upconversion pulses. Both possibilities are outlined in Fig. 6.15a and b.

In Fig. 6.15c an elegant way is shown to determine the phase relaxation time by a LIG. A short pulse, a, excites some species (e.g. excitons) in a sample. After the excitation pulse, the species will continue to maintain coherence for a period equal to the phase-relaxation time τ_2. If a second coherent pulse, b, hits the sample after pulse a, but before τ_2, it can form a LIG with this oscillation,

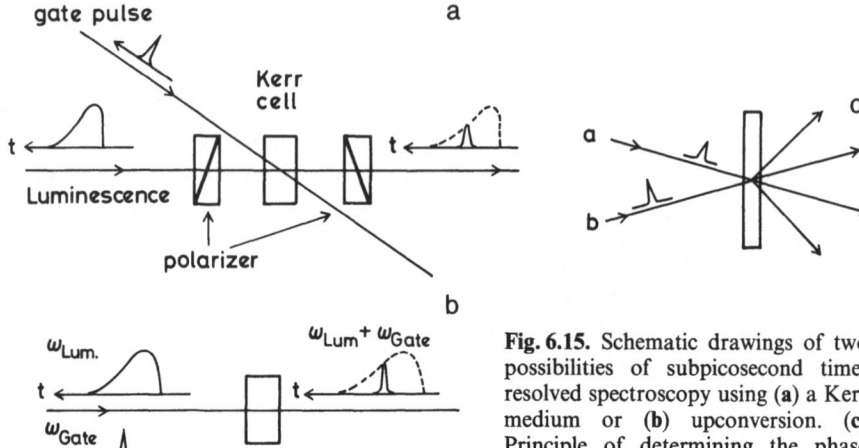

Fig. 6.15. Schematic drawings of two possibilities of subpicosecond time-resolved spectroscopy using (**a**) a Kerr medium or (**b**) upconversion. (**c**) Principle of determining the phase relaxation time by LIG

resulting in diffracted orders. The decay of the intensity of the diffracted orders with increasing time delay between pulses a and b contains information about τ_2.

Experimental techniques for laser spectroscopy are also discussed in [6.20, 26, 27, 28, 32, 70].

6.3 Examples of Laser Spectroscopy of Semiconductors

In this section we present some selected examples of laser spectroscopy in semiconductors. As mentioned in the introduction, this selection is somewhat arbitrary and far from exhaustive. Other recent reviews covering parts of this field are [6.20, 26–28, 32, 46, 51, 60, 69, 70, 73, 91].

6.3.1 Brillouin Scattering

Brillouin scattering describes the scattering from one polariton state to another by the emission or absorption of acoustic phonons. The process is shown schematically for backscattering and various incident laser-photon energies in Fig. 6.16a. Obviously the Stokes and anti-Stokes shifts change drastically around the resonance of the exciton polariton. Furthermore, the number of possible scattering processes increases for $\hbar\omega > E_{\rm L}$. In Fig. 6.16b we present experimental results for CdS from [6.95a] which verify the above statements. If the velocities of sound of the various phonon modes are known, it is possible to reproduce the dispersion relation of the exciton polariton very precisely from data as shown in Fig. 6.16b, i.e., to determine parameters like the transverse

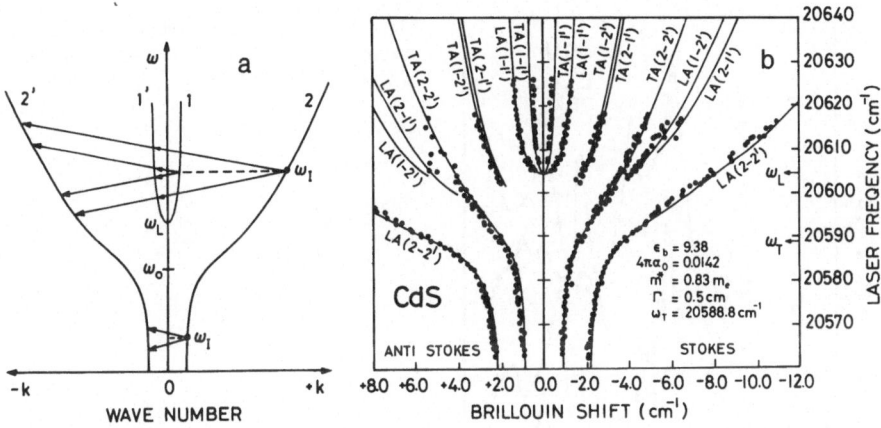

Fig. 6.16. Brillouin scattering around an exciton resonance; (**a**) schematic, (**b**) measured Brillouin shifts in CdS. (From [5.95a])

eigenenergy, the oscillator strength, the effective exciton masses, deviations from parabolicity, etc. The intensity of the Brillouin scattering as a function of $\hbar\omega_{exc}$ contains information about the phonon–polariton coupling and about the additional boundary conditions, i.e., about the fractions of the incident intensity propagating above $\hbar\omega_L$ on the lower and upper polariton branches, respectively. Resonant Brillouin scattering has been used to investigate the exciton polariton resonances, e.g., in GaAs, CdTe, ZnSe, CuBr, CdS or CdSe. More information about this process and references can be found in [6.20, 96].

6.3.2 Photoluminescence Excitation Spectroscopy of Bound Exciton Complexes

Our next example concerns photoluminescence excitation spectroscopy (PLES). We have chosen PLES of bound exciton complexes (BECs) in semiconductors. This technique can be and is partly performed at low excitation levels often with conventional light sources [6.97]. However, pulsed, tunable, high-intensity lasers are also used; such is the case if the relevant spectral range is more easily accessed by them than by cw lasers, as in ZnO, which has a gap in the near UV ($E_{A\Gamma_5}(T) = 3.375$ eV). High power pulsed lasers are also used since higher excitation intensities relax the selection rules so that forbidden transitions to excited states of BECs become weakly allowed and thus visible in PLES. We present in Fig. 6.17 PLES spectra of some of the prominent BECs in ZnO; these features are usually attributed to excitons bound to neutral acceptors and donors, respectively. Concerning the interpretation of the origin of the BECs see also [6.98]. Referring to Fig. 6.17, the resonances $R2$ have excitation energies close to the AB valence band splitting Δ_{AB} ($\Delta_{AB} = 5.4$ meV) and are attributed to BECs where a hole from the A valence band is replaced by one from the B band. In the case of acceptor BECs,

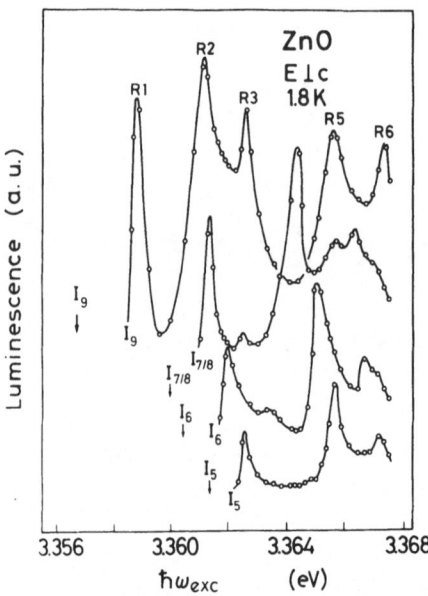

Fig. 6.17. Photoluminescence excitation spectra of various bound exciton complexes in ZnO. (From [5.98])

even two holes can be replaced resulting in resonance $R5$. The other resonances shown are considered as excited states of these complexes. These interpretations are supported by measurements under the influence of magnetic fields [6.99]. Similar experiments have been performed recently especially on various BECs in CdS [6.100–102].

6.3.3 High Excitation Luminescence

We proceed now to luminescence under high laser excitation. In Fig. 6.18 [6.103], we present two emission spectra of CdS recorded under band-to-band excitation at two different excitation intensities and geometries. In Fig. 6.18a the luminescence is recorded directly from the excitation spot in a backward scattering geometry and thus preferentially contains spontaneous emission. One observes luminescence lines due to the BECs at neutral donors (I_2) and acceptors (I_1) and on the low energy side the so-called M and P_M bands. The M band is attributed, as already mentioned, to biexciton recombination, and to recombination of the I_2 BEC under emission of acoustic phonons or under inelastic scattering by free carriers [6.32]. The P_M band had been thought to be due to inelastic biexciton-biexciton scattering but it is now predominantly interpreted as the acoustic phonon wing of the I_1 BEC. The spectrum in Fig. 18b has been recorded from the edge of the sample where mainly stimulated emission appears. One observes a weak M band. The main emission is centered in the region of the P bands (see Sect. 6.1). What is of special interest here is that the $I_2 - P_M$ feature of Fig. 6.18a appears with mirror symmetry as a

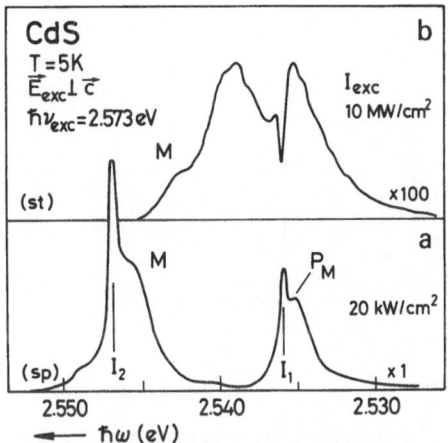

Fig. 6.18. Luminescence spectra of CdS obtained for different excitation intensities and geometries. (From [5.103])

reabsorption feature in Fig. 6.18b. This means, luminescence polaritons are absorbed while propagating through the excited and unexcited parts of the sample either by creating I_1 BECs directly or by emitting acoustic phonons. This symmetry of the I_1 BEC recombination and formation under emission of acoustic phonons supports the underlying model and is in agreement with photoluminescence excitation spectroscopy [6.100, 102, 104]. The P_M band luminescence lineshape and the ratio between recombination under emission and absorption of acoustic phonons has also been used as a phonon thermometer when the system is under laser excitation. Phonon temperatures as high as 14 K have been measured [6.105].

The inelastic scattering processes between excitons and excitons or free carriers have been presented in Sect. 6.1, together with some recent alternative interpretations. Both the experimental results, the theoretical models and the comparison between experiment and theory have been reviewed recently in [6.32] so we shall not go into more details here.

6.3.4 Spectroscopy of Biexcitons

Optical nonlinearities, which are connected with transitions to the biexciton, have been outlined in Sect. 6.1. Most of them (and some more) have been observed by nonlinear laser spectroscopy in a variety of materials, predominantly in copper halides and II–VI compounds. These results have been summarized recently in [6.20, 46] with emphasis on the copper halides. Therefore, we shall present here only results for the II–VI compounds and refer the reader to the above–mentioned reviews and the literature cited therein for details in the copper halides.

Luminescence spectroscopy under laser excitation did not give conclusive evidence for the existence of biexcitons in CdS and other II–VI compounds

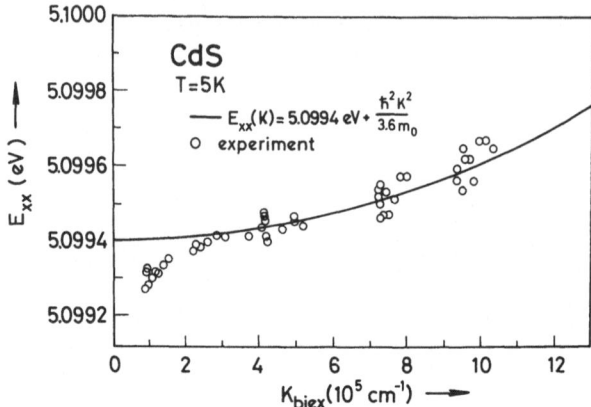

Fig. 6.19. The dispersion of the biexciton in CdS as determined by luminescence assisted two-polariton spectroscopy. (From [5.107])

[6.32] but led to some controversies concerning the eigenenergies of the biexcitons and the spectral shape of the expected emission bands [6.32, 90]. One of the problems which is characteristic for uniaxial (e.g., wurtzite C_{6v}) type samples is the following: In decay processes like the ones shown in Fig. 6.7e, the final state may not only be the lower polariton and the longitudinal exciton branches, but also the mixed mode polariton states [6.20, 32, 90]. The latter fill the energetic region between the former two limiting cases, forming a continuum which is a function of the angle $\star(K, c)$ for a polarization in the plane defined by K and c. These mixed mode polaritons are a peculiarity of uniaxial crystals and correspond to the extraordinary beam in classical birefringence.

Definite proof for the existence of biexcitons in II–VI compounds (mainly CdS and ZnO) came from the observation of the resonances in hyper-Raman scattering (HRS) [6.106], from luminescence assisted two-photon spectroscopy (LATS) [6.89, 90] or from spectroscopy with laser-induced gratings (LIGs) [6.93].

As an example, we show in Fig. 6.19 the dispersion of the biexciton as determined by LATS. By varying the orientation between the incident pump laser beam and the direction of observation of the luminescence, it is possible to vary the wave vector of the created biexcited from zero to 10^6 cm^{-1} [6.107]. The origin of the small discrepancy between the parabolic dispersion with twice the exciton mass (solid line) and the experimental points for small values of K is still not clear.

The diamagnetic shift of the biexciton in CdS has been determined by LATS to be slightly anisotropic and to have values around 5×10^{-6} eV T^{-2} as compared to 2×10^{-6} eV T^{-2} for the free A exciton [6.108].

LATS also revealed the correct level scheme of the biexcitons in ZnO. Due to the small splitting between A and B valence bands, Δ_{AB}, in ZnO as compared to the binding energy of the biexciton E_{xx} there exist three bound biexciton

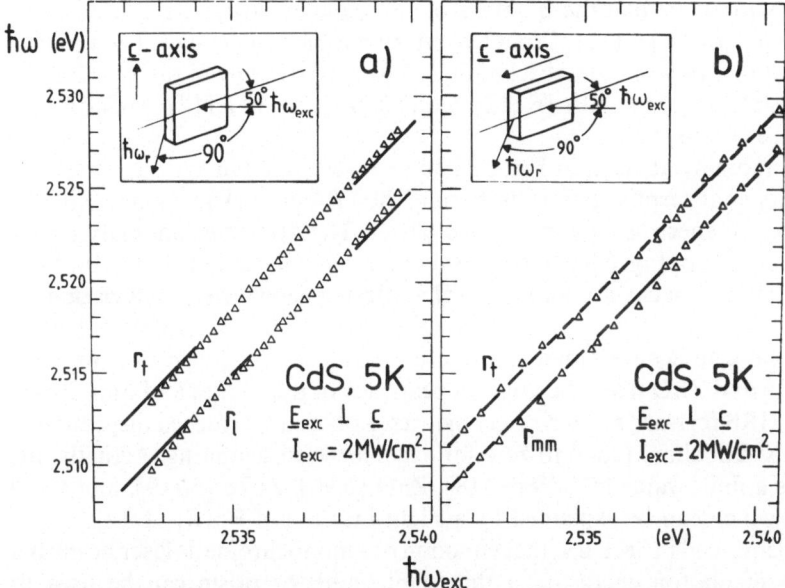

Fig. 6.20. The relation between $\hbar\omega_{exc}$ and $\hbar\omega_R$ in HRS experiments in CdS for two different scattering geometries. (From [5.109])

states involving two holes from the A VB, one from the A and one from the B VB or two B holes [6.90]. Due to the smaller exciton and biexciton radii in ZnO as compared to CdS, the values of the diamagnetic shift are below detection threshold. The corresponding (AB) and (BB) biexciton states are unbound in CdS due to the different relation between Δ_{AB} and E_{xx}^{b}. First hints for their existence have been found in LIG spectroscopy only recently [6.93].

High resolution spectroscopy experiments have been used for the determination of the dispersion of the exciton-polariton in the vicinity of the lowest free exciton resonances in the II–VI compounds CdS, ZnSe, ZnTe, and ZnO. For a recent review and detailed references, see [6.20]. In CdS it was possible to change the scattering geometry and polarization with respect to the crystallographic c axis [6.106, 109]. Figure 6.20 shows experimental results from [6.109] for a geometry close to backscattering. The slope predicted by (6.17) is shown by solid lines. In the case where all beams are polarized perpendicular to c and all wave vectors are perpendicular to c, one can identify the values of $\hbar\omega_f$ in (6.17) with the longitudinal eigenenergy for the wave vector given by the scattering geometry and a state on the LPB in the bottleneck region. In the scattering geometry where c lies in the scattering plane, one observes again the state in the bottleneck region but also a value for $\hbar\omega_f$ between the longitudinal and transverse eigenenergies, i.e., a mixed mode polariton.

HRS measurements in magnetic fields up to 20 T complemented by classical reflection and transmission spectroscopy have allowed the determination of

B-field parameters like the g values or the diamagnetic shifts in CdS, ZnO [6.110] and ZnTe [6.111]. Complementary measurements have been performed by direct two-photon absorption (TPA) spectroscopy of the $n_B = 2$ and 3 exciton states, e.g., in CuBr [6.112], CdS [6.113], ZnO [6.114], ZnSe [6.115] and ZnTe [6.116].

The anomaly at $\hbar\omega_{abs}$ in Fig. 6.8 can be investigated in various ways. The imaginary or absorptive part is the basis of the TPA or LATS measurements for the biexciton spectroscopy mentioned above. The dispersive anomaly on the right-hand side of Fig. 6.8 has also been observed by HRS in CdS [6.107, 117]. It leads, in both backward and forward scattering geometries, to deviations of the relation $\hbar\omega_R = f(\hbar\omega_{exc})$ from its usually smooth and monotonic behavior. Furthermore, indications for a self-renormalization at $\hbar\omega_{exc}$ have been found in both CdS and ZnO from the angle dependence of the Stokes and anti-Stokes shifts in HRS forward scattering geometries. For an unperturbed dispersion at $\hbar\omega_{exc}$ this shift should tend to zero for exact forward scattering. Actually one observes a finite value in CdS [6.118], ZnO [6.90], ZnTe [6.119], and CuCl [6.120], which can be explained by a slight increase of Re$\{K\}$ at $\hbar\omega_{exc}$.

As mentioned in Sect. 6.2, the refraction of a monochromatic laser beam as a function of photon energy in a thin semiconductor prism can be used to determine the polariton dispersion. This type of experiment revealed at higher incident intensities an anomaly around 2.5461 eV [6.121]. This anomaly is believed to be an induced transition from a high density of really excited $A\Gamma_5$ excitons ($\hbar\omega_T = 2.5523$ eV, $\hbar\omega_L = 2.5541$ eV) to the biexciton at 5.0995 eV [6.20].

6.3.5 Spectroscopy of the Electron–Hole Plasma

If the generation rate in semiconductors is raised to sufficiently high values, the electron–hole pair density will reach a point where the concept of excitons and biexcitons as individual quasiparticles breaks down and a new collective phase is formed, the electron–hole plasma. As a rough estimate, this transition is reached if the mutual distance of electron–hole pairs becomes comparable to the exciton diameter or if the Coulomb interaction is screened by Thomas-Fermi or Debye screening to an extent that no more bound states occur. It is evident that the transition density depends strongly on material parameters like the exciton binding energy and Bohr radius. The generation rate which is necessary to reach this density depends in addition on the lifetime of the created electron–hole pairs. As a consequence, it is possible under favorable circumstances to create an EHP in the indirect elemental semiconductor by illumination with a classical incandescent lamp [6.55], while pulsed lasers with excitation intensities in the GW/cm^2 range are necessary to create an EHP in the direct SC CuCl [6.122].

In Si and Ge the EHP undergoes a first-order phase transition into a liquid-like state (EHL) below a critical temperature T_c, as already mentioned in Sect. 6.1. This new phase manifests itself through a new emission band. From its analysis the phase diagram can be reconstructed [6.28, 55]. Under normal excitation conditions, the EHL appears in the form of small droplets (EHDs) with a diameter of a few micrometers surrounded by a gas of excitons and/or free carriers. The percentage of the volume occupied by EHDs increases with increasing pump rate but the density of the EHL is independent of the excitation intensity and decreases with temperature up to T_c. By applying inhomogeneous stress to the sample it is possible to minimize in the bulk of the sample the width of the forbidden gap E_g through the dependence of E_g on the stress. If the sample is strongly laser-excited at the surface, the flow of excitons and EHDs into the potential minimum and the formation of a big EHL drop (γ drop) can be visually observed. This process can be photographed by an IR sensitive TV system using the light produced by the recombination radiation [6.83]. Without stress, it is possible to monitor the expansion of an EHD cloud driven by the phonon wind. Since the phonon propagation is anisotropic in Si or Ge this expansion leads to highly aesthetical spatial patterns [6.83]. Above T_c one obtains an EHP with an excitation-dependent density. These topics have been reviewed in several places [6.55] and we shall not go into details here. The intermediate density regime, where excitonic processes are still predominant (see above) is less pronounced in Si and Ge as compared to the II–VI or I–VII compounds. Biexcitons have been observed under suitable excitation conditions [6.123] in Si and Ge as well as bound multiexciton complexes, which can be considered as nucleation centers for the EHDs [6.124]. A sophisticated way to produce electric power from laser light via an EHP has been investigated in [6.125].

Some experiments have been performed in the indirect III–V compound GaP revealing the existence of an EHP under laser excitation, perhaps even as an EHL below some T_c [6.56]. The parameters deduced by various authors from the analysis of the luminescence lineshape reveal, however, considerable disagreements concerning the values of variables such as the plasma density.

In the direct gap SC GaAs, which has a rather low exciton binding energy and hence a correspondingly large exciton radius [Table 6.1 and Eq. (6.8)], exciton–exciton or exciton–electron collisions have been reported. The biexciton binding energy E_{xx}^b is expected to be around 1 meV according to calculations of E_{xx}^b/E_x^b as a function of the ratio of electron and hole effective masses [6.32]. This small value makes verification of the biexciton very difficult and no successful attempt of observation is known to us, though this quasiparticle is also expected to exist as a bound state in this material. An EHP has definitely been observed in this material, however [6.126–128]. The plasma density varies strongly from experiment to experiment [6.126] and no clear indications for the appearance of a liquid phase have been found so far. There was a certain controversy about the spatial expansion of a degenerate EHP under inhomogeneous excitation conditions. Some authors proposed drift velocities, v_D, of the order of 10^7 cm/s [6.56c]. This value would equal or even exceed the Fermi velocities v_F resulting thus in an extremely anisotropic population of states in K space. Model calculations based on combined

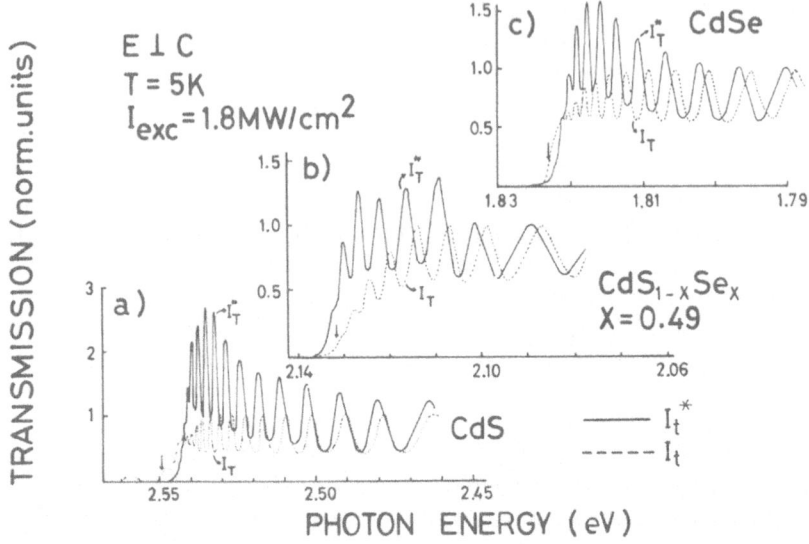

Fig. 6.21. Gain spectra of CdS, $CdS_{0.51}Se_{0.49}$, and CdSe observed under identical excitation conditions. (From [5.130])

transport equations for heat (phonons) and carriers (thermo-diffusion model) could not reproduce such high values but predicted values of v_D around 10^6 cm/s [6.128c]. More recent experiments in GaAs have obtained values of the drift length, $l_D < 25$ µm [6.128b] and thus $v_D < v_F$. We shall come back to this topic in connection with EHP measurements in CdS.

The II–VI compounds with band gaps in the range 1.5–2.5 eV occupy an intermediate position between the elementary and III–V semiconductors, where an EHP is easily reached under high excitation, and the Cu halides, where excitonic processes dominate and an EHP can be created under extreme excitation conditions only. In II–VI compounds, both regimes can be investigated in some detail, depending on the excitation conditions. This is one of the reasons why II–VI compounds, and especially CdS, are often used here as an example.

In Fig. 6.21 we show gain spectra of CdS, $CdS_{0.51}Se_{0.49}$, and CdSe obtained with the pump and probe beam technique under quasi-stationary excitation conditions in the way described in Sect. 6.2. (The results for the mixed crystal are discussed later). The gain spectra show the behavior presented qualitatively in Sect. 6.1 for low temperatures, i.e., a red shift of the absorption edge and optical gain between the chemical potential μ of the plasma and (roughly) the reduced gap, E_g'. The samples used in Fig. 6.21 are plane-parallel platelets with as-grown surfaces. They form thus a Fabry-Perot resonator resulting in the modulation of the transmission in the transparent part of the spectrum. The analysis of the gain spectra gives detailed information about the EHP parameters such as the plasma density ($n_p^{CdS} \approx 2 \ldots 3 \times 10^{18}$ cm^{-3}), the plasma temperature T_p ($T_p^{CdS} \approx 25$ K), and the band gap renormalization. Tabulations of the relevant EHP parameters for both pure materials are found in [6.87].

The fact that these gain spectra are truly related to an EHP has been proven by gain and reflection excitation spectroscopy [6.57]. It has been found that the plasma density is strongly excitation dependent at all temperatures. This means that no liquid phase is formed even below T_c. The reason is the short carrier lifetime ($\tau_p \approx 150$ ps) which prevents a spatial condensation of the carriers into droplets [6.57]. This finding has been confirmed by various authors [6.59, 131].

The transition from the exciton phase to the EHP is observed to occur gradually and continuously, in agreement with theoretical calculations [6.57, 62, 88].

The blue shift of the Fabry-Perot modes indicates a decrease of the refractive index under the creation of an EHP and is also in agreement with theory [6.32, 40, 132]. With increasing temperature, the experimental findings also behave in agreement with theory. Theory predicts that the gain decreases as the carrier population becomes less degenerate with increasing T_p and the red shift of the absorption edge changes over into a blue shift as shown schematically in Fig. 6.10. This change occurs experimentally at $T \approx 160$ K [6.133].

There are also some calculations which consider recombination of carrier pairs in the EHP under emission of plasmon-phonon mixed state quanta as shown in Fig. 6.3d [6.134, 135]. Both calculations show that for a degenerate EHP this contribution is one or two orders of magnitude smaller than the gain connected with direct recombination. In agreement with this finding, most gain spectra do not show the indirect process. The effect reported in [6.136] could be due to a rather high value of T_p caused by the large excess energy in the excitation. On the other hand, the variation of the excitation strip length used in the gain measurements of [6.136] is not very well suited to investigating EHP effects as has been discussed in Sect. 6.2. Excitonic gain processes may therefore be responsible for or contribute to the spectra reported there. The indirect process, together with stimulated recombination under LO phonon emission, is thought to form the so-called Q bands observed in some cases [6.137].

The expansion of the EHP under inhomogeneous excitation conditions has been investigated in CdS and CdSe by one or several of the techniques described in Sect. 6.2. Values of l_D of (6 ± 2) μm and (13 ± 7) μm have been found in CdS and CdSe under nanosecond excitation, respectively [6.87]. In both cases, the drift velocities are smaller than the Fermi velocities so that the anisotropy of the population in K-space does not contribute decisively to the gain lineshape under the excitation conditions of [6.87]. On the other hand, these values of l_D cannot be explained by classical diffusion. From $l_D = (D\tau)^{1/2}$ and a known carrier lifetime of 100–200 ps [6.87] values of D of the order of 10^3 cm^2 s^{-1} would follow. They are considerably above the generally accepted values around $D \approx 10$ cm^2 s^{-1} [6.59]. Therefore, we prefer to speak about drift length and expansion of the (degenerate) plasma and not about diffusion. The precise nature of the transport process involved here is to our knowledge not yet known, though there are promising attempts being made in connection with the thermodiffusion model [6.128].

A similar controversy about v_D and l_D exists for the EHP in Si above T_c [6.82, 83].

6.3.6 Spectroscopy with Laser-Induced Gratings

Some of the first detailed spectroscopic investigations using LIGs were done in Si and CuCl [6.138, 48]. We concentrate here on some recent experiments of LIG spectroscopy in CdS which have allowed the distinction of various nonlinearities and yielded data of the drift length of different excited species [6.130, 139].

The LIG was created by two coherent beams of $\hbar\omega = 2.637$ eV and a duration of several nanoseconds. The first value is so far above the exciton resonances at 2.553 eV that the absorption coefficient, and thus the excitation conditions, do not change with excitation even if an EHP is created. On the other hand, this value is close enough to the exciton resonance that no strong heating of the carriers is to be expected. Measurements in [6.57] indicate an increase of the carrier temperature above the lattice temperature for these conditions of less than 30 K. The LIG is read by a probe beam around 2.43 eV, which is situated in the transparent region of the sample and is also below the spectral region of optical gain; thus this beam sees only a phase grating. The angle between the pump beams can be varied thus altering the grating constant, Λ, of the LIG. The normalized efficiency is then recorded as a function of Λ^{-1} for various temperatures between 7 K and 300 K. In Fig. 6.22, we show the

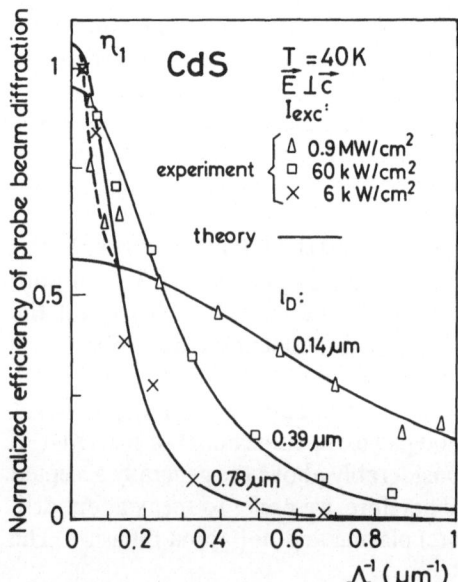

Fig. 6.22. Normalized diffraction efficiency of laser-induced, dynamic gratings in CdS as a function of the inverse grating constant Λ^{-1}. (From [5.139])

results for 40 K and different pump intensities. They are representative for the "low temperature" range, which extends up to ≈ 150 K. The solid lines are calculated curves from [6.139] based on a sinusoidally modulated generation rate and two-dimensional diffusion. Changes of the optical properties induced by the carrier-pair density n_p according to [6.61, 62, 88] are also included. Because of the arguments given in connection with the EHP expansion above, the theory is formulated so that l_D, and not the diffusion constant, enters as the relevant parameter. At the lowest excitation intensity, the data can be fitted with $l_D = 0.78$ μm. This value increases with decreasing temperature, e.g., to 1.1 μm at 7 K. This relation reflects increasing scattering of the excited species with phonons at increasing temperatures. At constant temperature, l_D decreases with increasing excitation intensity. See Fig. 6.22 for the case where $l_D \approx 0.78$ μm at 6 kW/cm^2 changes to $l_D \approx 0.39$ μm at 60 kW/cm^2. This reflects the increasing interparticle scattering with increasing excitation (Fig. 6.8) and/or scattering with phonons which are created by radiationless recombination of the electron-hole pairs. In this regime, the data are compatible with classical diffusion. With a low density lifetime of 1 ns a value of $l_D = 1$ μm leads to $D = 10$ cm^2/s.

In the above, the optical nonlinearities are of excitonic nature, like collision broadening of the exciton resonance or induced absorption from excitons to biexcitons. The situation changes drastically if we go to excitation intensities where an EHP is formed. When the intensity of the write beams equals 0.9 MW/cm^2 (Fig. 6.22), the experimental data separates into two branches. The "upper branch" with $\Lambda^{-1} < 0.2$ μm^{-1} drops rapidly with increasing Λ^{-1} (dashed line) and is compatible with values of $l_D \approx 5$ μm as determined by other pump and probe beam techniques [6.87]. We interpret the LIG produced under these conditions to be predominantly an incoherent (population) grating with the underlying optical nonlinearity being connected with the formation of an EHP. The contribution of the EHP is smeared out by drift processes and thus becomes negligible for $\Lambda^{-1} \approx 0.2$ μm^{-1}. For Λ^{-1}'s above this value, the efficiency is lower and decays only slightly with increasing Λ^{-1}. A fit to this "lower branch" reveals $l_D = 0.14$ μm, a length which is comparable with the penetration depth of the pump beams. Since this value is in contradiction with all l_D values for an EHP reported so far, we assume that the diffracted orders are due to coherent or $\chi^{(3)}$ effects. The diffracted orders have the same photon energy as the probe beam. Therefore, processes of the type $\chi^{(3)}$ ($\omega_{probe} : \omega_{exc}$, $\omega_{probe}, -\omega_{exc}$) seem to be the relevant ones. These processes, which have no drift length in the classical sense, are also found in nondegenerate four-wave mixing experiments in CdS [6.130, 140] where frequency-shifted diffracted orders can be easily observed by detuning the two pump beams by 3 meV and more. For these values a population grating with species of a lifetime of 100 ps would be completely smeared out by the motion of the grating.

At high temperatures ($T \approx 300$ K) one finds that the efficiency η as a function of Λ^{-1} is independent of the excitation intensity and l_D is a few tenths of a micrometer. Evidently the scattering with thermal phonons dominates over interparticle scattering through all values of I_{exc} used here ($I_{exc}^{max} \approx 1$ MW/cm^2). In agreement with this finding, we know from pump and probe beam experiments [6.133] that there is almost no more optical EHP gain, i.e., no degenerate carrier population at this temperature. Coherent processes are also suppressed, presumably due to the decreasing phase-relaxation times of the excited species [6.139].

6.3.7 Spectroscopy of Alloy Semiconductors

In the following section, we present some material on laser spectroscopy of alloy semiconductors. We chose the system $CdS_{1-x}Se_x$. In Fig. 6.23, we show (solid line) a typical low temperature luminescence spectrum of $CdS_{0.42}Se_{0.58}$ obtained under band-to-band excitation at low temperature. It has been found by excitation spectroscopy, which is site selective, or by polarization spectroscopy [6.23, 142, 143] that the broad band centered around 604 nm is due to the recombination of localized excitons and that the mobility edge coincides with the high energy edge of the luminescence band. The double structure between 608 and 616 nm is the LO phonon replica of the zero-phonon band at 604 nm. The system $CdS_{1-x}Se_x$ belongs with respect to the x dpendence of the phonons to the persistent-mode type, i.e., the two peaks in the LO-phonon satellite correspond in energy roughly to the CdS and CdSe LO phonon modes, respectively. The spectrum in Fig. 6.23 is a time-integrated one after excitation with pulses of 5 ps duration at $\hbar\omega_{exc} = 2.13$ eV from a dye laser synchronously pumped by a mode-locked Nd:YAG laser. It has been found by time-resolved measurements with a streak camera that the decay time of the luminescence increases with increasing localization energy, i.e., with increasing emission wavelength (see e.g. [6.142b]). One contribution to this phenomenon may be understood qualitatively by considering the overlap integral between the electron and hole wave functions in the localized exciton complex. This overlap and, consequently, the transition matrix element are maximal if the electron

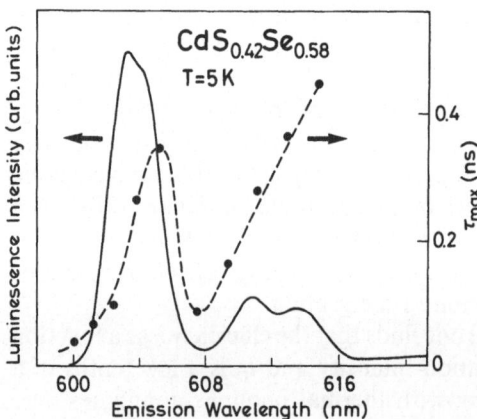

Fig. 6.23. Time-integrated luminescence spectrum of $CdS_{0.42}Se_{0.58}$ after picosecond excitation (—). Time delay between excitation pulse and luminescence maximum as a function of $\hbar\omega_{lum}$ (– – –). (Curve deduced from data in [6.141])

and hole are found in approximately identical volumes. This is the case for weakly bound holes close to the mobility edge or for cases where the exciton is localized in a potential well which is larger than its Bohr radius. In the system under consideration here, a free electron is bound to a localized hole at least for the deeper localized excitons. If the hole is localized in a volume smaller than the exciton or donor radius, the overlap integral between both wave functions decreases and the recombination time increases. Another reason for the increase of the decay time is the feeding of the population at lower energy from states at higher energy, for example by phonon-assisted tunneling [6.23, 142, 143].

It has been found in [6.141] that the dynamics of the onset of luminescence depend also on the localization depth. The carriers and/or excitons relax rapidly to the mibility edge in times shorter than or comparable to 20 ps, the time resolution of the setup. The relaxation into deeper localized states takes place via acoustic phonon emission and/or tunneling between various spatially separated and localized levels. This takes more time, the deeper localized states are. This is reflected by the dashed line in Fig. 6.23. It gives the time, τ_{max}, which elapses between the excitation pulse and the temporal maximum of the emission at the respective wavelength as indicated in the insert. The curve increases monotonically from a resolution limited value of 20 ps around the mobility edge to values around 400 ps for the deeper localized states. The same behavior is observed in the LO phonon replica.

Pump and probe experiments under nanosecond excitation show that the nonlinear optical properties of alloy semiconductors are markedly different from the ones of the pure materials (Fig. 6.21). Instead of a red shift of the absorption edge, one observes a blue shift. These findings can be understood under the assumption that all excited carrier pairs relax into the localized states [6.144]. Indeed, the estimated density of the localized states of 10^{18} cm^{-3} is sufficient to accommodate them. The bleaching of absorption and/or the appearance of gain is caused by the filling of the localized states and population inversion. The many-particle or renormalization effects caused by localized states are much less pronounced than for free ones. Consequently, one finds almost no reduction of the gap in the spectral region above the mobility edge [6.144].

6.3.8 Laser Spectroscopy of Quantum Well Structures

Time-resolved laser spectroscopy has been used in many cases, e.g., to investigate the decay dynamics of interband recombination processes of excitons, biexcitons or of the EHP. See, e.g., [6.27, 28, 46, 49b, 58, 59, 145, 146] and the literature cited therein. The intraband relaxation of hot carriers has been monitored by time-resolved pump and probe experiment [6.28]. Coherent, Raman or hyper-Raman type processes can be distinguished from incoherent luminescence by their respective temporal behavior [6.147].

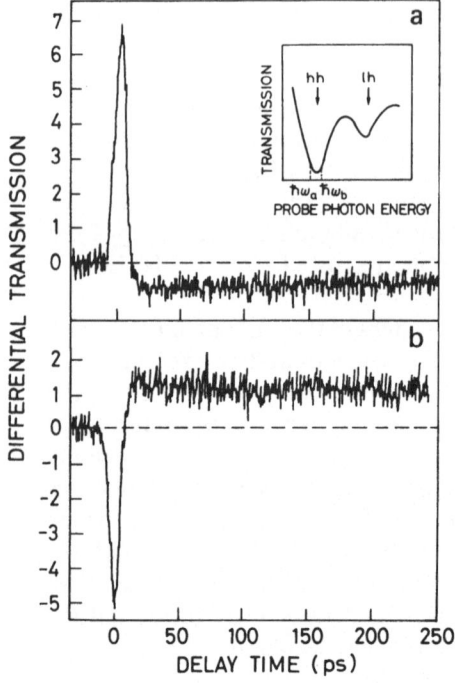

Fig. 6.24. Time-resolved changes of transmission in GaAs/GaAlAs MQWS under pulsed excitation below the lowest free exciton resonance. (From [5.74,77,159]. Copyright: Bell Telephone Laboratories, Inc., 1987; reprinted by permission)

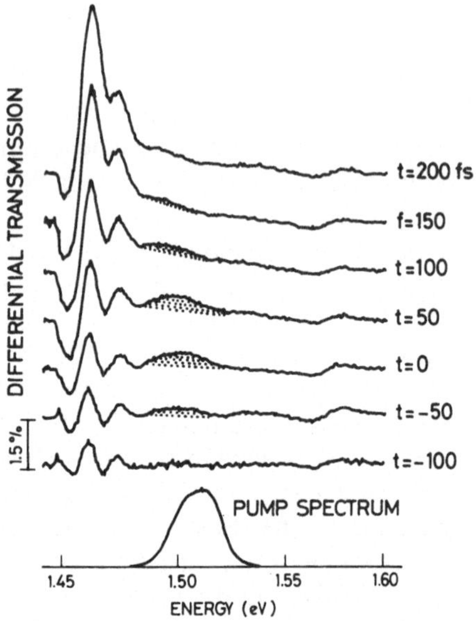

Fig. 6.25. Time-resolved changes to the transmission spectra of GaAs/GaAlAs MQW under pulsed band-to-band excitation (From [5.74,159]. Copyright: Bell Telephone Laboratories, Inc., 1987; reprinted by permission)

We shall use the rest of this chapter to outline two recent experiments on time-resolved spectroscopy on MQW structures. In Fig. 6.24 we show time-resolved changes of the transmission of a GaAs/GaAlAs MQW under pulsed excitation below the $n = 1hh$ exciton resonance [6.74, 77]. Trace (a) shows the change of transmission on the low energy flank of the $n = 1hh$ exciton, trace (b) on the high energy side. There are changes with relatively long time constants of the order of nanoseconds which are connected with the formation of an EHP. The fast features, which immediately follow the picosecond excitation pulse, indicate a transient blue shift of the exciton resonance. These shifts are attributed to the ac Stark effect outlined in Sect. 6.1, and are due to a "repulsion" between the exciton eigenenergy and the pump energy [6.76–78]. In Fig. 6.25, from [6.74], we show results in the same type of material where band-to-band excitation with pump pulses of 80 fs duration is employed. The change of transmission is measured by an 80 fs continuum at delay increments of 50 fs. There is some bleaching due to blocking of the band-to-band transition by the excited bunch of electrons (dotted areas). The temporal evolution of the intraband relaxation can be seen nicely. There is a decrease of excitonic absorption in the region of the $n = 2hh$ transition around 1.57 eV. This change is rather small and constant on the time scale shown here and it is due to direct screening of the Coulomb interaction in the exciton by the carriers. The screening is to a large extent independent of the energy or distribution of the free carriers in the bands. A similar behavior is found for the $n = 1hh$ and lh excitons in the three lowest traces. When the excited carriers relax down to the exciton levels, the change of transmission increases dramatically. This is attributed to phase-space filling and exchange interactions which become effective for these exciton resonances only after the excited species have relaxed down into the relevant states. Figure 6.25 demonstrates, thus, that direct screening of the Coulomb interaction is less efficient in quasi-two-dimensional systems as compared to exchange interaction and phase-space filling, as indicated in Sect. 6.1. This behavior differs significantly from that of three-dimensional semiconductors [6.32].

6.4 Applications and Outlook

Laser spectroscopy of semiconductors has contributed to our understanding of these systems since its beginning about a quarter of a century ago mainly in fundamental or basic research. Presently, however, we see that especially the nonlinear processes in SC are evolving rapidly towards application. We shall use this final section to outline one example, namely optical bistability (OB). Recent reviews of this topic are found in [6.73, 148, 149].

An optically bistable device has for a certain range of incident intensities I_0 two or more stable and reversible states of different transmission and/or reflection. The occurrence of one or the other state depends on the history, i.e., whether the bistable regime is reached from lower or higher intensities. Consequently, for example, a plot of the transmitted intensity as a function of the incident intensity shows hysteresis behavior. Necessary, but not sufficient preconditions are that an optical nonlinearity combined with a suitable feedback exist in the device.

The first SCs in which OB was found were InSb and GaAs, in 1979 [6.150]. Since then a large number of semiconductors have been investigated with respect to OB. The material which exhibits the largest variety of optical bistabilities is probably CdS. In Fig. 6.26, we show three hysteresis loops, which are all due to optical excitation in the electronic system, namely the formation of an EHP [6.62]. In Fig. 6.26a, we show OB caused by excitation-induced absorption. The effect occurs in the range of photon energies where the sample is transparent under weak excitation and becomes opaque under high excitation (Fig. 6.21). The positive feedback, which leads to an abrupt switch-down from high to low transmission, is based on the fact that increasing

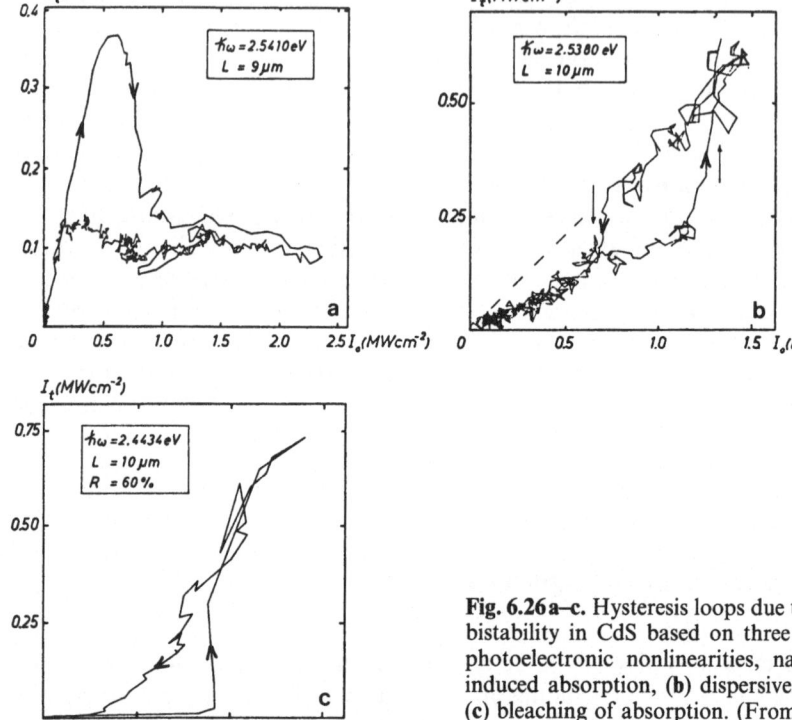

Fig. 6.26a–c. Hysteresis loops due to optical bistability in CdS based on three different photoelectronic nonlinearities, namely (**a**) induced absorption, (**b**) dispersive chances, (**c**) bleaching of absorption. (From [5.62])

absorption leads to an increasing generation rate of electron–hole pairs, which in turn leads to even higher absorption. Once the sample is in the highly absorbing state, I_0 can be lowered by some amount without switching back. The switching times are in the subnanosecond region for both transitions according to recent single shot experiments [6.139]. This type of OB caused by excitation-induced absorption was found in 1983 independently by various research groups. For a review, see [6.151].

Figure 6.26b shows OB due to excitation-induced changes of the real part of the refractive index (so-called dispersive OB). In this case the feedback is provided by a Fabry-Perot (FP) resonator containing the optically nonlinear medium. Here the FP resonator is simply formed by dielectric reflecting layers evaporated on the as-grown plane parallel surfaces of the CdS platelet-type samples. One starts at a transmission minimum, i.e., destructive interference of the partial waves and hence low intensity in the FP cavity. If this low intensity is sufficient to change n, positive feedback sets in because a finite Δn makes the interference in the FP less destructive and the subsequent increase of the intracavity intensity causes further increases in Δn. At a certain point, the sample switches to a state close to a transmission maximum; the high intracavity intensity allows the incident intensity to be reduced by a certain amount without leaving the high transmission branch. However, continued reduction results in reswitching to the lower stable state. The communication times between branches in Fig. 6.26b are as low as 300 ps [6.62].

The third example (Fig. 6.26c) shows OB due to excitation induced bleaching of absorption as it occurs above 160 K in CdS due to the EHP formation [6.62, 130]. Again a FP resonator is necessary; the resonator is however "switched off" as long as the material is absorbing. One tunes the incident laser to a position where one would have a transmission maximum if the sample were transparent. The feedback comes in this case from the increasingly efficient constructive interference in the FP with increasing bleaching. The switching times in Fig. 6.26c are in the nanosecond range.

Induced absorption OB has been found in CdS at 7 K also due to broadening of the exciton resonance for incident intensities as low as 300 W/cm^2 [6.152] and at 1.8 K due to an excitation induced thermal shift of a bound exciton complex absorption line [6.153]. Dispersive OB has been reported under similar conditions [6.154].

At room temperature both dispersive OB and induced absorption OB have been found under illumination with the green line of the Ar$^+$ ion laser [6.155, 156, 157]. Critical slowing down of the switching process has been observed as one approaches the critical switching intensity [6.156, 157]. Self-oscillations and the transition to deterministic optical chaos has been found, if the induced absorber is inserted in a hybrid (i.e., electro-optic) ring cavity [6.130, 156]. In Si, InP, and GaAs/GaAlAs MQWs, it has been possible to obtain optical bistability with hybrid self-electro-optci devices, known as SEEDs [6.151, 158]. However, this aspect leads to the field of electro-optics, which is beyond the scope of this review.

The technical interest in optical bistability is obvious. An OB device is a binary optical memory; the switching between various states can represent logic operations. Functions like AND, OR, and NOR have been implemented optically. This means one has all ingredients to build digital optical computers (DOCs), which would work with light pulses instead of current and/or voltage pulses. Since light beams do not interact with each other while propagating in vacuum, a DOC has the huge inherent advantage of parallel data processing. The present state of computer architecture for DOCs and other aspects of optical (digital) data handling are found in the proceedings of some recent conferences in this field [6.148]. Another aspect of future application of optical nonlinearities comes from LIGs, especially from the fact that some of the diffracted orders are phase conjugated with respect to the incident ones. This permits the construction of phase conjugate mirrors which allow, for example, the reconstruction of a wave front which is distorted and the focusing of laser beams on tiny targets or pellets. Recent reviews of this field including possible applications are to be found in [6.26, 27, 94].

To summarize, we can expect in the future the realization of novel concepts for data handling, image processing or computing which are based on linear and nonlinear (electro-)-optical properties of semiconductors and thus on the results of laser spectroscopy of semiconductors. It is to be hoped that these results will be used for the benefit of mankind. Everybody active in this field is asked to feel responsible for the outcome and the use of his research and to contribute so that the above-mentioned aim will be reached.

Note Added in Proof

Since the manuscript was finished in spring 1987 a lot of new results have been published in the field of laser spectroscopy of crystalline solids, indicating that this field is presently very active. In the following we want to mention briefly some of these new contributions.

The photoluminescence excitation spectroscopy of BEC has been extended to ZnS [6.160]. Biexciton states have been observed in II–VI compound MQW structures, showing a significant increase of the biexciton binding energy as compared to three-dimensional semiconductors [6.162]. The ac Stark effect has been observed with ps pulses and also calculated in CdS [6.163]. Furthermore it has become possible to observe this phenomenon for the first time also with ns pulses in GaAlAs MQW [6.164], thus complementing earlier ps measurements. The EHP has been investigated in GaAlAs MQW by pump- and probe-beam spectroscopy under quasi stationary excitation [6.165]. The renormalization of the fundamental gap has been analysed and found to be consistent with the theory. The renormalization of the higher sub-bands turned out to be significantly smaller than that of the fundamental gap and the disappearance of the exciton resonances is predominantly due to phase-space filling in these quasi-two-dimensional systems.

In the II–VI materials a lot of effort has been made to determine various relaxation times. In CdSe the following resonances have been investigated by various types of laser-induced grating spectroscopy: the exciton resonance, the induced absorption from the exciton to the biexciton level and the two photon absorption from the crystal ground state to the biexciton [6.166]. Phase relaxation times τ_2 up to 40 ps have been found at low excitation and temperature. These values decrease rapidly with increasing excitation or temperature.

The τ_2 values and the grating decay times have also been measured in the mixed crystal $CdS_{1-x}Se_x$ [6.167, 169]. It has been found that τ_2 can be as long as 80 ps under weak resonant excitation in the localized states. By varying the grating constant it has been found that the diffusion-length l_D in the localized states is significantly reduced as compared to the pure materials like CdS by a factor of about ten but it is definitely non-zero. Experiments on phase-conjugation and non-degenerate four wave mixing have recently been reported for CdS [6.170]. The proceedings of some recent conferences or workshops contain further information on new developments in this area [6.161,168,171,172].

Acknowledgements. Fruitful and stimulating discussions are acknowledged with colleagues from all over the world, e.g. Dr. D. S. Chemla, Dr. M. Dagenais, Prof. H. M. Gibbs, Prof. S. W. Koch, Dr. D. A. B. Miller, Dr. S. Schmitt-Rink (USA); Prof. J. B. Grun, Dr. B. Hönerlage, and Dr. R. Levy (France); Prof. S. D. Smith (U.K.); Prof. P. Mandel (Belgium); Prof. D. J. M. Hvam (Denmark); Prof. Xu Xurong and Prof. Rensong Dai (China); Dr. O. Gogoline, Dr. V. G. Lyssenko, Dr. S. Shevel, and Prof. Dr. V. B. Timofeev (USSR); Dr. F. Hennebergr (GDR); Prof. Dr. E. O. Göbel, Dr. Forchel, and Prof. Dr. H. Haug (FRG); and many others not mentioned here by name. I should like to thank all my co-workers, especially Dr. K. Bohnert, Dr. K. Kempf, Dr. H. Kalt, Dr. H. Schrey, Dr. M. Wegener, H. E. Swoboda, R. Renner, Ch. Weber, and F. A. Majumder for the beautiful results which they found during their Diploma and Ph.D. work and which could be presented in this review to only a small extent. Part of the work reported here has been supported by the Deutsche Forschungsgemeinschaft in the Sonderforschungsbereiche 65 „Festkörperspektroskopie" and 185 „Nichtlineare Dynamik" and by the Commission of the European Communities in the frame of the Stimulation Action. The careful and critical reading of the manuscript by Professor W. M. Yen is gratefully acknowledged.

References

6.1   H. Haken: *Light – Vol. II Laser Light Dynamics* (North-Holland, Amsterdam 1986)
6.2   F.P. Schäfer (ed.): *Dye Lasers*, Topics Appl. Phys., Vol. 1, 2nd ed. (Springer, Berlin, Heidelberg 1977)
6.3   S.L. Shapiro (ed.): *Ultrashort Light Pulses*, Topics Appl. Phys., Vol. 18, 2nd ed. (Springer, Berlin, Heidelberg 1984)
6.4   C.K. Rhodes (ed.): *Excimer Lasers*, Topics Appl. Phys., Vol. 30, 2nd ed. (Springer, Berlin, Heidelberg 1984)
6.5   O. Madelung: *Introduction to Solid-State Theory*, Springer Ser. Solid-State Sci., Vol. 2 (Springer, Berlin, Heidelberg 1978)
6.6   N.W. Ashcroft, N.D. Mermin: *Solid State Physics* (Holt-Saunders, New York 1976)
6.7   O. Madelung, M. Schulz, H. Weiss, (eds.): *Semiconductors*, Landolt-Börnstein, New Series, Group 3, Vol. 17 (Springer, Berlin, Heidelberg 1982–1985)
6.8   T. Brossat, F. Raymond: J. Cryst. Growth **72**, 280 (1985)
6.9   C.H. Aldrich, R.N. Silver: Phys. Rev. B **21**, 600 (1980)
6.10   W.G. Spitzer, H.Y. Fan: Phys. Rev. **106**, 882 (1957)
6.11   U. Rohrer, R. Claus: Phys. Rev. B **28**, 7048 (1983)
6.12   H. Nather, L.G. Quagliano: Solid State Commun. **50**, 75 (1984)
6.13   E. Hanamura, H. Haug: Phys. Rep. **33C**, 209 (1977)
6.14   J.P. Wolfe, A. Mysyrowicz: Sci. Am. **250–3**, 70 (1984)
6.15   C. Klingshirn: In *Energy Transfer Processes in Condensed Matter*, ed. by B. DiBartolo, NATO ASI Series, Vol. B **114** (Plenum, New York 1984) p. 285
6.16   M.M. Denisov, V.P. Makarov: Phys. Status Solidi B **56**, 9 (1973)
6.17   E.I. Rashba, M.D. Sturge (eds.): *Excitons*, Modern Problems in Condensed Matter Sciences, Vol. 2 (North-Holland, Amsterdam 1982)
6.18   K. Cho (ed.): *Excitons*, Topics Curr. Phys., Vol. 14 (Springer, Berlin, Heidelberg 1979)
6.19   B. DiBartolo (ed.): *Collective Excitations in Solids*, NATO ASI Series, Vol. B **88** (Plenum, New York 1982)
6.20   B. Honerlage, R. Levy, J.B. Grun, C. Klingshirn, K. Bohnert: Phys. Rep. **124**, 161 (1985)
6.21   M. Schreiber, Y. Toyozawa: J. Phys. Soc. (Jpn.) **51**, 1528, 1537, 1544 (1982); ibid. **52**, 318 (1983); also J. G. Liebler, S. Schmitt-Rink, H. Haug: J. Lumin. **34**, 1 (1985)

6.22   P.J. Dean, D.C. Herbert: In [6.17], p. 55
6.23   S. Permogorov, A. Reznitsky: *Proc. Int. Conf. "Excitons 84"*, Güstrow, GDR, p. 194
6.24   X.C. Zhang, A.V. Nurmikko: In *Proc. 17th Int. Conf. on the Physics of Semiconductors,*
       ed. by J.D. Chadi, W.A. Harrison (Springer, Berlin, Heidelberg 1985) p. 1443;
       see also A.V. Nurmikko: In [6.28], p. 355
6.25   D. Fröhlich: Adv. Solid State Phys. **XXI**, 77 (1981)
6.26   Y.R. Shen: *The Principles of Nonlinear Optics* (Wiley, New York 1984)
6.27   J.P. Eichler, P. Günter, D.W. Pohl (eds.): *Laser Induced Gratings*, Springer Ser. Opt.
       Sci., Vol. 50 (Springer, Berlin, Heidelberg 1986)
6.28   M.H. Pilkuhn (ed.): *High Excitation and Short Pulse Phenomena*, J. Lumin. **30** (1985)
6.29   C. Klingshirn, K. Bohnert, H. Kalt, V.G. Lyssenko, K. Kempf: In [6.28], p. 188
6.30   C. Benoit à la Guillaume, J.M. Debever, F. Salvan: Phys. Rev. **177**, 567 (1969)
6.31   S.G. Elkomoss, G. Munschy: J. Phys. Chem. Solids **42**, 1 (1981); ibid. **45**, 345 (1984)
6.32   C. Klingshirn, H. Haug: Phys. Rep. **70**, 315 (1981)
6.33   B.S. Razbirin, I.N. Ural'tsev, G.V. Mikhailov: Solid State Commun. **25**, 799 (1978)
6.34   I. Broser, J. Gutowski: J. Cryst. Growth **72**, 313 (1985)
6.35   R. Lindwurm, H. Haug: Z. Phys. B **53**, 281 (1983)
6.36   Y. Yoshikumi, H. Saito, S. Shionoya: Solid State Commun. **32**, 665 (1979)
6.37   J. Bille: Adv. Solid State Phys. **XIII**, 111 (1973)
6.38   C. Klingshirn: In *Spectroscopy of Solid State Laser Type Materials*, ed. by B.
       DiBartolo Ettore Majorana Intern. Science Series, Physical Sciences **30**, 485,
       Plenum, New York (1987)
6.39   H. Kalt, V.G. Lyssenko, R. Renner, C. Klingshirn: J. Opt. Soc. Am. B **2**, 1188 (1985)
6.40   H. Haug: Adv. Solid State Phys. **XXII**, 149 (1982)
6.41   H. Haug: In [6.28], p. 171
6.42   H. Haug, S.W. Koch, R. März, S. Schmitt-Rink: J. Lumin. **24/25**, 621 (1981)
6.43   M. Kuwata, T. Mita, N. Nagasawa: Solid State Commun. **40**, 911 (1981);
       T. Itoh, T. Katohno: J. Phys. Soc. (Jpn.) **51**, 707 (1982)
6.44   H.H. Kranz, H. Haug: Phys. Rev. A **34**, 2554 (1986)
6.45   J.M. Hvam, A. Bivas: Phys. Status Solidi B **101**, 363 (1980)
6.46   M. Ueta, H. Kazanki, K. Kobayashi, Y. Toyozawa, E. Hanimura: *Excitonic Processes
       in Solids*, Springer Ser. Solid State Sci., Vol. 60 (Springer, Berlin, Heidelberg 1986)
6.47   R. Baumert, I. Broser: Solid State Commun. **38**, 31 (1981)
6.48   A. Maruani, D.S. Chemla: Phys. Rev. B **23**, 841 (1981)
6.49   T. Mita, N. Nagasawa: Opt. Commun. **24**, 345 (1978);
       Y. Masumoto, S. Shionoya, T. Takagahara: Phys. Rev. Lett. **51**, 923 (1983)
6.50   H. Kalt, R. Renner, C. Klingshirn: IEEE J. QE-**22**, 1312 (1986)
6.51   H. Haug (ed.): *Optical Nonlinearities and Instabilities in Semiconductors* (Academic,
       New York, 1988)
6.52   C. Klingshirn: In [6.51], p. 13
6.53   R. Zimmermann, K. Kilimann, W.D. Kraeft, D. Kremp, R. Röpke: Phys. Status Solidi
       B **90**, 175 (1978)
6.54   H. Haug, S. Schmitt-Rink: Prog. Quantum Electron. **9**, 3 (1984)
6.55   T.M. Rice: Solid State Phys. **32**, 1 (1977);
       J.C. Hesnel, T.G. Philips, G.A. Thomas: Solid State Phys. **32**, 88 (1977);
       Ya. Prokowskii, V.B. Timofeev: Sov. Sci. Rev., A Phys. Rev. **1**, 191 (1979); see also,
       V.B. Timofeev: in [6.17], p. 349
6.56   D. Bimberg, M.S. Skolnick, L.M. Sander: Phys. Rev. B **9**, 2231 (1979);
       R. Schwabe, F. Thuselt, H. Weinert, R. Bindemann: Phys. Status Solidi B **95**, 571
       (1979);
       also, H. Schweizer, E. Zelinski: In Ref. [6.28], p. 37
6.57   K. Bohnert, M. Anselment, G. Kobbe, C. Klingshirn, H. Haug, S.W. Koch, S. Schmitt-
       Rink, F.F. Abraham: Z. Phys. B **42**, 1 (1981)
6.58   H. Yoshida, S. Shionoya: Phys. Status Solidi B **115**, 203 (1983);
       Y. Unuma, Y. Abe, Y. Masumoto, S. Shionoya: Phys. Status Solidi B **125**, 735 (1984)

Laser Spectroscopy of Crystalline Semiconductors   253

6.59   H. Saito: In [6.28], p. 303
6.60   A. Miller, D.A.B. Miller, S.D. Smith: Adv. Phys. **30**, 697 (1981)
6.61   L. Banyai, S.W. Koch: Z. Phys. B **63**, 283 (1986)
6.62   M. Wegener, C. Klingshirn, S.W. Koch, L. Banyai: Semicond. Sci. Technol. **1**, 366 (1986)
6.63   I. Balslev: Phys. Rev. B **30**, 3203 (1984); also in [6.28] p. 162
6.64   H. Schweizer, A. Forchel, A. Hangleiter, S. Schmitt-Rink, J.P. Löwenau, H. Haug: Phys. Rev. Lett. **51**, 698 (1983)
6.65   F. Keilmann, J. Kuhl: IEEE J. QE-**14**, 203 (1978)
6.66   A. Miller, G. Parry: Philos. Trans. R. Soc. London Ser. A **313**, 277 (1984)
6.67   K.A. Dmitrenko, S.G. Shevel, L.V. Marintchenko: Phys. Status Solidi B **134**, 605 (1986)
6.68   D.S. Chemla, D.A.B. Miller: J. Opt. Soc. Am. B **2**, 1155 (1985)
6.69   J. Hegarty, M.D. Sturge: J. Opt. Soc. Am. B **2**, 1143 (1985)
6.70   D.S. Chemla, A. Pinczuk (eds.): *Semiconductor Quantum Wells and Superlattices*, IEEE J. QE-**22**, No. 9 (1986)
6.71   D.S. Chemla: Private communication
6.72   S. Schmitt-Rink, C. Ell, H. Haug: Phys. Rev. B **33**, 1183 (1986);
       G. Tränkle, H. Leier, A. Forchel, H. Haug, C. Ell: Phys. Rev. Lett. **58**, 419 (1987)
6.73. D.S. Chemla (ed.): *Excitonic Optical Nonlinearities*, J. Opt. Soc. Am. B **2**, No. 7 (1985)
6.74   D.S. Chemla, S. Schmitt-Rink, D.A.B. Miller: In [6.51] p. 83
6.75   D. Fröhlich, A. Nöthe, K. Reimann: Phys. Rev. Lett. **55**, 1335 (1986)
6.76   A. Mysyrowicz, D. Hulin, A. Antonetti, A. Migus, W.T. Masselink, H. Morkoç: Phys. Rev. Lett. **56**, 2748 (1986)
6.77   A.V. Lehmen, D.S. Chemla, J.E. Zucker, J.P. Heritage: Opt. Lett. **11**, 609 (1986);
       S. Schmitt-Rink, D.S. Chemla: Phys. Rev. Lett. **57**, 2752 (1986)
6.78   B.R. Mollow: Phys. Rev. **188**, 1969 (1969); Phys. Rev. A **5**, 2217 (1972)
6.79   R. Claus: Phys. Status Solidi B **100**, 9 (1980)
6.80   I. Broser, R. Broser, E. Beckmann, E. Birkicht: Solid State Commun. **39**, 1209 (1981)
6.81   M.V. Lebedev, V.G. Lysenko, V.B. Timofeev: Sov. Phys. – JETP **86**, 2193 (1984);
       M.V. Lebedev, M.H. Strasnnikova, V.B. Timofeev, V.V. Czernyi: JETP Lett. **39**, 366 (1984)
6.82   A. Forchel, B. Laurich, H. Hollmer, G. Tränkle, M.H. Pilkuhn: In [6.28] p. 67
6.83   J.P. Wolfe: In [6.28] p. 82
6.84   H. Yoshida, H. Saito, S. Shionoga, V.B. Timofeev: Solid State Commun. **33**, 161 (1980)
6.85   K.L. Shaklee, R.F. Leheny: Appl. Phys. Lett. **18**, 475 (1971);
       K.L. Shaklee, R.E. Nahory, R.F. Leheny: J. Lumin. **7**, 284 (1973)
6.86   J.M. Hvam: J. Appl. Phys. **49**, 3124 (1978)
6.87   F.A. Majumder, H.E. Swoboda, K. Kempf, C. Klingshirn: Phys. Rev. B **32**, 2407 (1985)
6.88   K. Bohnert, F. Fidorra, C. Klingshirn: Z. Phys. B **57**, 263 (1984)
6.89   H. Schrey, V.G. Lysenko, C. Klingshirn: Solid State Commun. **32**, 897 (1979)
6.90   J.M. Hvam, G. Blattner, M. Reuscher, C. Klingshirn: Phys. Status Solidi B **118**, 179 (1983)
6.91   H.J. Eichler (ed.): *Dynamic Gratings and Four Wave Mixing*, IEEE J. QE-**22**, 1194ff (1986)
6.92   A.L. Amirl, T.F. Boggess, B.S. Wherrett, G.P. Perryman, A. Miller: Phys. Rev. Lett. **49**, 933 (1982)
6.93   H. Kalt, R. Renner, C. Klingshirn: In [6.91], p. 1312
6.94   V.V. Shkunov, B.Ya. Zel'dovich: Sci. Am. **251-12**, 40 (1986);
       D.M. Pepper: Sci. Am. **251-1**, 56 (1986)
6.95   J. Wicksted, M. Matsushita, H.Z. Cummins, T. Shigenari, X.Z. Yu: Phys. Rev. B **29**, 3350 (1984);
       M. Matsushita, J. Wicksted, H.Z. Cummins: Phys. Rev. B **29**, 3362 (1984);
       T. Shigenari, X.Z. Yu, H.Z. Cummins: Phys. Rev. B **30**, 1962 (1984)

254   C. Klingshirn

6.96   J.L. Birman: In [6.17], p. 27;
        also E.S. Koteles: In [6.17], p. 83
6.97   H. Malm, R.R. Haering: Can. J. Phys. **49**, 2432 (1971)
6.98   G. Blattner, C. Klingshirn, R. Helbig, R. Meinl: Phys. Status Solidi B **107**, 105 (1981)
6.99   M. Schilling, R. Helbig, G. Pensl: J. Lumin. **33**, 201 (1985)
6.100   R. Baumert, I. Broser, J. Gutowski, A. Hoffmann: Phys. Rev. B **27**, 6263 (1983);
        J. Gutowski, I. Broser: Phys. Rev. B **31**, 3611 (1985); also Solid State Commun. **58**, 523 (1986);
        J. Gutowski, J. Schott, I. Broser: Phys. Status Solidi B **138**, 673 (1986)
6.101   J. Puls, F. Henneberger, J. Voigt: Phys. Status Solidi B **119**, 291 (1983)
6.102   C. Klingshirn, W. Maier, G. Blattner, P.J. Dean, G. Kobbe: J. Cryst. Growth **59**, 352 (1982)
6.103   H. Schrey: Ph. D. Thesis, University of Karlsruhe (1979)
6.104   H. Schrey, C. Klingshirn: Phys. Status Solidi B **90**, 67 (1978)
6.105   J. Shah: Phys. Rev. B **9**, 562 (1974)
6.106   T. Itoh, Y. Nozue, M. Ueta: J. Phys. Soc. Jpn. **40**, 1791 (1976); ibid. **44**, 1305 (1978)
6.107   V.G. Lysenko, K. Kempf, K. Bohnert, G. Schmieder, C. Klingshirn, S. Schmitt-Rink: Solid State Commun. **42**, 401 (1982)
6.108   G. Kurtze, V.G. Lysenko, C. Klingshirn: Phys. Status Solidi B **110**, 103 (1982)
6.109   H. Schrey, V.G. Lysenko, C. Klingshirn: Solid State Commun. **31**, 299 (1979)
6.110   G. Blattner, G. Kurtze, G. Schmieder, C. Klingshirn: Phys. Rev. B **25**, 7413 (1982)
6.111   W. Maier, G. Schmieder, C. Klingshirn: Z. Phys. B **50**, 193 (1983)
6.112   H.J. Mattauch, Ch. Uihlein: Solid State Commun. **25**, 447 (1978); Phys. Status Solidi B **96**, 189 (1979)
6.113   D.G. Seiler, D. Heiman, R. Feigenblatt, R.L. Aggarwald, B. Lax: Phys. Rev. B **25**, 7666 (1982);
        D.G. Seiler, D. Heiman, B.S. Wherrett: Phys. Rev. B **27**, 2355 (1983)
6.114   R. Dinges, D. Fröhlich, B. Staginnus, W. Staude: Phys. Rev. Lett. **25**, 922 (1970);
        W. Kaule: Solid State Commun. **9**, 17 (1971);
        G. Pensl: Solid State Commun. **11**, 1277 (1972);
        R. Baltrameyumas, V. Gavryuskin, V. Kubertavichyus, R. Rachyukaitis: JETP Lett. **38**, 1 (1983)
6.115   H.W. Hölscher, A. Nöthe, Ch. Uihlein: Phys. Rev. B **31**, 2379 (1985)
6.116   D. Fröhlich, A. Nöthe, K. Reimann: Phys. Status Solidi B **125**, 653 (1984)
6.117   G. Kurtze, W. Maier, G. Blattner, C. Klingshirn: Z. Phys. B **37**, 9 (1980)
6.118   G. Schmieder, K. Kempf, K. Bohnert, G. Kobbe, V.G. Lysenko, A. Kreissl, C. Klingshirn, G. Kurtze, G. Blattner, W. Maier: J. Lumin. **24/25**, 613 (1981);
        K. Kempf, G. Schmieder, G. Kurtze, C. Klingshirn: Phys. Status Solidi B **107**, 297 (1981)
6.119   W. Maier: Ph.D. Thesis, University of Karlsruhe (1982)
6.120   Y. Masumoto, S. Shionoya: J. Phys. Soc. Jpn. **49**, 2236 (1980)
6.121   R. Baumert, I. Broser, K. Buschik: In [6.91], p. 1539
6.122   D. Hulin, A. Antonetti, L.L. Chase, J.L. Martin, A. Migus, A. Mysyrowicz, J.P. Löwenau, S. Schmitt-Rink, H. Haug: Phys. Rev. Lett. **52**, 779 (1984)
6.123   V.B. Timofeev: In [6.17], p. 349
6.124   M.L.W. Thewalt: In [6.17], p. 393
6.125   V.B. Timofeev, V.D. Kulakowskii, I.V. Kukushkin: Physica B **117/118**, 327 (1983)
6.126   O. Hildebrand, E.O. Göbel, K.M. Romanek, H. Weber, G. Mahler: Phys. Rev. B **17**, 4775 (1978)
6.127   S. Tanaka, H. Kobayashi, H. Saito, S. Shionoya: J. Phys. Soc. Jpn. **49**, 1051 (1980)
6.128   A. Selloni, S. Modesti, M. Capizzi: Phys. Rev. B **30**, 821 (1984);
        M. Capizzi, A. Frova, S. Modesti, A. Selloni, J.L. Staehli, M. Guzzi: Helv. Phys. Acta **58**, 272 (1985);
        also, G. Mahler, F. Fourikis: In [6.28], p. 18

6.129   O. Engstrom (ed.): *Proc. of the 18th Int. Conf. on the Physics of Semiconductors – 1986* (World Scientific, Singapore 1987)
6.130   C. Klingshirn, U. Becker, C. Dörnfeld, H. Kalt, M. Kunz, M. Lambsdorff, V.G. Lysenko, F.A. Majunder, R. Renner, S. Shevel, H.E. Swoboda, Ch. Weber, M. Wegener: In [6.129], p. 1667
6.131   H. Yoshida, S. Shionoya: Phys. Status Solidi B 115, 203 (1983); Y. Unuma, Y. Abe, Y. Masumoto, S. Shionoya: Phys. Status Solidi B 125, 735 (1984)
6.132   A. Kreissl, K. Bohnert, V.G. Lysenko, C. Klingshirn: Phys. Status Solidi B 114, 537 (1982)
6.133   H.E. Swoboda: Ph.D. Thesis, University of Kaiserslautern (1989)
6.134   R. Zimmermann, M. Rösler: Phys. Status Solidi B 75, 633 (1976)
6.135   H. Haug, J. Müller, R. Mewis: J. Lumin. 38, 239 (1987)
6.136   H. Saito: Solid State Commun. 39, 71 (1981)
6.137   H. Kuroda, S. Shionoya, H. Saito, E. Hanamura: J. Phys. Soc. Jpn. 35, 534 (1973); M. Hayashi, H. Saito, S. Shionoya: Solid State Commun. 24, 833, 837 (1977); J. Phys. Soc. Jpn. 44, 582 (1978)
6.138   J.P. Woerdman: Philips Res. Rep., Suppl. 7 (1971)
6.139   Ch. Weber: Ph.D. Thesis, University of Kaiserslautern (1989)
6.140   R. Renner: Ph.D. Thesis, University of Frankfurt (1989)
6.141   S. Shevel, R. Fischer, E.O. Göbel, G. Noll, P. Thomas, C. Klingshirn: J. Lumin., 37, 45 (1987); H.E. Swoboda, F.A. Majumder, C. Klingshirn, S. Shevel, R. Fischer, E.O. Göbel, G. Noll, P. Thomas, A. Reznitsky, S. Permogorov: J. Lumin. 38, 79 (1987)
6.142   E. Cohen, M.D. Sturge: Phys. Rev. B 25, 3828 (1982); J.A. Kash, A. Ron, E. Cohen: Phys. Rev. B 28, 6147 (1983)
6.143   S. Permogorov, A. Reznitskii, S. Verbin, G.O. Müller, P. Flögel, M. Nikiforova: Phys. Status Solidi B 113, 589 (1982); S.A. Permogorov, A.N. Reznitskii, S.Yu. Verbin, V.G. Lysenko: JETP Lett. 37, 462 (1983); S.A. Permogorov, A. Reznitskii, S. Verbin, A. Naumov, W. van der Osten, H. Stolz: J. de Phys. C 7, 173 (1985)
6.144   F.A. Majumder, S. Shevel, V.G. Lysenko, H.E. Swoboda, C. Klingshirn: Z. Phys. B 66, 409 (1987)
6.145   T. Amand, J. Collet, B.S. Razbirin: Phys. Rev. B 34, 2718 (1986)
6.146   E. Ostertag, J.B. Grun: Phys. Status Solidi B 82, 335 (1977)
6.147   Y. Masumoto, S. Shionoya, Y. Tanaka: Solid State Commun. 27, 1117 (1978); Y. Oka, K. Nakamura, H. Fujisaki: Phys. Rev. Lett. 57, 2857 (1986)
6.148a   B.S. Wherrett, S.D. Smith (eds.): *Optical Bistability, Dynamical Nonlinearity and Photonic Logic*, Philos. Trans. R. Soc. London, Ser. A 313 (1984)
6.148b   B.L. Dove (ed.): *Digital Optical Circuit Technology*, AGARD Conf. Proc. 362 (1985)
6.148c   H.M. Gibbs, P. Mandel, N. Peyghambarian, S.D. Smith (eds.): *Optical Bistability III*, Springer Proc. Phys., Vol. 8 (Springer, Berlin, Heidelberg 1986)
6.148d   H.M. Gibbs: *Optical Bistability: Controlling Light with Light* (Academic, New York 1985)
6.148e   M.E. Garmire (ed.): *Optical Bistability*, IEEE J. QE-21, No. 9 (1985)
6.148f   P. Mandel, S.D. Smith, B.S. Wherrett (eds.): From *Optical Bistability Towards Optical Computing* (North-Holland, Amsterdam 1987)
6.149   F. Henneberer: Phys. Status Solidi B 137, 371 (1986)
6.150   D.A.B. Miller, S.D. Smith: Opt. Commun. 31, 331 (1979); H.M. Gibbs, S.L. McCall, T.N.C. Verkatesan, A.C. Gossard, A. Passner, W. Wiegmann: Appl. Phys. Lett. 35, 451 (1979)
6.151   D.A.B. Miller: J. Opt. Soc. Am. B 1, 857 (1984)
6.152   C. Klingshirn, M. Wegener, C. Dörnfeld, M. Lambsdorff, J.Y. Bigot, F. Fidorra: In [6.148b], p. 129;

M. Wegener, C. Dornfeld, M. Lambsdorff, F. Fidorra, C. Klingshirn: *Proc. of the Quebec Int. Symp. on Optical Chaos,* SPIE Proc. **667**, 102 (1986)

6.153   M. Dagenais: Appl. Phys. Lett. **45**, 1267 (1984)
6.154   M. Dagenais, W.F. Sharfin: Appl. Phys. Lett. **46**, 230 (1985)
6.155   M. Lambsdorff, C. Dornfeld, C. Klingshirn: Z. Phys. B **64**, 409 (1986)
6.156   M. Wegener, C. Klingshirn: In [6.129], p. 1675; Phys. Rev. A **35**, 1740 and 4247 (1987), also, M. Wegener, C. Klingshirn, G. Muller-Vogt: Z. Phys., B **68**, 519 (1987)
6.157   I. Haddad, M. Kretzschmar, H. Rossmann, F. Henneberger: Phys. Status Solidi B **138**, 235 (1986)
6.158   D. Jäger, F. Forsmann, B. Wedding: In [6.148], p. 1453;
        D. Jäger, F. Forsmann: Solid-State Electron. **30**, 67 (1987)
6.159   D.S. Chemla: In [6.129], p. 513
6.160   J. Gutowski, I. Broser, G. Kudlek: *Spectroscopy of Bound Excitons in Cubic ZnS at Moderate to High Excitation Densities* Phys. Rev. B in press
6.161   H. Haug, L. Banyai (eds.): NATO *Workshop on Optical Switching in Low Dimensional Systems, Marbella Oct (1988)* to be published by Plenum in NATO ASI series (1989)
6.162   A. Mysyrowicz: *Evidence for Biexciton ZnSe Quantum Wells* in [6.161]
6.163   H. Haug, C. Ell, J.F. Müller, K. EL Sayed: *Stationary Solutions for the Excitonic Optical Stark Effect in Two- and Three-Dimensional Semiconductors* in [6.161]
6.164   Ch. Weber: private communication
6.165   Ch. Weber, C. Klingshirn, D.S. Chemla, D.A.B. Miller, J.E. Cunningham, C. Ell: *Gain Measurements and Band-Gap Renormalization in GaAs/Al$_x$Ga$_{1-x}$As Multiple Quantum Well Structures* Phys. Rev. B, Rapid Commun., **38**, 12748 (1988)
        C. Klingshirn, Ch. Weber, D.S. Chemla, D.A.B. Miller, J.E. Cunningham, C. Ell, H. Haug: *The Electron-Hole Plasma in Quasi Two-Dimensional and Three-Dimensional Semiconductors* in [6.161]
6.166   C. Dörnfeld, J.M. Hvam: *Optical Nonlinearities and Phase Coherence in CdSe Studied by Transient Four Wave Mixing* to be published in IEEE J. Quantum Electron;
        J.M. Hvam, C. Dörnfeld: *Nonlinearities, Coherence and Dephasing in Layered GaSe and in CdSe Surface Layer* in [6.161]
6.167   H.-E. Swoboda, F.A. Majumder, Ch. Weber, R. Renner, C. Klingshirn, G. Noll, E.O. Göbel, S. Permogorov, A. Reznitzky: *The Influence of Localized Excitons on the Optical Behaviour of CdS$_{1-x}$Se$_x$* Proc. 19th Intern. Conf. Phys. Semicond., Warschau, Aug (1988) in press;
        H.-E. Swoboda, F.A. Majumder, R. Renner, Ch. Weber, M. Sence, Lu Jie, G. Noll, E.O. Göbel, J. Vaitkus, C. Klingshirn: *Nonlinear Optical Properties of the System CdS$_{1-x}$Se$_x$* invited contribution to [6.168], p. 749
6.168   Proc. Intern. Conf. on Optical Nonlinearity and Bistability of Semiconductors, Berlin Aug (1988), to be published in Phys. Status Solidi B **150** (1988)
6.169   J.M. Hvam, C. Dörnfeld, H. Schwab: *Optical Nonlinearity and Phase Coherence in CdSe and CdS$_{1-x}$Se$_x$* in [6.168], p. 387
6.170   C. Klingshirn, M. Kunz, F.A. Majumder, D. Oberhauser, R. Renner, M. Rinker, H.-E. Swoboda, A. Uhrig, Ch. Weber: Review of Nonlinear Optical Processes in Wide Gap II–VI *Compounds* invited contribution to [6.171]
6.171   R. Gebhardt: NATO Workshop on II–VI Semiconductors, Regensburg, Aug (1988) to be published by Plenum, NATO ASI series
6.172   SPIE Intern. Congress on Optical Science and Engineering, Nonlinear Optical Materials, Hamburg, Sept (1988), to be published

7. Laser Spectroscopy of Amorphous Semiconductors

P. Craig Taylor

With 32 Figures

Amorphous semiconductors are often loosely classified into two different groups, the tetrahedrally coordinated amorphous materials which are primarily based upon group IV atoms and the amorphous materials based upon group VI atoms, which usually have lower average coordination numbers. Although this distinction has no strong theoretical justification, there are enough empirical differences between these two classes of amorphous semiconductors that the separation has proved useful.

Of the tetrahedrally coordinated amorphous semiconductors the prototype material is amorphous silicon (a-Si). Often hydrogen is added to a-Si to remove defects such as silicon "dangling bonds" from the material. The alloy which contains hydrogen is called hydrogenated amorphous silicon (a-Si:H) and is currently the material upon which most of the important electronic devices, such as photovoltaic cells and thin film transistors, are based. Also important in this class of amorphous semiconductors are the alloys based upon a-Si:H such as hydrogenated silicon germanium and hydrogenated silicon carbon alloys. These materials can only be made in thin film form by some kind of deposition technique which provides very rapid cooling of individual atoms or molecules onto a substrate.

The prototype amorphous semiconductors which contain group VI atoms are the so-called chalcogenide glasses. These amorphous semiconductors, which typically contain one of the elements sulfur, selenium, or tellurium, can usually be made in bulk form by quenching from the liquid phase. Hence these amorphous semiconductors are glasses. The prototype chalcogenide glasses are Se, As_2Se_3, As_2S_3, $GeSe_2$, and GeS_2. Although the chalcogenide glasses are not as commercially important as the tetrahedrally bonded amorphous semiconductors, these materials do find use in some applications such as filters, modulators and detectors of infrared radiation.

7.1 Introductory Comments

Laser Spectroscopy is used in many different ways in the study of electronic, optical and structural properties of amorphous semiconductors. Laser excitation is often used for ordinary absorption and reflection spectroscopy, for photoluminescence and light scattering experiments, and for various optical modulation experiments. There are also several varieties of electronic transport experiments which use laser excitation. In the chalcogenide glasses lasers have been employed to produce changes in the optical and structural properties. Although not much work has been done on the nonlinear optical properties of amorphous semiconductors, this is another area where laser spectroscopy will be of considerable importance in the future. Finally, lasers have been employed to probe the low frequency vibrational properties of amorphous semiconductors.

7.2 Optical Properties of Amorphous Semiconductors

The amorphous semiconductors exhibit electrical conductivities which are analogous to those observed in intrinsic crystalline semiconductors. In particular, the electrical conductivity is thermally activated with an activation energy which is half the energy of the optical band gap. In fact, this property more than any other is what gives these amorphous solids their name. Beyond this general experimental feature, the optical properties of the two general classes of amorphous semiconductors are often very different. For this reason we will discuss the two classes separately.

The one general feature of the optical properties of amorphous semi-conductors which is common to both the tetrahedrally coordinated amorphous semiconductors and the chalcogenide glasses is the general shape of the optical absorption edge. A schematic representation of the absorption edge is shown in Fig. 7.1. In this figure there are three general regions denoted as A, B, and C. The region at highest energy (region A) can be used to extract an estimate of the

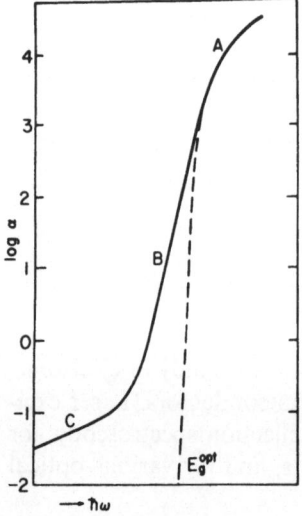

Fig. 7.1. Schematic diagram of the absorption edge in an amorphous semiconductor. Region *A* can be used to determine an estimate of the optical band gap. Region *B* is the Urbach tail and region *C* is the absorption well below the gap. (After [7.6])

optical energy gap for an amorphous solid. Following a procedure first suggested by *Tauc* [7.1] one can assume that the matrix element for the optical absorption process is constant and derive the following expression for the shape of the optical absorption edge:

$$\alpha \omega n = C(\hbar\omega - E_g)^2 , \qquad (7.1)$$

where α is the absorption coefficient, ω the angular frequency, n the real part of the index of refraction (which may depend on ω), C a constant and E_g the optical gap energy, which in this model is $E_v - E_c$, the difference between the valence and conduction band localization edges. Although other functional forms have been proposed, which are based upon slightly different starting assumptions [7.2], the form shown in (7.1) is conventionally used to define the band gap energies (E_g) of amorphous semiconductors.

The second region (B) in Fig. 7.1 is called the Urbach edge region in analogy with the presence in crystalline solids of a similar region originally discovered by *Urbach* [7.3]. In this region the optical absorption coefficient depends exponentially on the energy as shown in the following equation:

$$\alpha(\omega) \sim \exp[\sigma(\hbar\omega - \hbar\omega_0)] , \qquad (7.2)$$

where σ is a constant whose interpretation depends on the details of the model used to interpret the data [7.4, 5].

A microscopic picture for the origin of the absorption in the Urbach "tail" region of the spectrum is still not entirely clear, but the most commonly accepted interpretation is due to *Dow* and *Redfield* [7.4, 5]. The Dow-Redfield model suggests that the Urbach tail results from ionization of excitons caused by local electric fields. The source of the ionizing electric fields is supposed to be phonons. The shape of the absorption edge is insensitive to the details of these electric fields. The major difficulty with this model, in addition to the fact that it is not a microscopic model, is that the Urbach edges in the amorphous semiconductors are so broad that the internal electric fields must be quite large.

A more qualitative picture of the origin of the Urbach tails in amorphous semiconductors has been presented many times in the literature. This picture suggests that the absorption in this spectral region is due to "strained bonds". The idea is that the amorphous network must be strained locally in order to accommodate the atoms in a nonperiodic array. These strained regions will create near the band edges electronic states which generate the Urbach tail in the absorption spectrum. Although this qualitative picture is appealing, it is difficult to make predictions based solely upon it.

The third region (C) in Fig. 7.1 is a region where the absorption depends less dramatically on the energy than in the Urbach region [7.6]. This region extends well below the optical band gap and the absorption in this region may result from different processes in different amorphous semiconductors. In the chalcogenide glasses the absorption in this region is probably due to the presence of impurities. In the tetrahedrally coordinated amorphous semiconductors the absorption in this region is due to the presence of unsatisfied bonds.

Although the shapes of the absorption edges in the chalcogenide glasses and the tetrahedrally coordinated amorphous semiconductors are essentially the same, there is one significant difference. In the chalcogenide glasses the optical absorption edges are highly reproducible and insensitive to preparation conditions. On the other hand, the tetrahedrally coordinated amorphous semiconductors, most of which can only be made in thin film form, exhibit optical absorption edges which often depend dramatically on preparation conditions.

7.2.1 The Chalcogenide Glasses

In the chalcogenide glasses the optical properties are dominated by strong electron–lattice interactions. This strong electron–lattice coupling is manifested in many experimental situations, including a metastable optically induced absorption which extends well below the band gap and a photoluminescence (PL) which is strongly Stokes-shifted.

Besides the absorption regions shown in Fig. 7.1 there are several other absorption processes which are common to essentially all chalcogenide glasses. In addition to the metastable optical absorption mentioned above, these processes include a transient optically induced absorption, which extends well into the gap, and a "free carrier" absorption, which is remarkably frequency independent but is only observed at room temperature in the most highly conducting glasses.

We illustrate the general features of the optical absorption in the chalcogenide glasses using glassy As_2S_3 as an example. This glass is perhaps the most studied of all of the chalcogenide glasses. In Fig. 7.2 are displayed data compiled [7.7] from several sources [7.6, 8–22]. There are two regions where the absorption is low in Fig. 7.2. These two regions occur where the reduced frequency $\bar{v}$ (where $\bar{v}$ is given by the inverse of the wavelength) is either in the range $\bar{v} \leq 10 \text{ cm}^{-1}$ or $10^3 \leq \bar{v} \leq 10^4 \text{ cm}^{-1}$. These two transparent regions are separated by a region of strong vibrational absorption in which the strongest peak occurs around 300 cm^{-1}. The series of peaks above the major one is generated by multiphonon processes. Impurity absorption peaks sometimes also exist in this region. The characteristic rise in the absorption above about 10^4 cm^{-1} is the band edge absorption which was shown in more detail in Fig. 7.1.

Above 10^4 cm^{-1} the absorption is due to interband electronic transitions which are similar to those which occur in crystalline solids. Because these are amorphous materials, however, there is no structure due to critical points in the Brillouin zone. Using visible and ultraviolet reflectivity and electro-reflectivity data the interband absorption spectra have been extracted for a great many chalcogenide glasses. We again use glassy As_2S_3 as an illustrative example. The interband spectra for glassy As_2S_3 (and also for glassy As_2Se_3) are shown in Fig. 7.3.

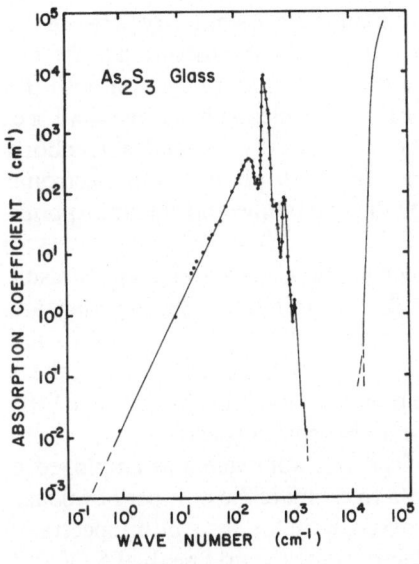

Fig. 7.2. Absorption coefficient (in $[cm^{-1}]$) in glassy As_2S_3 at 300 K. (After [7.7] as originally compiled from [7.6, 8–22])

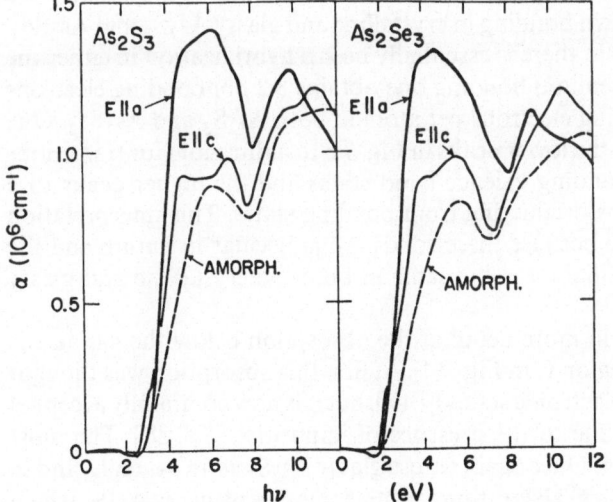

Fig. 7.3. Interband absorption spectra for glassy and crystalline As_2S_3 and As_2Se_3 at 300 K. (After [7.8])

In both of these materials the interband absorption exhibits two peaks, one near 6 eV and one near 10 eV [7.8, 22–27]. These two peaks are separated by a broad minimum near 7 eV. As can be seen from Fig. 7.3, the glassy spectra are similar to those of the corresponding crystals except that the anisotropy and the sharp structure are missing from the spectra of the glasses. A further important difference between the crystalline and glassy spectra is an overall broadening of the general two-peak structure. This broadening is attributed to distortions in the glassy network.

In spite of these two differences, there remains a close similarity between the spectra of the glasses and the crystals. This similarity is an illustration of the fact that the disorder in the glasses has only a relatively minor effect on both the electronic densities of states and the optical matrix elements. As we shall see, this situation will not be the case for the tetrahedrally bonded amorphous semiconductors where the matrix elements for both phonon and electronic absorption processes are very different for the crystalline and the amorphous solids.

The similarities in the case of the chalcogenide glasses and crystals result from the fact that these solids are very molecular in nature. Because they are molecular the interband (and also the vibrational) absorption spectra are determined in large part by the "molecules" and their excited states. In these cases the extra symmetries implied by the periodic long range order have little effect on the gross features of the interband absorption spectra.

Drews et al. [7.8] have invoked the electronic sum rule and employed a Kramers-Kronig analysis of reflectivity data to estimate the number of valence electrons which contribute to each of the two peaks in the absorption spectra of Fig. 7.3. They obtain between 3 and 3.5 electrons for the first peak at 5 eV and less than 3 electrons for the second peak at 10 eV. These two estimates are consistent with the known bonding in crystalline and glassy As_2S_3 and As_2Se_3. With the assumption that there is essentially no s-p hybridization in either the arsenic or sulfur (or selenium) bonding one obtains 3.2 nonbonding electrons per atom and 2.4 bonding electrons per atom in both As_2S_3 and As_2Se_3. One may therefore attribute the lower peaks in Fig. 7.3 to a threshold for transitions originating from nonbonding valence band states and the higher peaks to a threshold for transitions originating from bonding states. This interpretation makes qualitative sense because these solids are molecular in nature and the nearest-neighbor coordination is the same in both the crystalline and glassy phases.

We shall now look in more detail at the absorption below the gap in the chalcogenide glasses (region C in Fig. 7.1). At first this absorption was thought to be due to "intrinsic" electronic states [7.17], but it is now commonly accepted that the absorption is due to the presence of impurities [7.6, 28]. The most common impurity is iron. Once again we use glassy As_2S_3 as an example, and in Fig. 7.4 we show the optical absorption for three samples of glassy As_2S_3 which differ in the amount of iron impurities they contain [7.28]. The lowest absorption is shown by the sample which is nominally undoped, but chemical analysis has indicated that the residual iron in this sample is approximately 5 ppm. In all chalcogenide glasses prepared with existing technologies iron exists at about this level. By comparisons with ESR measurements, *Tauc* et al. [7.28] have determined that the absorption is due to charge transfer transitions of the form $Fe^{2+} \rightarrow Fe^{3+} + e$.

So far we have discussed optical absorption in the chalcogenide glasses where the light which is used to measure the absorption does not itself produce any changes in the absorption. There are, however, several optically induced

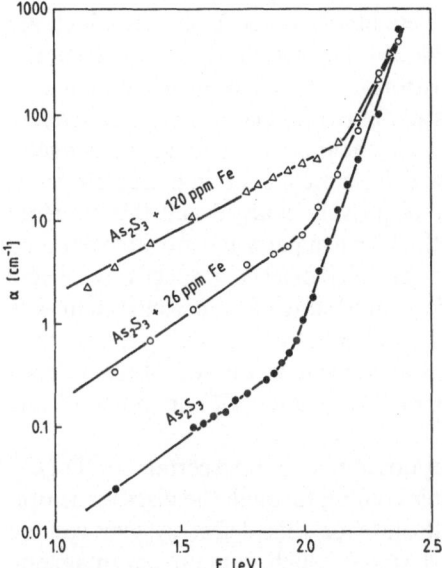

Fig. 7.4. Absorption coefficient at 300 K in glassy As_2S_3 with varying amounts of iron. The "pure" As_2S_3 is estimated to have about 5 ppm Fe. (After [7.28])

changes in the absorption spectra in the chalcogenide glasses. The interpretation of these spectra depends upon an understanding of the strong role that lattice relaxation plays in determining the optical properties of the chalcogenide glasses. We therefore briefly describe the current models within which the optically induced changes in the optical properties can be understood.

It has been known for some time that when disorder is introduced into an otherwise periodic system the electronic states near the edges of the bands become localized [7.29]. It is these localized electronic states which are of great importance in determining the optical properties of amorphous semiconductors in general and the chalcogenide glasses in particular. Early models for the electronic densities of states in the chalcogenide glasses predicted "tails" of localized one-electron states which extended well into the gap from both the valence and conduction bands [7.30, 31]. The overlap of these bands near the middle of the gap was the mechanism for the pinning of the Fermi level and the inability to dope the semiconductors. Unfortunately, these models also predicted a strong paramagnetism which was never observed [7.32].

The resolution to this dilemma, which was first suggested by *Anderson* [7.33], is that a strong electron lattice interaction leads to the pairing of all electronic states, even those which exist within the gap. In Anderson's original formulation of the model there is no gap in the paired-electron density of states, and the observed optical gap comes from the fact that it still requires a finite energy to remove one electron from the pairs. This model is very general and not especially useful in determining the microscopic properties of the localized electronic states which are most important in determining the optical properties. For this reason a specific model has evolved, which although based upon Anderson's original idea, nonetheless invokes specific defects which are very strongly chemically motivated [7.34–36]. The specific defects are basically under- and over-coordinated chalcogen atoms. The chalcogen atoms in at least the most common semiconducting glasses are twofold coordinated. Therefore the important defects are threefold and onefold coordinated chalcogens. Because of Anderson's hypothesis, these defects must contain paired electrons which means that the onefold defect is negatively charged and the threefold defect is positively charged. The defects thus form in pairs and are often denoted as C_3^+ and C_1^- for the threefold and onefold defects, respectively. These configurations make up the diamagnetic ground state of the defect system. The changes in coordination are facilitated by utilizing the lone-pair p electrons of the chalcogen atoms. These electronic states occur at the top of the valence band (C_1^-) or the bottom of the conduction band (C_3^+) in most of the chalcogenide glasses.

The chalcogenide glasses are thus supposed to contain a certain level of C_3^+ (and C_1^-) defects which are frozen-in after cooling through the glass transition temperature. The diamagnetic ground state of this defect system can be perturbed optically to produce excited states which contain paramagnetic forms of the defects. These paramagnetic states are produced by transferring charge between the two charged defects, and this new defect is often denoted as C_1^0 or C_3^0 depending on whether the new defect is singly or triply coordinated. The exact microscopic description of these paramagnetic excited states (C_1^0 or C_3^0) is still somewhat controversial [7.35–37], but the description which appears to be most consistent with the electron spin resonance (ESR) data is the C_1^0 description. The C_1^0 state is often called a chalcogen "dangling bond".

Thus the chalcogenide glasses are diamagnetic in the ground state, but pairs of charged, diamagnetic defects exist in these amorphous semiconductors which can, if the temperature is low enough, be rendered paramagnetic after the appropriate optical excitation. At temperatures below about 100 K, irradiation with band gap light (where α is greater than or equal to $100\,\text{cm}^{-1}$) produces metastable C^0 states which show an ESR signal [7.37] and an optical absorption [7.37–39].

A typical optically induced absorption spectrum is shown for glassy As_2S_3 in Fig. 7.5. Also shown in this figure are comparable data for glassy As_2Se_3. This absorption can be induced with intensities on the order of a few mW cm^{-2}. The absorption has an onset at an energy which is roughly half that of the optical gap, and the absorption is relatively flat up to the band gap. The common interpretation of this low temperature absorption is that it is due to the presence of C^0 states in the middle of the gap.

This mid-gap absorption can be optically bleached by light with energies below the band gap. Light energies between half of the optical gap energy and an energy where α is less than about 10 cm^{-1} are all effective in bleaching this optically induced mid-gap absorption. This bleaching process also eliminates the optically induced paramagnetism and returns the glass to its diamagnetic ground state configuration. As a practical matter it is difficult to bleach the

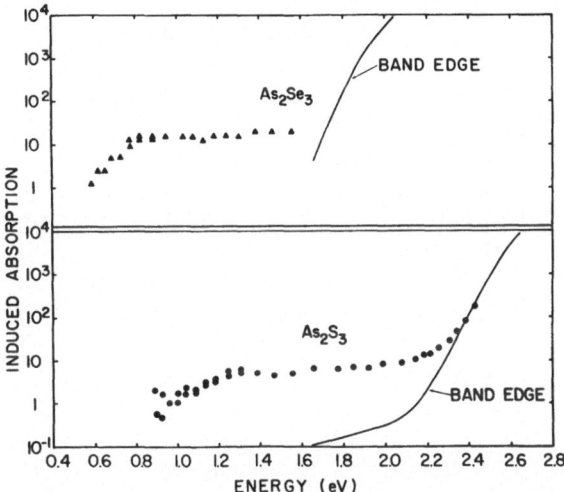

Fig. 7.5. Absorption coefficient in glassy As_2S_3 (*bottom*) and As_2Se_3 (*top*) at 6 K. Solid lines represent absorption edges in the absence of optical excitation. The data points indicate the optically induced absorption as described in the text. (After [7.37])

samples completely because the absorption goes down upon bleaching so that further bleaching takes a progressively longer time. Bleaching can also be initiated by thermal cycling. Temperatures above about 200 K result in essentially complete bleaching of the optically induced mid-gap absorption and the optically induced ESR.

The optical absorption discussed so far has been either stable or metastable. There also exist in the chalcogenide glasses optically induced absorption processes which decay in time. Experiments have been performed on both nanosecond [7.40–43] and picosecond [7.44, 45] time scales. The results of *Orenstein* and *Kastner* [7.42] for glassy As_2Se_3 are shown in Fig. 7.6. In part (a) of this figure are shown the spectral dependence of the optically induced absorption curves for various delay times after pulsed laser excitation with light whose energy is greater than the band gap. These data, which were all taken at room temperature, have been normalized to unity at the highest energy (1.5 eV).

One general trend exhibited by the data of Fig. 7.6a is that the absorption shifts to higher energies with time after pulsed excitation. A second trend, which is not so obvious since the data have been normalized, is that the magnitude of the absorption decreases with the time after pulsed laser excitation. At the shortest times recorded in Fig. 7.6a, the absorption near 1.5 eV is approximately $1 \, \text{cm}^{-1}$. At the longer delay times the absorption near 1.5 eV has decreased by factors of 1.6, 2.7, and 9.3 for delay times of 10^{-4}, 10^{-3}, and 10^{-2} s, respectively.

The spectral shape of the transient optically induced absorption shown at 300 K in Fig. 7.6a is very different from the metastable optically induced absorption seen at low temperatures (Fig. 7.5). The absorption induced at low

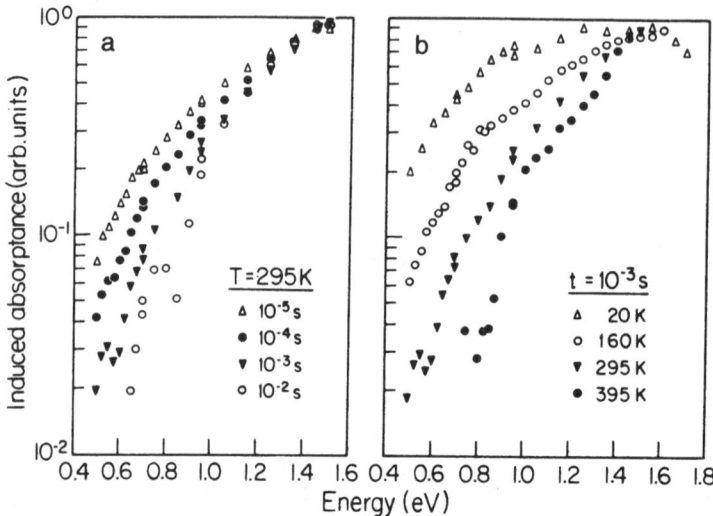

Fig. 7.6 a, b. Transient optically induced absorption in glassy As_2Se_3. (**a**) Absorption at 300 K with delay time after pulsed laser excitation as a variable parameter. (**b**) Absorption after a delay of 1 ms following laser excitation with temperature as a variable parameter. (After [7.42])

temperatures rises at energies which are near the middle of the gap, and above this onset this absorption is essentially constant in magnitude up to the band edge. Figure 7.6b shows that as the temperature is lowered there is a general transition from absorption which increases rapidly with energy to absorption which has an onset and is relatively constant in energy above this onset. In Fig. 7.6b the transient optically induced absorption at the longest delay time measured (1 ms) is shown as a function of temperature. The data taken at 20 K are very similar to those shown for glassy As_2Se_3 in Fig. 7.5.

It is apparent that the transient optically induced absorption at long delay times and low temperatures approaches the spectral shape of the metastable optically induced absorption. This trend shows that there is at least consistency between the two different sets of measurements. The mechanisms for these absorption spectra are not well characterized. *Orenstein* and *Kastner* [7.40] have suggested that the same mechanism applies at all temperatures, but there is as yet no definitive proof of this assertion.

On shorter time scales, such as the picosecond regime, for technical reasons there have been no measurements of the spectral dependences of the optically induced absorption in the chalcogenide glasses. There have been, however, pump-and-probe experiments where the excitation and measurement energies are identical. Typical data for glassy As_2Se_3 are shown in Fig. 7.7.

Measurements such as those shown in Fig. 7.7 indicate that the optically excited carriers thermalize on a time scale less than 1 ps [7.44, 45]. There are at least two decay processes which occur on the picosecond time scale. At high

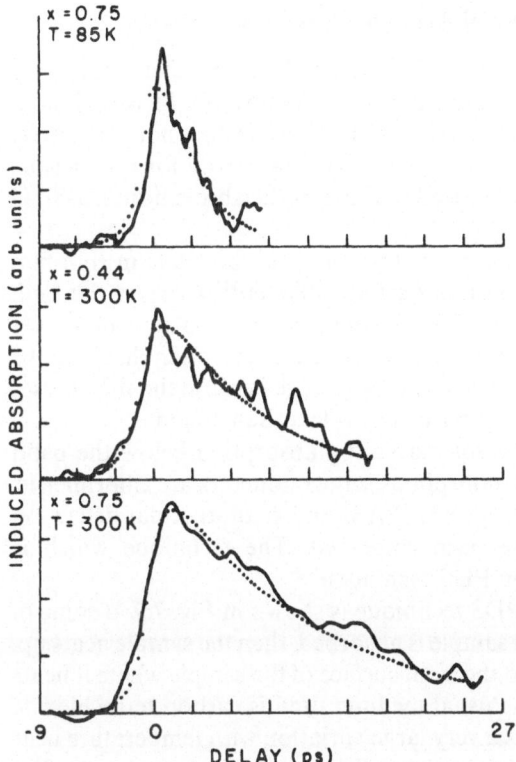

Fig. 7.7. Optically induced component of the absorption, $\Delta\alpha$, as a function of time after pulsed laser excitation at 6150 Å in glassy $As_2S_{3-x}Se_x$ at 85 K and 300 K. Full lines are experimental data and dotted lines are exponential fits to the data. (After [7.45])

excitation intensities, a long-lived optically induced absorption is observed [7.44]. This absorption, which is excited by a two-photon process, takes more than 300 ps to decay. At low excitation intensities, the optically induced absorption decays rapidly on the time scale of a few picoseconds. The rapid decay of this absorption can be explained as due to geminate recombination of the thermalized carriers [7.45].

There are additional optically induced, metastable, or in some cases irreversible, changes in the optical absorption of the chalcogenide glasses. Two of the most important of these will be discussed in Sect. 7.3. The first of these is called photodarkening, which is a metastable shift of the optical absorption edge to lower energies after laser irradiation with light of energy greater than or equal to the band gap energy. The second optically induced change is an irreversible change in the structure of chalcogenide glass films, called the photostructural effect. In this effect the laser light produces a photopolymerization of the film whereby the structure changes from a predominantly molecular structure to a polymeric one.

268 P. C. Taylor

7.2.2 The Tetrahedrally Coordinated Amorphous Semiconductors

The prototype tetrahedrally coordinated or tetrahedrally bonded amorphous semiconductor is hydrogenated amorphous silicon (a-Si:H). In this class of amorphous semiconductor the electron-lattice interaction does not play nearly as strong a role as it does in the chalcogenide glasses. Partly for this reason, there are fewer kinds of optically induced changes in the absorption in a-Si:H and related amorphous alloys.

Because these amorphous semiconductors can only be made in thin-film form, the optical properties are not as reproducible as they are in the bulk chalcogenide glasses. Some examples [7.46] from early data on amorphous Ge are shown in Fig. 7.8. In addition, different techniques often have to be employed to measure absorption below the band edge because the thicknesses of films are limited for all practical purposes to less than 10 μm.

The most common techniques for measuring absorption below the band edge in the tetrahedrally bonded amorphous semiconductors are calorimetric. Both photoacoustic spectroscopy [7.47] (PAS) and photothermal deflection spectroscopy [7.48] (PDS) have been employed. The technique which is currently used most widely is the PDS technique.

A schematic diagram of the PDS technique is shown in Fig. 7.9. If some of the light incident on the thin-film sample is absorbed, then the sample heats up. Some of the heat is transported to the front surface of the sample where it heats up the surrounding medium. The usual medium used is carbon tetrachloride because its index of refraction has a very large variation with temperature near room temperature. A laser beam is aimed parallel to the front surface of the film as close to the front surface as possible. As the CCl$_4$ heats up, a gradient in the index of refraction is established, which causes the laser beam to deflect. The

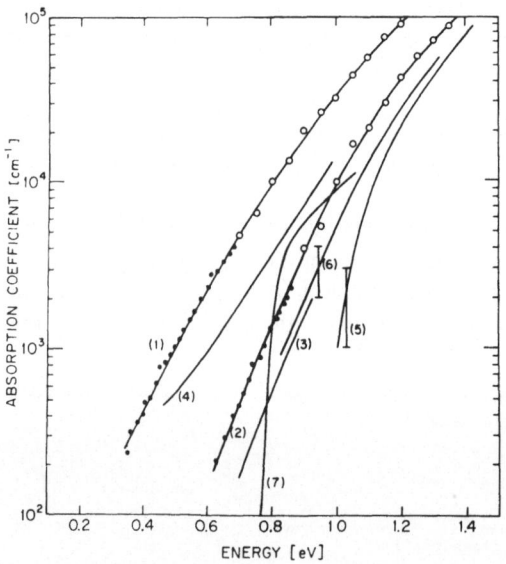

Fig. 7.8. Absorption coefficient for amorphous (curves *1* through *6*) and crystalline (curve *7*) germanium at 300 K. In the amorphous films the substrate temperature T_S is a parameter. (*1*) $T_S = 25\,°C$, sputtered film; (*2*) T_S 350 °C, sputtered film; (*3*) $T_S = 25\,°C$ followed by annealing at 150 °C for 100 h, sputtered film; (*4*) $T_S = 25\,°C$, evaporated film; (*5*) $T_S = 25\,°C$ followed by annealing at 300 °C, evaporated film; (*6*) $T_S = 300\,°C$, evaporated film; (*7*) crystalline Ge. (After [7.37])

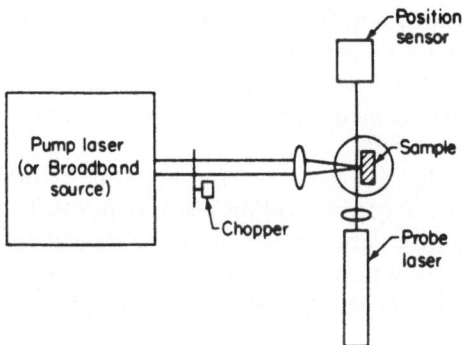

Fig. 7.9. Experimental arrangement for measuring optical absorption by photothermal deflection spectroscopy (PDS). The amplitude of the deflection on the position-sensitive detector is measured using a phase-sensitive detection scheme and the system is calibrated using a blackbody absorber. (After [7.48 b])

deflection is measured with a position sensitive detector (Fig. 7.9). The system is calibrated using a black absorber.

With care, very low absorption can be measured using the PDS technique, although measurements must be made near room temperature to achieve the best sensitivity. At 300 K, values of αd (where α is the absorption coefficient and d is the sample thickness) as low as 10^{-5} can be easily measured. The PDS technique has the advantage that it does not measure scattering as absorption. One potential disadvantage is that any absorbed light which is reradiated as photoluminescence will not be measured as absorption.

Examples [7.2, 49] of absorption spectra measured on a-Si:H using the PDS technique are presented in Fig. 7.10. It is clear from this figure that the "scatter" in the measurements for different films is not nearly as great as that in Fig. 7.8. Recently there has been considerable progress in improving the reproducibility in making films of a-Si:H from laboratory to laboratory and from film to film.

Note from Fig. 7.10 that there is some absorption well below the Urbach region in a-Si:H. This absorption looks very much like the absorption in glassy As_2S_3 (Fig. 7.4) which was attributed to the presence of iron impurities. In the

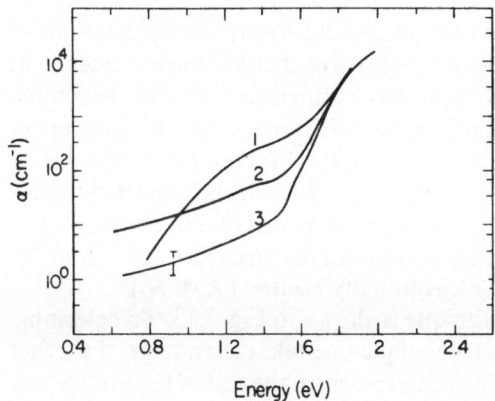

Fig. 7.10. Optical absorption at energies below the optical energy gap in a-Si:H at 300 K as measured by the PDS technique. The parameter plotted is the substrate temperature used during sample deposition: (1) 100°C, (2) 330°C, (3) 230°C. (After [7.48c])

case of a-Si:H, the absorption is probably not due to impurities but rather to unsatisfied Si bonds ("dangling bonds"). The major reason for putting hydrogen in a-Si is to reduce the dangling bond density. It appears that the hydrogen, although very effective in removing dangling Si bonds, nonetheless always leaves some silicon bonds unsatisfied.

Irradiation of films of a-Si:H with light of energy greater than that of the band gap produces an increase in the below-gap absorption, just as is the case for the chalcogenide glasses. However, there are many important differences between these two systems. The optically induced absorption in a-Si:H is metastable at 300 K whereas in the chalcogenide glasses the absorption is metastable only below about 150 K.

Optically induced metastabilities in a-Si:H are all called by a single name, the Staebler-Wronski effect [7.50], even though there are undoubtedly several different processes going on in general. Staebler and Wronski were the first to observe optically induced metastabilities in a-Si:H when they reported changes in the dark and photoconductivities after irradiation with band-gap light.

An example [7.51] of the increases in absorption which occur in a-Si:H after optical excitation is given in Fig. 7.11. This increase in absorption can be annealed away by heating the films to approximately 200°C for about 20 min. The absorption is probably due to silicon dangling bonds because of a close parallel between the magnitude of the electron spin resonance (ESR) spin density and the integrated below-gap absorption [7.51].

One can also change the below-gap absorption by rapid quenching of the films of a-Si:H from elevated temperatures. An example [7.52] of such changes is presented in Fig. 7.12. In Fig. 7.12 it is apparent that the below-gap absorption is decreased after rapid quenching from 200°C. After the film has remained at room temperature for a few weeks, the absorption has increased. Also in this case there is a direct correlation between changes in the below-gap absorption and changes in the ESR signal. The cause of these thermal metastabilities is not well known, but the mechanism may involve the diffusion of hydrogen in these films [7.53]. The inset in Fig. 7.12 shows the background absorption from the substrate. The peak near 0.9 eV is due to OH$^-$ ions in the quartz substrate.

Although most of the techniques for measuring absorption in thin films of a-Si:H and related semiconducting alloys involve thermal measurements, at least one ordinary absorption technique has been employed. This technique involves the use of a broad photoluminescence (PL) spectrum which occurs in all of these alloys. The general idea is to excite the PL with laser light whose energy is near the optical band gap. Some of the PL which is emitted travels down the plane of the film in a waveguide mode (provided that the substrate has a lower index of refraction than the film). One can then measure absorption by varying the position at which the PL is initially excited [7.54–56].

A schematic diagram of this technique is shown in Fig. 7.13. To determine the absorption coefficient using this technique one takes advantage of the fact that the intensity of the PL light which emerges from the end of the sample can

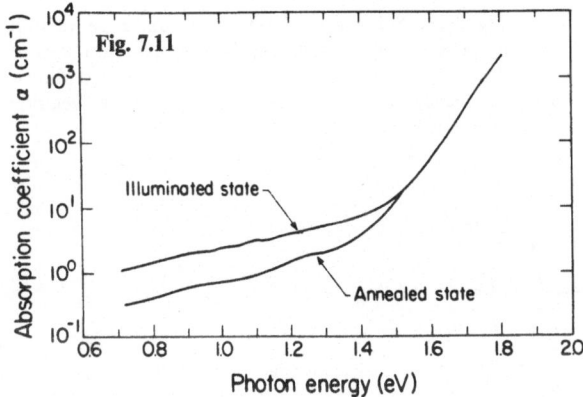

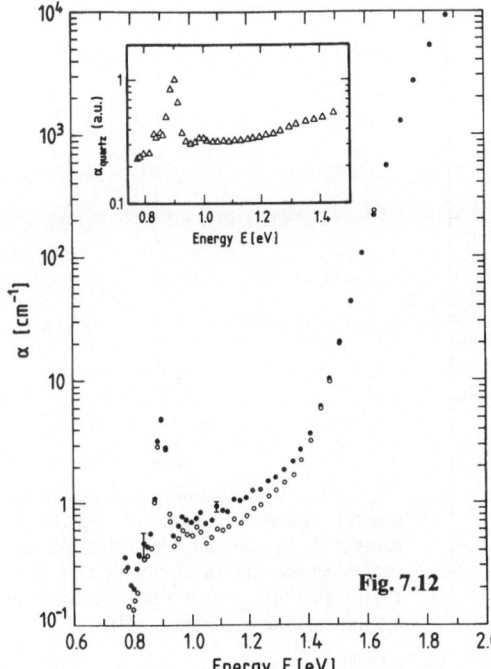

Fig. 7.11. Optical absorption at 300 K in a-Si:H as measured by the PDS technique. The increase in the absorption well below the optical energy gap is caused by laser irradiation of energy greater than that of the band gap. See text for details. (After [7.51])

Fig. 7.12. Optical absorption in a-Si:H at 300 K as measured by the PDS technique. Open circles represent data taken immediately after rapid quenching from 200 °C to room temperature. The filled circles represent data taken after 1000 h of storage in the dark at 300 K. There is a noticeable increase in the absorption after the sample has "annealed" at 300 K. The inset shows the background absorption in the vitreous quartz substrate. The peak near 0.9 eV is due to OH⁻ ions in the substrate

be approximated by

$$I = I_0 e^{-\alpha L}, \tag{7.3}$$

where I is the transmitted intensity, I_0 is the intensity in the absence of absorption, α is the absorption coefficient, and L is the length of the sample through which the light has traveled. Because path lengths between about a millimeter and about a centimeter can be easily obtained, absorption coefficients in the range $0.1 \leq \alpha \leq 10 \text{ cm}^{-1}$ can be measured.

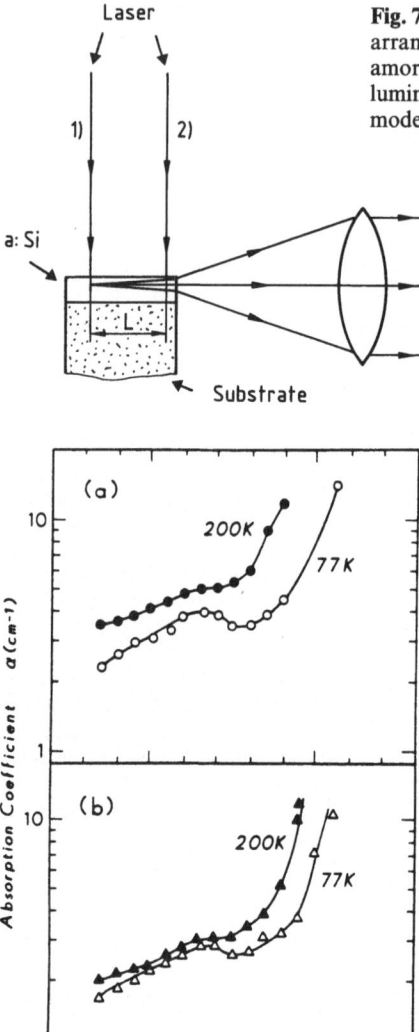

Fig. 7.13. Schematic diagram of an experimental arrangement for measuring absorption in thin films of amorphous semiconductors using laser excited photoluminescence traveling down the film in a waveguide mode. (After [7.54])

Fig. 7.14a, b. Absorption well below the optical absorption edge in a-Si:H as measured by the PLAS technique described in the text. (a) A film of a-Si:H of 1.5 μm thickness. (b) A film of a-Si:H of 0.5 μm thickness. (After [7.55])

As with all techniques which measure absorption directly, this one has the disadvantage that scattering will be measured as absorption unless special precautions in the detection procedure are taken. There are two advantages to this technique over the thermal measurements (PDS or PAS). The first advantage is that absorption which is reradiated is measured as true absorption. The second advantage is that one may measure the absorption at low temperatures where both PDS and PAS have difficulties.

A typical absorption spectrum using PL absorption spectroscopy (PLAS) is shown in Fig. 7.14. This spectrum, which was measured on a thin film of a-Si:H, is very similar to those measured by PDS (Figs 7.10 and 7.11). The only difference between the results obtained using these two different techniques is the appearance of a peak in the absorption at 1.15 eV in the low temperature PLAS measurements. Because the band gap shifts to lower energy with increasing temperature, this absorption peak becomes masked above about 250 K by the absorption in the Urbach tail region. It is possible that the thermal measurements will also yield a peak when measurements are made at low temperatures. It is also possible that this peak represents an absorption which efficiently excites PL at lower energies and hence will not be observed in thermal experiments.

Transient, optically induced absorption also exists in the tetrahedrally bonded amorphous semiconductors [7.57–60]. In the nanosecond regime the transient optically induced absorption is qualitatively similar to the spectra observed in the chalcogenide glasses (Fig. 7.6) except that the slope of the absorption at any given temperature and time delay is smaller in the case of a-Si : H and most other tetrahedrally coordinated amorphous semiconductors. There are also differences from film to film in the tetrahedrally coordinated amorphous semiconductors depending on the method of preparation (sputtering, evaporation, glow discharge, and so forth) and on the deposition parameters (substrate temperature, growth rate, and so forth).

The time dependence of the transient optically induced absorption is also very similar to that observed in the chalcogenide glasses. Most films exhibit power law decays of the optically induced absorption and the temperature dependence of the exponent in the power law decay [7.59] is very similar to that observed in the chalcogenide glasses [7.40–43].

On a picosecond time scale pump-and-probe experiments on films of tetrahedrally coordinated amorphous semiconductors also exhibit qualitative

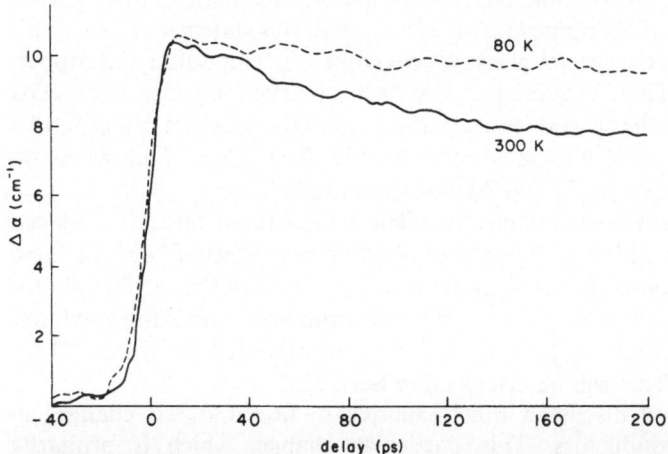

Fig. 7.15. Optically induced absorption in a-Si : H on a picosecond time scale. The dashed and solid curves represent data taken at 80 and 300 K, respectively. (After [7.58])

similarities with those measurements made on bulk samples of the chalcogenide glasses. The interpretations of the data are, however, very different due to the fact that in one case the excitation is below the band gap (Fig. 7.7 for a chalcogenide glass) and in the other case [7.57, 58] the excitation is at energies above the band gap.

At short times there is a rapid decay of the optically induced absorption on the scale of a few picoseconds in a-Si : H. This decay has been interpreted as due to the thermalization of the hot carriers which are produced by the exciting laser light [7.57]. When the thermalization is completed there exists a slower decay of the optically induced absorption, which is attributed to geminate recombination of trapped carriers, although this interpretation is still somewhat controversial [7.57].

Data for a-Si : H [7.58] in this longer time regime are presented in Fig. 7.15. Note that at 80 K the decay is slower than at 300 K. This trend could be explained as due to the temperature dependence of hopping or multiple trapping of the carriers provided that the mobility is great enough. As mentioned, there may be some difficulties with the assumption that the mobility is great enough. For the present these interpretations must be considered as tentative.

7.3 Laser-Induced Structural and Electronic Changes

One of the most universal and prominent optical effects in the chalcogenide glasses is the so-called photodarkening (PD) effect. This effect, which is a shift of the optical absorption edge to lower energies after irradiation with band-gap light, is observed to varying degrees in almost all chalcogenide glasses [7.61–66]. To date there is one known exception to this statement [7.67]. This exception occurs in the ternary glass systems copper-arsenic-sulfur and copper-arsenic-selenium. The PD effect has also been observed in some oxide glass systems [7.68], but the effect does appear to require the presence of a group VI atom in the glass [7.69]. We shall discuss the PD effect in Sect. 7.3.1, where we once again use glassy As_2S_3 and As_2Se_3 as examples.

Thin films of chalcogenide glasses often exhibit gross optically induced structural changes which are termed photostructural effects [7.62]. In these films laser excitation with band-gap light can produce an irreversible photopolymerization which changes the structure from one consisting predominantly of small molecules to one consisting of extended polymers. These photostructural effects will be discussed in Sect. 7.3.2.

In Sect. 7.3.3 we discuss a third example of laser-induced changes in amorphous semiconductors. This particular change, which is primarily electronic in origin, occurs in both glassy chalcogenide films and in films of tetrahedrally coordinated amorphous semiconductors. The change consists of a modulation of the index of refraction in the amorphous thin films. The modulation occurs as a result of a thermal modulation of the band gap. Since all solids have band gaps which vary with temperature, this effect is certainly not limited to amorphous semiconducting films, but as will be seen in Sect. 7.3.3, the amorphous semiconductors have certain advantages over other solids.

7.3.1 Photodarkening

A typical example [7.70] of the photodarkening effect in glassy As_2S_3 is shown in Fig. 7.16. Although this effect occurs in both well-annealed bulk and in thin-film samples, we shall discuss only the bulk samples where the effect is not complicated by the presence of gross photostructural changes. Photodarkening is a subtle effect which is created by photons of energy roughly equal to or greater than the band gap and which can be bleached by photons of less than band-gap energies [7.63]. The bleaching process is inefficient for two reasons. First, the absorption below the gap is weak and not many photons are absorbed. Second, as the absorption below the gap is bleached by below-gap light fewer and fewer photons are absorbed per unit time. Photodarkening can also be bleached by annealing at elevated temperatures near the glass transition temperature.

Several experiments have shown that PD is a subtle effect which does not involve gross bonding rearrangements. Infrared absorption, Raman scattering and ^{75}As nuclear quadrupole resonance (NQR) measurements [7.66] in glassy As_2Se_3 suggest that the changes important for the PD effect primarily involve

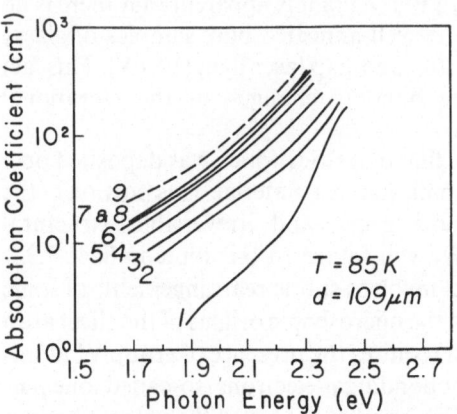

Fig. 7.16. Photodarkening in glassy As_2S_3 at 85 K. The sample thickness is 109 μm. Curve *1* represents the optical absorption before laser irradiation at 2.41 eV. Curves *2, 3, 4, 5, 6, 7* and *8* represent the optical absorption after the sample was irradiated for 60, 150, 400, 1 000, 4 600, 10 400, and 17 500 s, respectively, with 150 mW cm^{-2} intensity. Note that curves 7 and 8 are identical within the experimental accuracy. This apparent saturation is discussed in the text. Curve *9* represents one additional 13 200 s irradiation from the other side of the sample. (After [7.70])

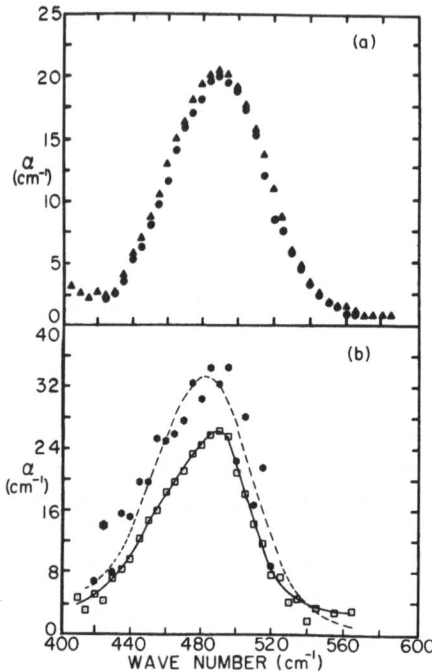

Fig. 7.17 a, b. Infrared absorption in glassy As$_2$Se$_3$ in the two-phonon spectral region. (a) Absorption coefficient in bulk, glassy As$_2$Se$_3$ (500 μm thick sample) before (▲) and after (●) laser irradiation at 77 K for 40 min at 6764 Å. (b) Absorption coefficient in an evaporated film of As$_2$Se$_3$ (150 μm thick and made with the substrate at 300 K) as deposited (□) and after laser irradiation at 300 K for 40 min at 6764 Å. The dashed line is a fit to data taken on the bulk sample of part (a) at 300 K. See text for details. (After [7.66])

the nonbonding p electrons. Figure 7.17 shows the two-phonon infrared absorption in glassy As$_2$Se$_3$ at 77 K. In Fig. 7.17a it is apparent that there is no change in the two-phonon spectrum in well-annealed bulk samples of glassy As$_2$Se$_3$ before and after irradiation with band-gap laser light (1.8 eV). This fact implies that there are no significant bonding changes, or the vibrational spectrum would certainly change.

Figure 7.17b shows data for a thin film of As$_2$Se$_3$ which was deposited on a 300 K substrate by evaporation from bulk starting material. This portion of the figure clearly demonstrates that more gross, and irreversible, structural rearrangements can occur in films. We will return to this topic in Sect. 7.3.2.

Because the photodarkening effect involves subtle rearrangements of some of the electrons and atoms in the glass, the microscopic origins of the effect are a matter of some debate. Since the effect requires the presence of group VI atoms where the valence band consists of nonbonding p-electrons (so-called lone pair electrons), the necessity of a nonbonding band at the top of the valence band is commonly accepted [7.64, 65, 71–78]. Beyond this one essential feature, however, the proposed models take very different forms. The possibilities which have been suggested include singly coordinated chalcogen atoms (dangling bonds) [7.79], under- and over-coordinated chalcogen atoms [7.72], tunneling chalcogen atoms [7.64], and changes in the overlap between nonbonding chalcogen wave functions [7.80].

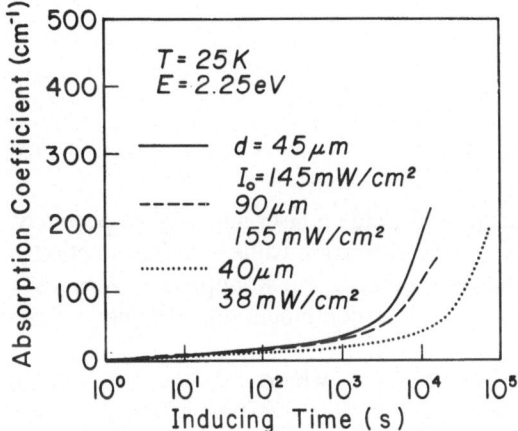

Fig. 7.18. Photodarkening in bulk, glassy As_2S_3 at 25 K. The laser irradiation was for ~ 40 min at 2.4 eV. The solid curve is for a sample 45 μm thick irradiated at a power density of 145 mW cm^{-2}. The dashed curve is for a sample 90 μm thick irradiated at a power density of 155 mW cm^{-2}. The dotted curve is for a sample 40 μm thick irradiated at a power density of 38 mW cm^{-2}. As described in the text, within experimental error the solid and dotted curves fall on top of one another when scaled in the logarithm of the time. (After [7.70])

With the PD effect even the experiments themselves are sometimes difficult to interpret. This situation occurs because the PD effect is inherently nonlinear. The quantity which is measured is the average absorption coefficient, but after photodarkening the actual absorption coefficient varies dramatically across the thin sample.

One example of the difficulty in interpreting some experimental results pertaining to the PD effect is shown in Fig. 7.16. This figure shows an apparent saturation of the PD effect after about 10000 s of irradiation with 150 mW cm^{-2} intensity of 2.41 eV laser light. The fact that this saturation is an apparent one, and not real, is demonstrated by curve 9 in Fig. 7.16, where the thin sample was irradiated from the opposite side following saturation of the effect on the first side. The apparent saturation involves the collapse of the penetration depth of the exciting laser light as the absorption coefficient near the irradiated surface increases dramatically [7.70].

Although the kinetics of the PD process are not well understood, one important feature can be explained phenomenologically. For samples of the same thickness, the dependence of the average absorption coefficient on the logarithm of the time scales with irradiation intensity. This feature is illustrated in Fig. 7.18. In this figure the solid and dotted curves were obtained on samples of essentially the same thickness, and the two curves lie on top of one another when translated along the time axis by an amount equal to the ratio of the inducing laser light intensities. This fact, and others [7.70, 81], can be explained with the following phenomenological description. Assume that the PD effect obeys

$$\frac{\partial \alpha(x,t)}{\partial t} = k(\omega)(I\alpha)^\beta f(\alpha,\alpha_s), \quad \text{where} \tag{7.4}$$

$$I(x,t) = (1-R)I_0 \exp\left[-\int_0^x \alpha(x',t)dx'\right] \tag{7.5}$$

and where ω and I_0 are the frequency and incident laser intensity, respectively, $k(\omega)$ is an efficiency function, and α_s is the saturation value of α. The function f, which ensures saturation of the PD process, must approach zero as t approaches infinity. When (7.4) and (7.5) are combined, one obtains

$$\frac{\partial \alpha(x,t)}{\partial t'} = \alpha^\beta f(\alpha,\alpha_s) \exp\left[-\beta \int_0^x \alpha(x',t)dx'\right], \quad \text{where} \tag{7.6}$$

$$t' = k(\omega)[(1-R)I_0]^\beta t. \tag{7.7}$$

Equations (7.6) and (7.7) suggest that plots of $\bar{\alpha}(t) - \alpha_0$ as a function of $\ln(t)$ for various values of incident laser intensities I_0 will be shifted with respect to each other but otherwise identical. (The quantity $\bar{\alpha}$ is the measured, or average, value of the absorption coefficient and the quantity α_0 is the value of α before laser irradiation.) This is exactly what occurs in Fig. 7.18 where the data are best fit with β equal to unity.

Several additional features of the photodarkening effect which are harder to explain are a more rapid reinducing of the effect after partial thermal annealing at a slightly higher temperature than the inducing temperature, a component of the PD which decays isothermally at the inducing temperature, an optical anisotropy when the effect is induced with polarized light, and an apparent increase in the volume of the sample after photodarkening. Some of these features can be explained at least qualitatively by the wide distributions of inducing and annealing times.

Recent results on PD in glassy As_2S_3 and As_2Se_3 alloyed with Cu [7.67] have the potential for unraveling some of the mysteries surrounding this effect. For the first time, chalcogenide glass compositions have been found for which the PD effect is absent. As shown in Fig. 7.19, the PD effect has essentially disappeared for glassy copper-arsenic-selenide with 5 at.% Cu. This fact is illustrated by the curves in the left-hand portion of Fig. 7.19. The curves in the right-hand portion of this figure indicate the effect in films of copper-arsenic-selenide with nominally 5 at.% Cu. (The actual Cu concentration in these films is near zero for reasons discussed elsewhere [7.81].) The optical absorption edge in glassy As_2Se_3 is also indicated in this figure for comparison. In the copper-arsenic-sulfide system the effect is even more dramatic, and the PD effectively disappears at 1 at.% Cu.

A structural model for these systems [7.67, 82] indicates that the copper always goes into these glasses tetrahedrally coordinated and that in order to accommodate the copper, some of the sulfur or selenium atoms also become tetrahedrally coordinated. When the chalcogen atoms become tetrahedrally

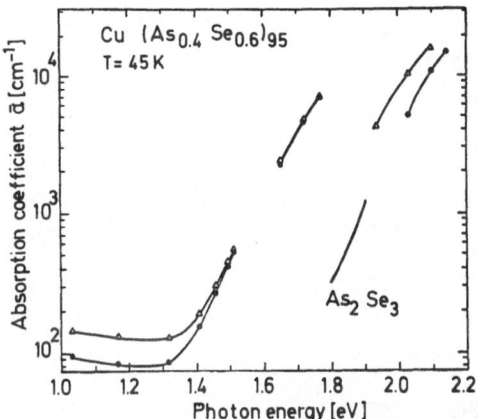

Fig. 7.19. Optical absorption in glassy $Cu_5(As_{0.4}Se_{0.6})_{95}$ before (●) and after (Δ) irradiation with band-gap light ($\sim 250\,mW\,cm^{-2}$) for 50 min at 45 K. The data at the left-hand side are for bulk samples (69 μm thick for lower values of α and 4.1 μm thick for higher values of α). The data at the right-hand side are for an evaporated film (1.85 μm thick) of nominally the same composition. Note the difference in the optical band gap between the bulk samples and the film. Note also that photodarkening only occurs in the evaporated film sample. Typical data for bulk glassy As_2Se_3 at this temperature are shown as the solid line. (After [7.81])

coordinated, the lone pair (or nonbonding) p electrons are no longer present, and therefore do not contribute to the PD process. The surprising fact is that very few tetrahedrally coordinated chalcogens are required to destroy the PD effect.

7.3.2 Photostructural Effects

When thin films of the chalcogenide glasses are made by sputtering or by evaporation, much more dramatic, and irreversible, laser-induced effects accompany the photodarkening. The initial structures of these films tend to consist of small molecules imbedded in a matrix which may have a polymeric character. Laser irradiation of these films produces a more polymeric structure by breaking up the small molecular structures. These photopolymerization processes are often referred to as photostructural effects [7.62].

An example of a photostructural effect in a film of As_2Se_3 is shown in Fig. 7.17b where the infrared absorption in the two-phonon region is shown. As mentioned in Sect. 7.3.1, part (a) of this figure shows that there is no change in the two-phonon spectrum in bulk As_2Se_3 after laser irradiation with band-gap light. The dashed line in Fig. 7.17b indicates the spectrum observed in the bulk material. It is clear that laser irradiation of the film produces a two-phonon spectrum which approaches that of the bulk.

Other experiments can also be used to monitor photostructural effects. In addition to infrared optical absorption [7.83–85], these experiments include x-ray scattering [7.61, 62, 86], nuclear quadrupole resonance [7.80, 87], and

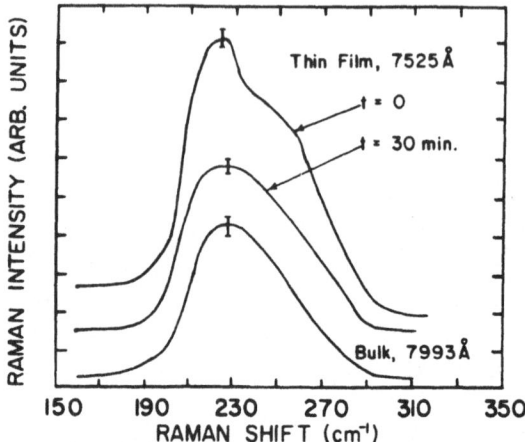

Fig. 7.20. One-phonon Raman spectrum for glassy As_2Se_3 at 300 K. The two upper traces are for a film evaporated onto a 300 K substrate before ($t=0$) and after ($t=30$ min) irradiation with laser light at 7525 Å. The bottom curve is for a bulk sample of glassy As_2Se_3. Excitation of the bulk sample with laser light (7993 Å) produced no measurable change in the Raman spectrum. (After [7.66])

Raman scattering. Although Raman scattering will be discussed in more detail in Sect. 7.5, we illustrate the application of this technique to monitor photostructural changes in Fig. 7.20. This figure shows Raman spectra for both thin-film and bulk samples of As_2Se_3. As in the case of the two-phonon infrared absorption spectra of Fig. 7.17b, the one-phonon Raman spectra of Fig. 7.20 show that the structure of the film approaches that of the well-annealed bulk sample after laser irradiation with band-gap light. The top curve in Fig. 7.20 shows the Raman spectrum for the unirradiated film and the middle curve the spectrum for the film after irradiation for 30 min with light of 7525 Å from a Kr^+ laser. The bottom trace shows the spectrum for the bulk glass. In this case also, the Raman spectrum of the bulk sample is unaffected by laser irradiation.

Photostructural effects in the chalcogenide glass films are of more than academic interest because of one accompanying feature which we have not yet discussed. This feature is a difference in etch rates which occurs before and after the photopolymerization process. Because of this difference in etch rates these films have been suggested as candidates for inorganic photoresists to be used in the manufacture of small scale electronic devices. The resolution of these "photographic films" is excellent because the changes occur on essentially an atomic scale; however, the contrast is not yet sufficient to make the process commercially viable.

7.3.3 Thermal Modulation

A final example of laser-induced changes in amorphous semiconductors is a transient thermal modulation of the index of refraction in thin bulk or film samples. This effect, which occurs in both tetrahedrally coordinated and chalcogenide amorphous semiconductors, is the result of a thermal modulation of the optical band gap by the absorbed laser light [7.88].

The experiment consists of measuring either the transmission or reflection of a thin sample while modulating with a pump laser beam whose energy is near the optical band gap. The modulated reflection or transmission is measured using phase-sensitive detection with the pump laser as the reference signal.

A typical modulated reflection spectrum below the optical gap in a-Si:H is shown in Fig. 7.21. The fringes are the result of thermal modulation of the real part of the index of refraction which modulates the optical path length in the film. Strictly speaking the effect does not require a laser because any heating

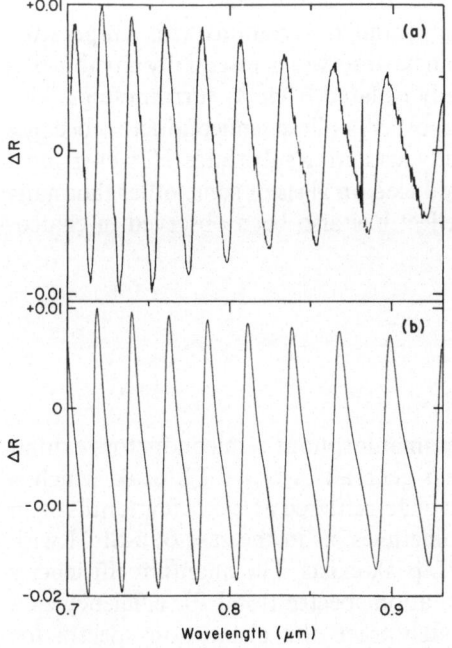

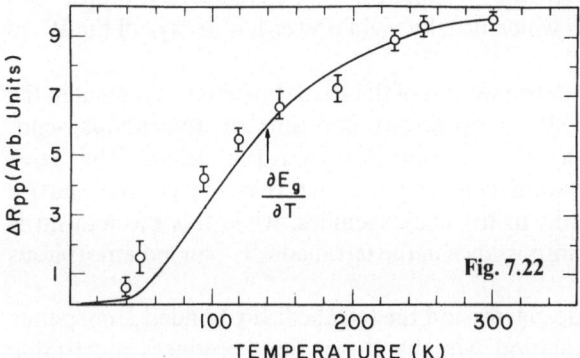

◀ **Fig. 7.21. (a)** Modulated 300 K reflectivity spectrum at 40 Hz modulation frequency in a 3 μm thick film of a-Si:H on a glassy substrate. **(b)** Calculated spectrum with $n = 3.5$ as described in the text. (After [7.88])

Fig. 7.22. Temperature dependence of the peak-to-peak amplitude of the modulated reflectivity at 40 Hz modulation frequency in a 3 μm thick sample of a-Si:H on a glass substrate (○). The solid line is the derivative of the band gap with temperature scaled to match the data at 300 K. (After [7.88])

mechanism can be employed, but the laser provides a very convenient source which may find practical application in optical data processing and storage.

The bottom portion of Fig. 7.21 shows a fit to the data assuming only a modulation of the real part of the index, n. Proof that the effect predominantly involves a modulation of n is shown in Fig. 7.22 where the magnitude of the photoreflection is plotted as a function of temperature for low incident laser power (where there is no dc component to the heating). Also shown in Fig. 7.22 is the known [7.2] derivative of the band gap with temperature scaled to match the data at 300 K. In many semiconductors the real part of the index below the band gap is proportional to a power of the optical gap [7.89]. The obvious agreement between the rate of change of the band gap with temperature and the magnitude of the modulated effect establishes convincingly that the modulation is indeed due to a modulation of the real part of the index.

Because most semiconductors exhibit similar variations with temperature for their optical band gaps, this effect can be observed in essentially any thin film provided that the film can be thermally isolated from its surroundings. This latter restriction effectively eliminates most crystalline semiconductors because they are difficult to produce on thermally insulating substrates. However, most amorphous semiconductors can be produced on glass or some other thermally insulating substrate, and in fact the effect has also been observed in chalcogenide glass films [7.88].

7.4 Photoluminescence

The photoluminescence (PL) processes in amorphous semiconductors exhibit several fairly universal features. There is generally a broad PL peak, which is typically a few tenths of an electron volt wide, centered at least a few tenths of an electron volt below the band edge. Sometimes, as in the case of a-Si:H with many defects, more than one broad PL peak exists. The quantum efficiencies can often approach unity and the excitation spectra (total PL efficiency as a function of excitation wavelength) often track the absorption spectra for energies below the band-gap energy. Time-resolved PL spectra exhibit broad distributions of decay rates which usually yield power law decays of the PL in time.

Although the detailed interpretations of the PL data are very different in the chalcogenide glasses and the tetrahedrally coordinated amorphous semiconductors, in both cases the PL is attributed to "intrinsic" defects. The major difference in the interpretation concerns the fact that the electron–lattice interaction is very important in the chalcogenides, while this mechanism is thought to be of much less importance in the tetrahedrally bonded amorphous semiconductors.

In both the chalcogenide glasses and the tetrahedrally bonded amorphous semiconductors, laser irradiation with band-gap light produces metastable changes in the PL efficiencies and sometimes also in the PL lineshapes. These metastabilities are in both systems attributed to rearrangements of both atoms and electrons.

7.4.1 The Chalcogenide Glasses

Photoluminescence in the chalcogenide glasses was first observed by *Kolomiets* et al. [7.90]. A typical spectrum [7.38] for glassy As_2Se_3 is shown in Fig. 7.23. Also shown in Fig. 7.23 are the excitation spectrum (on a linear scale) and the absorption edge (on a log scale). Note the large energy difference between the peak of the excitation spectrum and the PL peak. This large Stokes shift is due to the strong electron lattice interaction.

This PL spectrum is usually interpreted as due to the recombination of an electron with a trapped hole whose energy is lowered to near mid-gap by a lattice distortion [7.34–36]. One model associates the PL with the existence of charged radiative recombination centers which are derived from under- and over-coordinated chalcogen atoms [7.35, 91]. Another model associates the PL with uncharged, dipolar nearest-neighbor pairs of under- and over-coordinated chalcogen atoms [7.36]. Yet a third model suggests that small polarons are responsible for the PL [7.92].

Just as occurs in the optical absorption, there are metastable changes in the PL after optical excitation with band-gap light. In particular, the excitation of PL produces a metastable decrease in the PL efficiency [7.93–95]. This decrease, which is called fatiguing, can be restored either by annealing at higher temperatures or by irradiation with light of energy below the band gap. These restoration processes parallel those discussed in Sect. 7.2.1 for the mid-gap absorption. Although there are parallels between the optically induced absorption and the PL, these two effects are probably not due to precisely the same defect centers [7.37, 96].

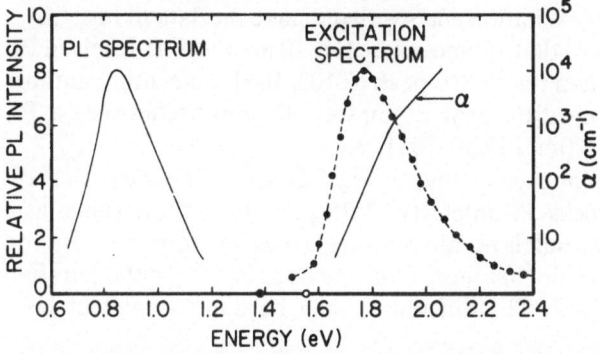

Fig. 7.23. Photoluminescence (PL) and PL excitation spectra for bulk, glassy As_2Se_3 at ~ 1.6 K. The excitation spectrum has been normalized to the number of incident photons. The solid line on the right is the absorption coefficient on a log scale (*right-hand side*). (After [7.38])

284   P. C. Taylor

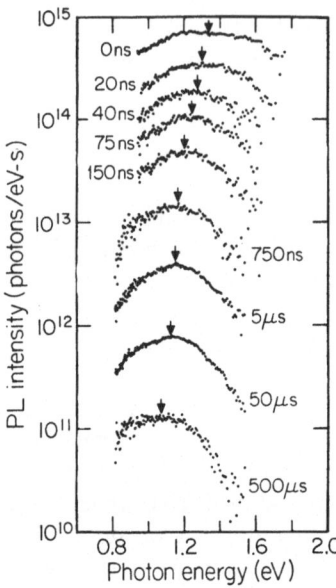

Fig. 7.24. Photoluminescence (PL) spectra in bulk glassy As$_2$S$_3$ at ~2 K. The parameter plotted is the delay time after pulsed laser excitation at 2.5 eV (~0.2 MW cm^{-2}) for ~10 ns. Arrows indicate an estimate of the mean PL energy at each delay time. (After [7.102])

The time decay of the PL has also been investigated in some detail in several chalcogenide glasses [7.97–105]. For technical reasons most of these measurements have been performed on glassy As$_2$S$_3$, which is a relatively wide-band-gap glass. At short times (10 ns) some authors [7.98, 99] have observed two PL peaks while others [7.100, 101] observe only one broad peak. All groups observe the general trend in which the peak (or peaks) shift down in energy with increasing delay time after laser excitation.

Figure 7.24 shows time-resolved PL spectra for glassy As$_2$S$_3$ at 2 K [7.102]. These spectra show only one broad peak, which shifts to lower energy with increasing delay time. The arrows are a rough indication of the peak energy at each delay time. The laser excitation energy used to take the data in Fig. 7.24 is 2.5 eV. Some authors report that at times exceeding 10 ms, the time-resolved PL peak is at lower energy than the cw PL peak [7.103, 104], while other authors report that the time-resolved PL peak asymptotically approaches the cw PL peak at the longest delay times [7.99, 105].

If one takes from a plot such as that in Fig. 7.24 either the integrated PL intensity [7.102] or the peak PL intensity [7.106] and plots these values as a function of time, then the result is usually a power law decay. Some authors find that the decay on a log-log plot has more than one slope [7.106], but others find only a single slope [7.102]. The differences may be due to different laser excitation energies.

The polarization of the emitted PL after excitation with polarized laser light has also been measured in glassy As$_2$S$_3$ [7.107, 108]. At short times (less than about 10^{-4} s) the emitted PL remains polarized, but at longer times this polarization rapidly disappears.

7.4.2 The Tetrahedrally Coordinated Amorphous Semiconductors

In undoped a-Si:H the PL spectrum generally has contributions from at least two different bands, one peaked somewhere between 1.2 and 1.4 eV and one peaked somewhere between 0.7 and 0.9 eV [7.109]. The higher energy peak dominates the PL at all temperatures in the best quality films. Films of the "best quality" are those with the lowest ESR spin densities, a fact which indicates the lowest number of paramagnetic defects located deep within the energy gap. In films of a-Si:H with much greater ESR spin densities (and also greater below-gap absorption), the lower energy PL band dominates at all temperatures. For spin densities in between these two extremes, the higher energy PL usually dominates at low temperatures ($T < 100$ K), while at higher temperatures the lower energy PL peak usually dominates.

The PL spectra in a-Si:H and in a series of a-Si$_{1-x}$Ge$_x$:H alloys are shown in Fig. 7.25 [7.110]. Note that these spectra are presented with the PL intensity plotted on a logarithmic scale. The low energy portions of these spectra are all exponential with essentially the same slope. This is a characteristic feature of the PL lineshapes in the tetrahedrally bonded amorphous semiconductors.

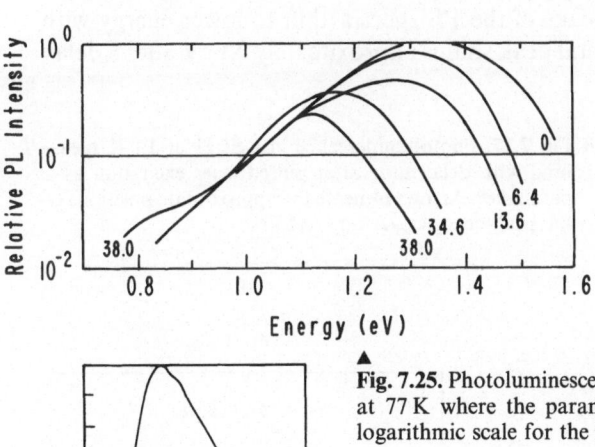

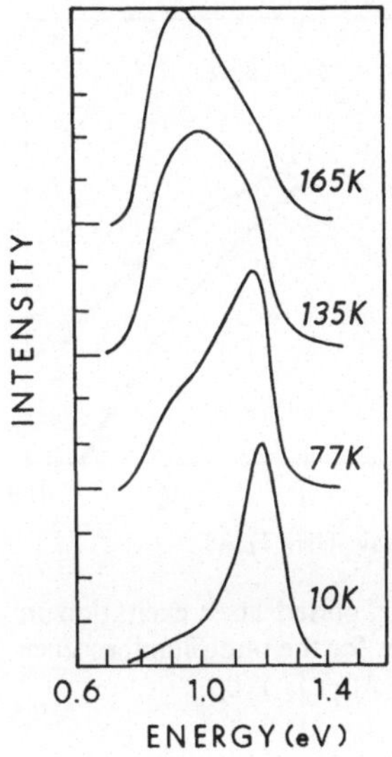

Fig. 7.25. Photoluminescence (PL) in a-Si$_{1-x}$Ge$_x$:H alloys at 77 K where the parameter plotted is $100x$. Note the logarithmic scale for the PL intensity. (After [7.110])

Fig. 7.26. Photoluminescence (PL) spectra in a-Si$_{62}$Ge$_{38}$:H. The parameter plotted is the temperature. At low temperatures the PL is dominated by a peak near 1.2 eV, while at high temperatures the PL is dominated by a peak near 0.85 eV. (After [7.111])

At higher temperatures, or in poorer quality films, the lower energy PL peak dominates the PL spectra. A typical example [7.111] is shown in Fig. 7.26 where spectra are shown for a silicon-germanium alloy (a-Si$_{1-x}$Ge$_x$:H with $x = 0.38$) as a function of temperature. At 10 K the PL spectrum is dominated by the higher energy PL, which for this particular alloy is peaked around 1.2 eV. As the temperature increases the PL intensity shifts to lower energy until at about 165 K only the lower energy PL, which peaks near 0.85 eV, is important.

The standard interpretation of the higher energy PL in a-Si:H is in terms of radiative transitions between localized electronic states at the bottom of the conduction band (conduction band-tail states) and localized states at the top of the valence band (valence band-tail states). The lower energy PL is usually interpreted in terms of radiative transitions between electrons in the conduction band and the silicon dangling bond states which lie near the middle of the gap [7.112].

The decay rates for the PL processes in a-Si:H and related alloys are very similar to those observed in the chalcogenide glasses. Typical PL spectra [7.113] as a function of delay time after pulsed laser excitation are shown in Fig. 7.27. A comparison with Fig. 7.24 indicates the parallels with the chalcogenide glasses. The peaks of the PL spectra shift to lower energy with increasing delay time. The total peak shift is approximately 0.1 eV after a delay of about 1 ms.

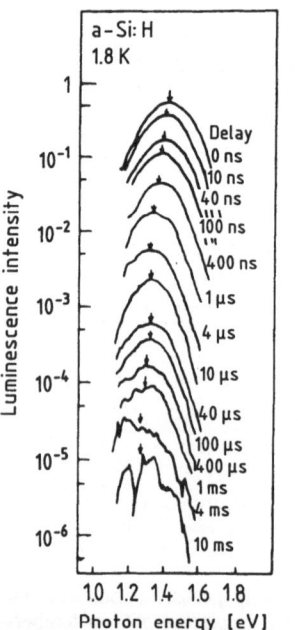

◀ **Fig. 7.27.** Photoluminescence in a-Si:H at 1.8 K measured with delay time after pulsed laser excitation as a parameter. Arrows indicate the approximate positions of the peak energies. (After [7.113])

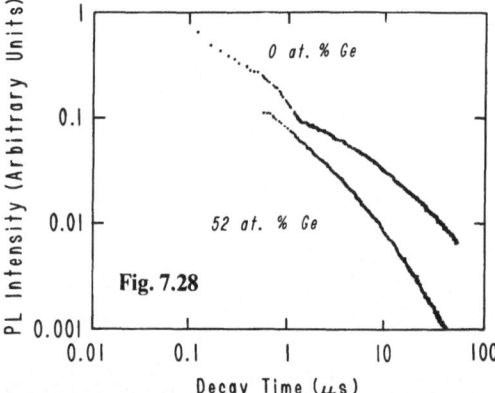

Fig. 7.28. Photoluminescence as a function of delay time after pulsed laser excitation in a-Si$_{1-x}$Ge$_x$:H alloys. Note the approximate power law behavior for the photoluminescence decay and the more rapid decay in the germanium-rich alloy. (After [7.110])

When the total PL intensity or the peak PL intensity is plotted as a function of time on a log-log scale, the results typically show power law behavior [7.110, 114–116]. Figure 7.28 illustrates this power law behavior [7.110] in a-Si:H and in the alloy system a-Si$_{1-x}$Ge$_x$:H. Although the decay is faster in the alloy containing germanium, the general features are similar in the two decay curves displayed in Fig. 7.28. Similar behavior has been observed for other tetrahedrally coordinated alloy systems such as a-Si$_{1-x}$C$_x$:H. The interpretation of these power law decays is the same as that discussed for the chalcogenide glasses in Sect. 7.4.1. A broad distribution of decay rates is the most commonly accepted explanation.

There are also optically induced metastabilities in the PL spectra of the tetrahedrally coordinated amorphous semiconductors [7.117], but the magnitudes of these effects are not nearly as great as in the chalcogenide glasses. After prolonged irradiation with laser light whose energy is greater than the optical band gap, there is a decrease of a few percent in the amplitude of the higher energy PL and a concomitant increase in the amplitude of the lower energy PL. These changes are very similar to the increase in the below-gap absorption and the ESR discussed in Sect. 7.2.2 and to the changes in photoconductivity which are now universally termed the Staebler-Wronski effect [7.50].

7.5 Raman Scattering

Unlike the case of PL, the Raman scattering results in the chalcogenide glasses and the tetrahedrally coordinated amorphous semiconductors are not even qualitatively similar [7.118]. In the chalcogenide glasses the spectra are often dominated by the local "molecular" structure in the glasses, while in the tetrahedrally coordinated amorphous semiconductors the spectra are to first order an approximation to the phonon densities of states. Thus the chalcogenide glasses tend to exhibit relatively sharp peaks in their Raman spectra (although not nearly as sharp as in crystalline solids), and the tetrahedrally coordinated amorphous semiconductors tend to exhibit very broad Raman features.

One of the major problems in interpreting Raman or infrared spectra in amorphous solids is the difficulty in evaluating the matrix elements for the various transitions. In Raman scattering the appropriate matrix elements are contained in a second rank tensor which expresses the displacement-dependent polarizability. Often these matrix elements are assumed to be constant within a given vibrational band [7.119]. This approximation, which is often not justified, was originally suggested by *Shuker* and *Gammon* [7.119], and the procedure is still called the Shuker-Gammon reduction scheme. Although this procedure may be a close approximation to the truth for the case of narrow vibrational bands which are well separated in energy, as sometimes occurs in the chalcogenide glasses, the procedure is particularly questionable when applied to all of the vibrational "bands" together in either the tetrahedrally coordinated amorphous semiconductors or the chalcogenide glasses.

Within the Shuker-Gammon approximation one can define a reduced Raman spectrum for a given vibrational "band" as

$$I_\varrho^R(\omega) = \omega(\omega_L - \omega)^{-4}[1 + n(\omega, T)]^{-1} I_\varrho(\omega), \tag{7.8}$$

where $I_\varrho(\omega)$ is the observed Raman scattering intensity, $n(\omega, T)$ is the Bose-Einstein distribution function, and ω_L is the angular frequency of the laser excitation. This expression is written for the case where the scattered light is observed at a lower frequency due to the creation of a phonon (Stokes spectrum). For the anti-Stokes spectrum the factors $[1 + n(\omega, T)]$ and $(\omega_L - \omega)$ are replaced by $n(\omega, T)$ and $(\omega_L + \omega)$, respectively.

Because there are in general no preferred orientations for the local environments in amorphous solids, there are also difficulties in defining the polarization properties of the Raman spectra. The random orientations of the individual scatterers mean that the symmetry properties of any given vibrational mode can be characterized experimentally only by a depolarization ratio, which is defined as the ratio of the intensity of the scattered laser light polarized in the scattering plane to that polarized perpendicular to the scattering plane [7.118]. Because amorphous solids tend to have Raman spectra which are essentially continuous in energy over some limited range, the depolarization ratio is a continuous function of ω.

7.5.1 The Chalcogenide Glasses

Because the vibrational spectra of many chalcogenide glasses are dominated by the local "molecular" structure, one can often obtain reasonable facsimiles of these spectra by comparisons with relevant crystalline phases [7.120]. This procedure, for which there is little theoretical justification, is only useful in those glass compositions for which there exists a stoichiometric crystalline compound. Glassy As_2S_3 is one such composition where the vibrational properties of the crystalline layered compound have been studied in some detail [7.121].

Figure 7.29 shows the polarized Raman spectra taken at 300 K in glassy As_2S_3. On the left side of the figure is the spectrum where the scattered and incident laser light are of the same polarization (H–H), and on the right side of the figure is the spectrum where the polarizations of the incident and scattered laser intensities are perpendicular (H–V). The arrows represent the positions and relative strengths of the Raman scattering peaks observed when the same geometries are imposed in the crystalline case. These arrows represent a suitable average over all orientations of the crystalline axes with respect to the incident laser beam [7.29].

The qualitative conclusions to be drawn from the plot in Fig. 7.29, and another like it for the infrared absorption spectrum, are that the vibrational spectra for crystalline and glassy As_2S_3 are very similar, that the matrix elements are not slowly varying functions of the frequency as would be

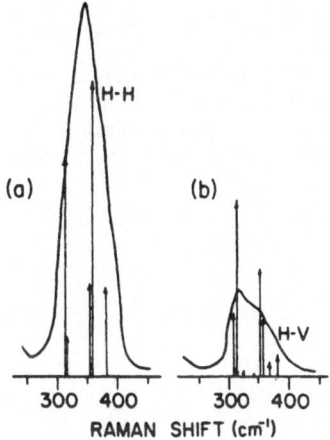

Fig. 7.29. Polarized (H–H) one-phonon Raman spectrum in glassy As_2S_3 (*left-hand side*). Depolarized (H–V) one-phonon Raman spectrum in glassy As_2S_3 (*right-hand side*). The arrows represent a histogram of the average of the polarized (or depolarized) Raman modes in crystalline As_2S_3. (After [7.120])

(a)   (b)

H-H

H-V

300   400   300   400
RAMAN SHIFT (cm⁻¹)

suggested by the Shuker-Gammon approximation, and that the dominant features in the Raman and infrared spectra are controlled by the local molecular structure. In the case of As_2S_3 the local order consists of AsS_3 pyramidal units linked together in a topology which is predominantly two dimensional.

7.5.2 The Tetrahedrally Coordinated Amorphous Semiconductors

The infrared absorption and Raman spectra for the typical tetrahedrally coordinated amorphous semiconductor are very different from the spectra described in the previous section for the chalcogenide glasses. As an example we present the Raman spectrum of a-Si [7.122] in Fig. 7.30. The Raman spectrum of a-Si:H, the hydrogenated alloy of amorphous silicon which is commercially useful, is very similar to the spectrum shown in Fig. 7.30 for a-Si [7.123].

Both the Raman and the infrared spectra in the tetrahedrally coordinated amorphous semiconductors reflect all of the main features of the phonon densities of states [7.122, 124, 125]. Therefore, the Shuker-Gammon approximation [7.119] although not strictly valid is reasonably accurate in this class of amorphous semiconductor. In addition, the Raman and infrared spectra, and hence the phonon densities of states, appear to be remarkably similar to those of the crystalline forms. This result is apparent by comparing the top and bottom portions of Fig. 7.30 which contain the infrared and Raman spectra for a-Si and the crystalline Si density of states, respectively.

One might at first question how the densities of states could be so similar in the crystalline and amorphous forms because the crystals exhibit the diamond cubic structure and the amorphous forms exhibit a continuous random network [7.126] of Si atoms. There are several reasons for this similarity. First, model calculations [7.126] indicate that the gross features of the phonon density of states are independent of the details of the structure. Second, the density of states seems to be dominated by the local tetrahedral order (four nearest neighbors) which is largely preserved in the amorphous form [7.127].

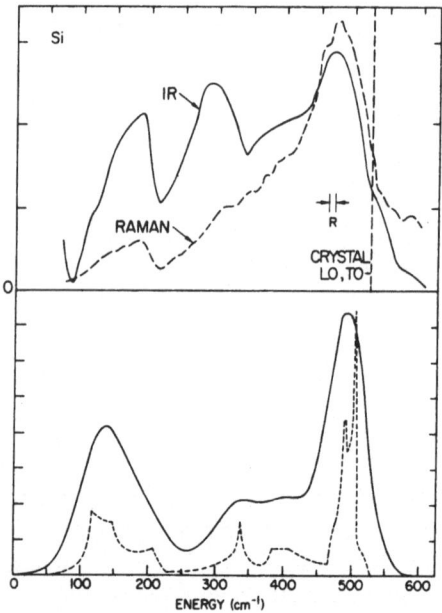

Fig. 7.30. *Top:* Infrared absorption (—) and reduced Raman spectrum (– – –) for a-Si at 300 K. *Bottom:* Density of states in crystalline Si as derived from inelastic neutron scattering experiments (– – –). Broadened density of states (—). (After [7.118, 124, 125])

Because the phonon density of states appears to be so insensitive to the details of the structural order, one cannot use the Raman and infrared spectra to make any detailed structural inferences. Specifically, one cannot infer details of the random network model, such as the number of rings of bonds of various sizes, merely from comparisons with the Raman and infrared spectra.

7.6 Far Infrared Laser Spectroscopy

We conclude this chapter with a discussion of the application of laser spectroscopy in the far infrared spectral region (approximately $1-100 \, \text{cm}^{-1}$) and its application to amorphous semiconductors. Because the chalcogenide glasses and the tetrahedrally coordinated amorphous semiconductors both display similar properties is this spectral range, we shall treat them together in this section.

There are two important questions to be answered from spectroscopic measurements in the $1-100 \, \text{cm}^{-1}$ range. First, what is the asymptotic form of the infrared and Raman spectra for amorphous semiconductors at very low frequencies? Second, what is the role in the infrared and Raman spectra at low frequencies of the very anharmonic tunneling modes [7.128, 129] which have been observed in specific heat [7.130, 131] and ultrasonic attenuation [7.132–134] measurements?

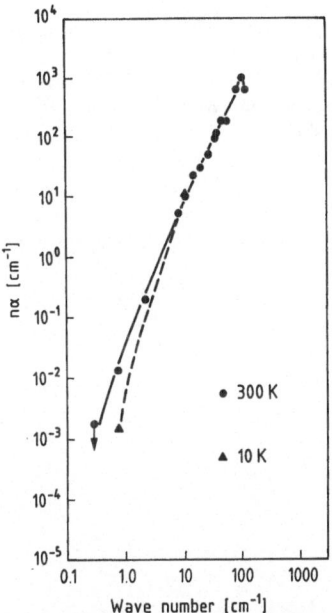

Fig. 7.31. Index of refraction n times absorption coefficient α at 300 K (—) and 10 K (– – –) in glassy As_2Se_3. (After [7.137])

The collection of Raman data at low frequencies requires special care to avoid stray background light and the Rayleigh tail of the exciting laser light. Triple monochromators are often necessary in this spectral range. The infrared absorption data are typically recorded using a molecular gas laser pumped by a CO_2 laser [7.135]. By changing organic gasses in the molecular gas laser, a nearly continuous (every few wave numbers in cm^{-1}) spectrum can be obtained from a few cm^{-1} to greater than 100 cm^{-1}.

Figure 7.31 shows far infrared absorption data for glassy As_2Se_3 at 300 K and 10 K. These data have been compiled from several sources [7.136–139]. Below about 10 cm^{-1}, the infrared absorption data [7.136, 137] become temperature dependent, but above this reduced frequency the data are essentially independent of temperature.

The Raman (data not shown) and infrared spectra exhibit similar trends in glassy As_2Se_3. These data are fairly characteristic of amorphous semiconductors, and indeed of amorphous solids, in general. The first general feature which emerges is the occurrence of a spectral range (usually above about 10 cm^{-1}) where the infrared absorption and the reduced Raman spectrum are temperature independent and scale roughly as the frequency squared. A second general feature is the occurrence of a temperature-dependent region (usually below about 10 cm^{-1}) where the reduced Raman scattering and the infrared absorption vary more rapidly than the square of the frequency.

In a number of the amorphous semiconductors there appears to be a correlation between the magnitude of the far infrared absorption, in the region where the absorption is temperature independent, and the magnitude of the

292   P. C. Taylor

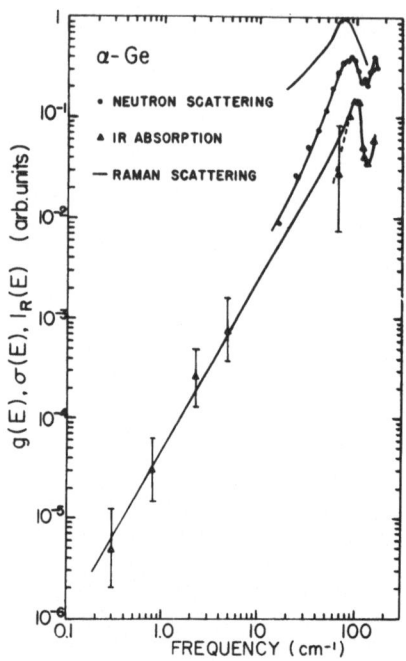

Fig. 7.32. Far infrared optical absorption (△), reduced Raman spectrum (—) and density of states as measured by inelastic neutron scattering (●) in a-Ge at 300 K. (After [7.140])

density of states as determined by specific heat measurements [7.136, 137]. Because the frequency dependence is the same (ω^2) in both the density of states and the far infrared absorption, the matrix elements which couple the photons to phonons must be only slowly varying with frequency in this range. There is also evidence [7.136] that these matrix elements are remarkably constant from amorphous semiconductor to amorphous semiconductor.

As mentioned above, similar far infrared absorption is observed in the tetrahedrally coordinated amorphous semiconductors. Figure 7.32 shows far infrared absorption [7.140] and reduced Raman [7.141] spectra for a-Ge at 300 K. Also shown in this figure for comparison is the density of states as determined from inelastic neutron scattering experiments [7.142]. Note that both the reduced Raman and the far infrared absorption intensities scale roughly as ω^2 in this region, as does the phonon density of states. A comparison of Figs. 7.31 and 7.32 shows that the low frequency features are essentially the same in the two main types of amorphous semiconductor. The only apparent difference between these two classes is the fact that the magnitude of the far infrared absorption in the tetrahedrally bonded amorphous semiconductors appears to vary with the sample preparation conditions [7.140] while no such variation has been observed in the chalcogenide glasses.

7.7 Summary

Laser spectroscopy is an important tool for studying the optical, electronic, and vibrational properties of amorphous semiconductors. Throughout this chapter we have distinguished between the tetrahedrally coordinated amorphous semiconductors, which are primarily based upon group IV atoms, and the vitreous chalcogenide semiconductors (chalcogenide glasses), which are based upon group VI atoms. Although this distinction is primarily experimental, the division into these two groups is not without some justification. For example, the electronic properties of the chalcogenide glasses are dominated by a strong electron–lattice interaction, while this interaction is not important to first order in the tetrahedrally coordinated amorphous semiconductors.

In this chapter we have concentrated on glassy As_2S_3, As_2Se_3, and Se as the prototype chalcogenide glasses and on amorphous silicon (Si), hydrogenated amorphous Si (a-Si:H) and hydrogenated amorphous silicon–germanium alloys (a-$Si_{1-x}Ge_x$:H) as the prototype tetrahedrally coordinated amorphous semiconductors. The examples which have been presented are those which illustrate features common to most, if not all, of the amorphous semiconductors in these two respective groups.

The optical properties of the amorphous semiconductors are dominated by three important and universal features of the absorption in the vicinity of the optical energy gap. There is a region with high absorption coefficient $(\alpha \sim 10^4\,\mathrm{cm}^{-1})$ which can be used to estimate the optical energy gap. At lower energies is a region where α depends exponentially on energy (the Urbach tail region), which is related to localized electronic states near the edges of the valence and conduction bands. At still lower energies is a region where the absorption is slowly varying with energy and is due to either intrinsic defects or impurities. The absorption in this region extends down to energies which are approximately half of the optical band gap energy.

In the chalcogenide glasses there are several other important absorption processes in addition to the three just mentioned. There is a metastable, optically induced absorption which occurs primarily at low temperatures and relates to movement of electrons in intrinsic defects such as over- and under-coordinated chalcogen atoms. There are also transient contributions to the optically induced absorption in the chalcogenide glasses whose origins are not as well understood. In order of roughly increasing time scales, these transient effects involve thermalization of hot carriers, geminate recombination of carriers, hopping and multiple trapping of carriers in band tails, and creation and annihilation of defects.

In the tetrahedrally coordinated amorphous semiconductors, many of the absorption processes parallel those observed in the chalcogenide glasses. There exists a metastable component to the optically induced absorption below the gap in these amorphous semiconductors. In contrast to the chalcogenide glasses, this component, which is one manifestation of the Staebler-Wronski effect, exists even at room temperature. This absorption is annealed at temperatures of about 200 °C in many samples. The transient contributions to the optically induced absorption are qualitatively similar to those which occur in the chalcogenide glasses, and the proposed mechanisms are the same as those listed for the chalcogenide glasses.

In the chalcogenide glasses, particularly in films of these amorphous semiconductors, more permanent optically induced changes can occur. One of the most universal and subtle of these changes is the so-called photodarkening effect, or the metastable shift of the band edge to lower energies after laser irradiation with band gap light. In films of chalcogenide glasses, gross photostructural changes, which are irreversible, can also occur. These changes involve primarily a photoinduced polymerization of the glass structure from small isolated molecules to large extended polymers.

Both the chalcogenide glasses and the tetrahedrally coordinated amorphous semiconductors exhibit a modulation of the real part of the index of refraction when they are irradiated with band-gap light. This modulation results from a thermal modulation of the optical band gap. This effect may have practical importance for optical switching devices.

Photoluminescence is another important application of lasers in the spectroscopy of amorphous semiconductors. In essentially all amorphous semiconductors there are several important features which dominate the PL spectra. There is generally a broad (a few tenths of an electron volt) PL peak which is several tenths of an electron volt below the optical band gap. Sometimes more than one PL peak exists. Quantum efficiencies for these PL processes often approach unity at low temperatures, and the excitation spectra tend to follow the absorption coefficient at the band edge. The time-resolved spectra exhibit broad distributions of decay rates which yield power-law decays of the PL with time after laser excitation.

The detailed interpretations of the PL processes in the chalcogenide glasses and the tetrahedrally coordinated amorphous semiconductors are very different because of major differences in the relative importance of the electron–lattice interaction in these two classes of amorphous semiconductors. In both of these materials, however, the dominant PL processes are thought to be due to intrinsic defects and not to impurities.

Laser irradiation with band-gap light produces metastable changes in the PL spectra in both the chalcogenide glasses and the tetrahedrally coordinated amorphous semiconductors. The PL efficiencies either increase or decrease with time. In both amorphous semiconducting systems, these changes are attributed to rearrangements of the electrons and atoms in various defect structures.

Raman scattering in amorphous semiconductors is very different for the chalcogenide glasses and the tetrahedrally coordinated films. In the chalcogenide glasses the Raman spectra are usually dominated by the local molecular or polymeric structure of the glasses so that certain vibrational modes couple much more strongly than others to changes in the polarizability with displacement. In the tetrahedrally coordinated amorphous semiconductors, the spectra are a close approximation to the phonon densities of states and thus exhibit in general much broader Raman spectra than the chalcogenide glasses.

In the far infrared region of the electromagnetic spectrum (approximately $1-100 \, \text{cm}^{-1}$), lasers can be employed to study low energy vibrational excitations of amorphous semiconductors. Both classes of amorphous semiconductors display similar features in this spectral range. Both the infrared absorption and the reduced Raman spectra show a frequency dependence which is proportional to ω^2. This frequency dependence implies that the matrix elements are essentially independent of energy in this spectral region, and that the spectra are proportional to the phonon densities of states.

From the far infrared to the ultraviolet, laser spectroscopy provides a very important tool to be used in understanding the electronic, optical and vibrational properties of amorphous semiconductors. These important tools provide information on both steady-state processes and on transient processes which occur on time scales as short as $10-100 \, \text{fs}$.

Acknowledgements. Much of the research discussed in this chapter was supported by the National Science Foundation under grant number DMR-86-15217, the Department of Energy through the Solar Energy Research Institute under subcontract number XM-5-05009-2, and the Office of Naval Research under contracts number N00014-83-K-0535 and number N00014-86-K-0710.

References

7.1   J. Tauc: "Optical Properties of Amorphous Semiconductors", in *Amorphous and Liquid Semiconductors*, ed. by J. Tauc (Plenum, New York 1974) p. 159
7.2   G.D. Cody: "The Optical Absorption Edge of Amorphous Silicon Hydride", in *Semiconductors and Semimetals*, Vol. 21B, ed. by J.I. Pankove (Academic, New York 1984) p. 11
7.3   R. Urbach: Phys. Rev. **92**, 1324 (1953)
7.4   J.D. Dow, D. Redfield: Phys. Rev. B **1**, 3358 (1970)
7.5   J.D. Dow, D. Redfield: Phys. Rev. B **5**, 594 (1972)
7.6   D.L. Wood, J. Tauc: Phys. Rev. B **5**, 3144 (1972)
7.7   P.C. Taylor: "Optical Properties of Non-Crystalline Semiconductors", in *Amorphous Semiconductors*, ed. by M. Pollak (CRC, New York 1987) p. 19
7.8   R.E. Drews, R.L. Emerald, M.L. Slade, R. Zallen: Solid State Commun. **10**, 293 (1972)
7.9   J.A. Savage, S. Nielsen: Infrared Phys. **5**, 195 (1965)
7.10   M. Tanaka, T. Minani: Jpn. J. Appl. Phys. **4**, 1023 (1965)
7.11   M.S. Maklad, R.K. Mohr, R.E. Howard, P.B. Macedo, C.T. Moynihan: Solid State Commun. **15**, 855 (1974)
7.12   M. Onomichi, T. Arai, K. Kudo: J. Non-Cryst. Solids **6**, 362 (1971)
7.13   P.A. Young: J. Phys. C **4**, 93 (1971)
7.14   J.T. Edmond, M.W. Redfearn: Proc. Phys. Soc., London **81**, 380 (1963)
7.15   S. Tsuchihashi, Y. Kawamoto: J. Non-Cryst. Solids **5**, 286 (1971)
7.16   D. Treacy, P.C. Taylor: In *Optical Properties of Highly Transparent Solids*, ed. by S.S. Mitra, B. Bendow (Plenum, New York 1975) p. 261
7.17   J. Tauc, A. Menth, D.L. Wood: Phys. Rev. Lett. **25**, 749 (1970)

296   P. C. Taylor

7.18   G. Lucovsky: Phys. Rev. B **6**, 1480 (1972)
7.19   J.P. Mathieu, D. Poulet: Bull. Soc. for Mineral Crystallogr. **22**, 532 (1970)
7.20   P.C. Taylor, S.G. Bishop, D.L. Mitchell, D. Treacy: In *Amorphous and Liquid Semiconductors*, ed. by J. Stuke, W. Brenig (Taylor and Francis, London 1974) p. 1267
7.21   P.B. Klein, P.C. Taylor, D.J. Treacy: Phys. Rev. B **16**, 4511 (1977)
7.22   R. Zallen, R.E. Drews, R.L. Emerald, M.L. Slade: Phys. Rev. Lett. **26**, 1564 (1971)
7.23   L.B. Zlatkin, E.K. Ivanov: J. Phys. Chem. Solids **32**, 1733 (1971)
7.24   M.L. Belle, B.T. Kolomiets, B.V. Pavlov: Fiz. Tekh. Poluprovodn. **2**, 1448 (1968)
7.25   A.G. Leiga: J. Opt. Soc. Am. **58**, 1441 (1968)
7.26   G. Weiser, J. Stuke: Phys. Status Solidi **35**, 747 (1969)
7.27   J. Stuke: J. Non-Cryst. Solids **4**, 1 (1970)
7.28   J. Tauc, F.J. DiSalvo, G.E. Peterson, D.L. Wood: In *Amorphous Magnetism*, ed. by H.O. Hooper, A.M. de Graaf (Plenum, New York 1973) p. 119
7.29   P.W. Anderson: Phys. Rev. **109**, 1492 (1958)
7.30   M.H. Cohen, H. Fritzsche, S.R. Ovshinsky: Phys. Rev. Lett. **22**, 1065 (1969)
7.31   N.F. Mott: Philos. Mag. **24**, 935 (1971)
7.32   S.C. Agarwal: Phys. Rev. B **7**, 685 (1973)
7.33   P.W. Anderson: Phys. Rev. Lett. **34**, 953 (1975)
7.34   R.A. Street, N.F. Mott: Phys. Rev. Lett. **35**, 1293 (1975)
7.35   N.F. Mott, E.A. Davis, R.A. Street: Philos. Mag. **32**, 961 (1975)
7.36   M. Kastner, D. Adler, H. Fritzsche: Phys. Rev. Lett. **37**, 1504 (1976)
7.37   S.G. Bishop, U. Strom, P.C. Taylor: Phys. Rev. B **15**, 2278 (1977)
7.38   S.G. Bishop, U. Strom, C.S. Guenzer: In *Amorphous and Liquid Semiconductors*, ed. by J. Stuke, W. Brenig (Taylor and Francis, London 1974) p. 963
7.39   J. Cernogora, F. Mollot, C. Benoit à la Guillaume: In *The Physics of Semiconductors*, ed. by M.H. Pilkuhn (Teubner, Stuttgart 1974) p. 1027
7.40   J. Orenstein, M. Kastner: Phys. Rev. Lett. **43**, 161 (1979)
7.41   M. Kastner, J. Orenstein: In *Physics of Semiconductors – 1978*, ed. by B.L.H. Wilson (Institute of Physics, London 1979) p. 1301
7.42   J. Orenstein, M.A. Kastner: Phys. Rev. Lett. **46**, 1421 (1981)
7.43   J. Orenstein, M.A. Kastner, V. Vaninov: Philos. Mag. B **46**, 23 (1982)
7.44   R.L. Fork, C.V. Shank, A.M. Glass, M. Migus, M.A. Bösch, J. Shah: Phys. Rev. Lett. **43**, 394 (1979)
7.45   D.E. Ackley, J. Tauc, W. Paul: Phys. Rev. Lett. **43**, 715 (1979)
7.46   R.J. Temkin, G.A.N. Connell, W. Paul: In *Amorphous and Liquid Semiconductors*, ed. by J. Stuke, W. Brenig (Taylor and Francis, London 1974) p. 133
7.47   S. Yamasaki, N. Nakagawa, H. Yamamoto, A. Matsuda, H. Okushi, K. Tanaka: In *Tetrahedrally Bonded Amorphous Semiconductors*, ed. by R.A. Street, D.K. Biegelsen, J.C. Knights, AIP Conf. Proc. No. 73 (American Institute of Physics, New York 1981) p. 258
7.48   N.M. Amer, W.B. Jackson: In *Semiconductors and Semimetals*, Vol. 21B, ed. by J.I. Pankove (Academic, New York 1984) p. 83;
       W.B. Jackson, N.M. Amer, A.C. Boccara, D. Fournier: Appl. Opt. **20**, 1333 (1981);
       W.B. Jackson, N.M. Amer: In *Tetrahedrally Bonded Amorphous Semiconductors*, ed. by R.A. Street, D.K. Biegelsen, J.C. Knights, AIP Conf. Proc. No. 73 (American Institute of Physics, New York 1981) p. 263
7.49   G.D. Cody, B.G. Brooks, B. Abeles: Solar Energy Mater. **4**, 231 (1982)
7.50   D.L. Stabler, C.R. Wronski: Appl. Phys. Lett. **31**, 292 (1977)
7.51   N.M. Amer, A. Skumanich, W.B. Jackson: Physica (Utrecht) **117B+118B**, 897 (1983)
7.52   J.M. Viner, C. Lee, P.C. Taylor: Unpublished
7.53   R.A. Street, J. Kakalios, C.C. Tsai, T.M. Hayes: Phys. Rev. B **35**, 1316 (1988)
7.54   M. Gal, R. Ranganathan, P.C. Taylor: J. Non-Cryst. Solids **77+78**, 543 (1985)
7.55   R. Ranganathan, P.C. Taylor: J. Non-Cryst. Solids **97–98**, 707 (1987)
7.56   R. Ranganathan, M. Gal, P.C. Taylor: Solar Cells **24**, 257 (1988)

7.57   J. Tauc: "Time Resolved Spectroscopy", in *Semiconductors and Semimetals*, Vol. 21B, ed. by J.I. Pankove (Academic, New York 1984) p. 299

7.58   Z. Vardeny, J. Strait, J. Tauc: Appl. Phys. Lett. **42**, 580 (1983)

7.59   S. Ray, Z. Vardeny, J. Tauc, T. Moustakas, B. Abeles: In *Tetrahedrally Bonded Amorphous Semiconductors*, ed. by R.A. Street, D.K. Biegelsen, J.C. Knights, AIP Conf. Proc. No. 73 (American Institute of Physics, New York 1981) p. 253

7.60   Z. Vardeny, P. O'Connor, S. Ray, J. Tauc: Phys. Rev. Lett. **44**, 1267 (1980)

7.61   K. Tanaka, M. Kikuchi: Solid State Commun. **11**, 1311 (1972)

7.62   J.P. deNeufville: "Photostructural Transformations in Amorphous Solids", in *Optical Properties of Solids, New Developments*, ed. by B.O. Seraphin (North-Holland, Amsterdam 1976)

7.63   K. Tanaka: "Reversible Photo-Induced Structural Changes in Chalcogenide Glasses", in *Structure and Excitations of Amorphous Solids*, ed. by G. Lucovsky, F.L. Galeener, AIP Conf. Proc. No. 31 (American Institute of Physics, New York 1976) p. 148

7.64   Ke. Tanaka: Solid State Commun. **34**, 201 (1980)

7.65   Ke. Tanaka: Phys. Rev. B **30**, 4549 (1984)

7.66   D.J. Treacy, P.C. Taylor, P.B. Klein: Solid State Commun. **32**, 423 (1979)

7.67   J.Z. Liu, P.C. Taylor: Phys. Rev. Lett. **59**, 1938 (1987)

7.68   W.M. Pontuschka, P.C. Taylor: Solid State Commun. **38**, 573 (1981)

7.69   E. Mytilineou, P.C. Taylor, E.A. Davis: Solid State Commun. **35**, 497 (1980)

7.70   J.Z. Liu, P.C. Taylor: "Photodarkening in Bulk Glassy As_2S_3", in *Amorphous and Liquid Semiconductors*, ed. by F. Evangelisti, J. Stuke (North-Holland, Amsterdam 1985) p. 1195

7.71   N.F. Mott, E.A. Davis, R.A. Street: Philos. Mag. **32**, 961 (1975)

7.72   M. Kastner, D. Adler, H. Fritzsche: Phys. Rev. Lett. **37**, 1504 (1975)

7.73   D. Emin: "Small Polaron Model of the Low-Temperature Optically Induced Properties of Chalcogenide Glasses", in *Amorphous and Liquid Semiconductors*, ed. by W.E. Spear (University of Edinburgh, Edinburgh 1977) p. 261

7.74   K.L. Ngai, P.C. Taylor: Philos. Mag. B **37**, 175 (1978)

7.75   S.R. Ovshinsky: Phys. Rev. Lett. **36**, 1469 (1976)

7.76   K.L. Ngai, T.L. Reinecke, E.N. Enconomou: Phys. Rev. B **17**, 790 (1978)

7.77   D.P. Jones, N. Thomas, W.A. Phillips: Philos. Mag. B **38**, 271 (1978)

7.78   K. Tanaka: J. Non-Cryst. Solids **35+36**, 1023 (1980)

7.79   D.K. Biegelsen, R.A. Street: Phys. Rev. Lett. **44**, 803 (1980)

7.80   D.J. Treacy, U. Strom, P.B. Klein, P.C. Taylor, T.P. Martin: J. Non-Cryst. Solids **35+36**, 1035 (1980)

7.81   Z. Liu: *Metastable Photoinduced Absorption in Chalcogenide Glasses*, Ph. D. Thesis, University of Utah (1987) unpublished

7.82   J.Z. Liu, P.C. Taylor: J. Non-Cryst. Solids **97+98**, 1123 (1987)

7.83   R.J. Nemanich, G.A.N. Connell, T.M. Hayes, R.A. Street: Phys. Rev. B **18**, 6900 (1978)

7.84   U. Strom, T.P. Martin: Solid State Commun. **29**, 527 (1979)

7.85   S.A. Keneman, J. Bordogna, J.N. Zemel: J. Appl. Phys. **49**, 4663 (1978)

7.86   J.P. de Neufville, S.C. Moss, S.R. Ovshinsky: J. Non-Cryst. Solids **13**, 19 (1974)

7.87   P.C. Taylor, J.Z. Liu: "Defects and the Photodarkening Process in the Chalcogenide Glasses", in *Defects in Glasses*, Vol. 61, ed. by F.L. Galeener, D.L. Griscom, M.J. Weber (Materials Research Society, Pittsburgh 1986) p. 223

7.88   R. Ranganathan, M. Gal, P.C. Taylor: Phys. Rev. B **37**, 1021 (1988)

7.89   J.I. Pankove: *Optical Processes in Semiconductors* (Dover, New York 1975)

7.90   B.T. Kolomiets, T.M. Mamantova, A.A. Babeev: Phys. Status Solidi **27**, K15 (1968)

7.91   R.A. Street: Adv. Phys. **25**, 397 (1976)

7.92   D. Emin: Adv. Phys. **24**, 305 (1975)

7.93   F. Mollot, J. Cernogora, C. Benoit à la Guillaume: Phys. Status Solidi A **21**, 281 (1974)

7.94   J. Cernogora, F. Mollot, C. Benoit à la Guillaume: Phys. Status Solidi A **15**, 401 (1973)

7.95   R.A. Street, T.M. Searle, I.G. Austin: J. Phys. C **6**, 1830 (1973)

7.96   P.C. Taylor, U. Strom, S.G. Bishop: Solar Energy Mater. **8**, 23 (1982)
7.97   R.A. Street, T.M. Searle, I.G. Austin: In *Amorphous and Liquid Semiconductors*, ed. by J. Stuke, W. Brenig (Taylor and Francis, London 1974) p. 953
7.98   K. Murayama, T. Ninomiya, H. Suzuki, K. Morigaki: Solid State Commun. **24**, 197 (1977)
7.99   M.A. Bösch, J. Shah: Phys. Rev. Lett. **42**, 118 (1979)
7.100 G.S. Higashi, M. Kastner: J. Phys. C **12**, L821 (1979)
7.101 G.S. Higashi, M. Kastner: J. Non-Cryst. Solids **35+36**, 921 (1980)
7.102 G.S. Higashi, M. Kastner: Phys. Rev. B **24**, 2295 (1981)
7.103 R.A. Street: Solid State Commun. **34**, 157 (1980)
7.104 J. Shah: Phys. Rev. B **21**, 4751 (1980)
7.105 G.S. Higashi, M. Kastner: Phys. Rev. Lett. **47**, 124 (1981)
7.106 K. Murayama, T. Ninomiya: Jpn. J. Appl. Phys. **21**, L512 (1982)
7.107 K. Murayama, H. Suzuki, T. Ninomiya: J. Non-Cryst. Solids **35+36**, 915 (1980)
7.108 K. Murayama, K. Kimura, T. Ninomiya: Solid State Commun. **36**, 349 (1980)
7.109 R.A. Street: Adv. Phys. **30**, 593 (1981)
7.110 M. Gal, J.M. Viner, P.C. Taylor, R.D. Wieting: Phys. Rev. B **31**, 4060 (1985)
7.111 R. Ranganathan, M. Gal, J.M. Viner, P.C. Taylor: Phys. Rev. B **35**, 9222 (1987)
7.112 R.A. Street, D.K. Biegelsen, R.L. Weisfield: Phys. Rev. B **30**, 5861 (1984)
7.113 I. Hirabayashi, K. Morigaki, S. Nitta: J. de Phys. C **4-42**, C4-587 (1981)
7.114 R.W. Collins, W. Paul: J. de Phys. C **4-42**, C4-591 (1981)
7.115 H. Oheda, S. Yamasaki, A. Matsuda, K. Tanaka: J. Non-Cryst. Solids **59+60**, 373 (1983)
7.116 D.J. Dunstan, S.P. Depinna, B.C. Cavenett: J. Phys. C **30**, L425 (1982)
7.117 J.I. Pankove, J.E. Berkeyheiser: Appl. Phys. Lett. **37**, 705 (1980)
7.118 D. Weaire, P.C. Taylor: "Vibrational Properties of Amorphous Solids", in *Dynamical Properties of Solids*, Vol. 4, ed. by G.K. Horton, A.A. Maradudin (North-Holland, New York 1980) p. 1
7.119 R. Shuker, R.W. Gammon: Phys. Rev. Lett. **25**, 222 (1970)
7.120 P.B. Klein, P.C. Taylor, D.J. Treacy: Phys. Rev. B **16**, 4511 (1977)
7.121 R. Zallen, M.L. Slade, A.T. Ward: Phys. Rev. B **3**, 4257 (1971)
7.122 M.H. Brodsky, A. Lurio: Phys. Rev. B **9**, 1646 (1974)
7.123 M.H. Brodsky, M. Cardona, J.J. Cuomo: Phys. Rev. B **16**, 3556 (1977)
7.124 J.E. Smith, Jr., M.H. Brodsky, B.L. Crowder, M.I. Nathan, A. Pinczek: Phys. Rev. Lett. **26**, 642 (1971)
7.125 G. Dolling, R.A. Cowley: Proc. Phys. Soc., London **88**, 463 (1966)
7.126 D. Weaire, R. Alben: Phys. Rev. Lett. **29**, 1505 (1972)
7.127 M.F. Thorpe: Phys. Rev. B **8**, 5352 (1973)
7.128 P.W. Anderson, B.I. Halperin, C. Varma: Philos. Mag. **25**, 1 (1972)
7.129 W.A. Phillips: J. Low Temp. Phys. **7**, 351 (1972)
7.130 R.C. Zeller, R.O. Pohl: Phys. Rev. B **4**, 2029 (1971)
7.131 R.B. Stephens: Phys. Rev. B **8**, 2896 (1973)
7.132 S. Hunklinger, W. Arnold, S. Stein, R. Nava, K. Dransfeld: Phys. Lett. **42**A, 253 (1972)
7.133 S. Hunklinger, W. Arnold, S. Stein: Phys. Lett. **45**A, 311 (1973)
7.134 B. Golding, J.E. Graebner, B.I. Halperin, R.J. Schultz: Phys. Rev. Lett. **30**, 223 (1973)
7.135 R.J. Wagner, A.J. Zelano, L.H. Ngai: Opt. Commun. **8**, 46 (1973)
7.136 U. Strom, J.R. Hendrickson, R.J. Wagner, P.C. Taylor: Solid State Commun. **15**, 1871 (1974)
7.137 U. Strom, P.C. Taylor: Phys. Rev. B **16**, 5512 (1977)
7.138 M. Gorman, S.A. Solin: Solid State Commun. **18**, 1401 (1976)
7.139 R.J. Nemanich: Phys. Rev. B **16**, 1655 (1977)
7.140 P.C. Taylor, U. Strom, J.R. Hendrickson, S.K. Bahl: Phys. Rev. B **13**, 1711 (1976)
7.141 J.S. Lannin: Solid State Commun. **12**, 947 (1973)
7.142 J.D. Axe, D.T. Keating, G.S. Cargill, R. Alben: AIP Conf. Proc. **20**, 279 (1974)

Subject Index

Topics in Applied Physics Founded by Helmut K. V. Lotsch